About the Authors

Patricia Barnes-Svarney has traveled worldwide as a nonfiction science and science fiction writer for more than 18 years. She has a bachelor's degree in geology and a master's degree in geography/geomorphology, and has worked professionally as a geomorphologist and oceanographer. Barnes-Svarney has had more than 350 articles published in magazines and journals and is the author or coauthor of nearly 30 books, including the award-winning *New York Public Library Science Desk Reference* and *Asteroid: Earth Destroyer or New Frontier?,* and several internationally best-selling children's books. In her spare time, she volunteers at a local animal shelter and runs an organic herb farm.

Thomas E. Svarney brings extensive scientific training and experience, a love of nature, and creative artistry to his various projects. With Barnes-Svarney, he has written extensively about the natural world, including paleontology (*The Handy Dinosaur Answer Book*), oceanography (*The Handy Ocean Answer Book*), weather (*Skies of Fury: Weather Weirdness around the World*), natural hazards (*A Paranoid's Ultimate Survival Guide*), and reference (*The Oryx Guide to Natural History*). His passions include martial arts, Zen, *Felis catus,* and nature.

Recently, as participants in the National Science Foundation's Antarctica Artists and Writers Program, the authors conducted research in Antarctica for a book on that icy continent's extreme weather. When they aren't traveling, the authors reside in the Finger Lakes region of upstate New York with their cats, Fluffernutter, Worf, and Pabu.

Also from Visible Ink Press

THE
HANDY
GEOLOGY
ANSWER
BOOK

THE HANDY GEOLOGY ANSWER BOOK

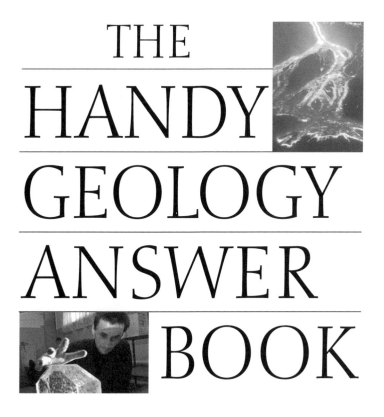

Patricia Barnes-Svarney and
Thomas E. Svarney

VISIBLE
INK
PRESS

Detroit

THE HANDY GEOLOGY ANSWER BOOK

Visible Ink Press®
43311 Joy Rd. #414
Canton, MI 48187–2075

Visible Ink Press is a registered trademark of Visible Ink Press LLC.

Most Visible Ink Press books are available at special quantity discounts when purchased in bulk by corporations, organizations, or groups. Customized printings, special imprints, messages, and excerpts can be produced to meet your needs. For more information, contact Special Markets Director, Visible Ink Press, at www.visibleink.com or (734) 667–3211.

Cover image of sand dune courtesy Photo Disk. All other cover and back cover images reprinted by permission of AP/Wide World Photos.

Art Director: Mary Claire Krzewinski
Typesetting: Graphix Group
ISBN 1–57859–156–2

Cataloging-in-Publication Data is on file with the Library of Congress.

Printed in the United States of America
All rights reserved

10 9 8 7 6 5 4 3 2 1

Contents

Introduction

Courtesy of *The Lord of the Rings* trilogy, movie audiences have been drawn to a fantasy place called "Middle Earth," an ever-changing landscape that seems almost alien. *The Lord of the Rings* may have come from the imagination of two great storytellers, author J. R. R. Tolkien and director Peter Jackson, but the landscape chosen for the movies was real. This enthralling fantasy world was actually New Zealand, a land formed and shaped by two of the most awe-inspiring, yet destructive, forces known to humans: volcanoes and plate tectonics. This was one instance where geology made a movie even more fascinating.

But geology is more than just amazing landscapes. It is rocks, minerals, fossils, processes, cycles, the physical characteristics and features of our amazing planet. We are surrounded by geology—it is in everyone's backyard; it is at everyone's feet. Geology describes how our environment became what it is, in effect providing a history of the planet and its universe. Ranging from the backyard to outer space, *The Handy Geology Answer Book* answers nearly 1,100 fundamental questions about this most fascinating science.

With more than 100 photographs and illustrations, *The Handy Geology Answer Book* takes you on a tour of our world. It answers questions on topics ranging from the microscopic formation of crystals to the titanic, eons-long processes that form islands, mountains, continents, and even planets. You'll be taken back in time to uncover the mysteries of dinosaur fossils and then catapulted toward the forefront of science, where you will learn how artificial gems are synthesized and why glaciologists fret over the effects of global warming on Earth's massive ice shelves. Along the way, you will get to know some of the famous geologists who cleared a path for future generations of scientists as they explore the mysteries of caves, mine the planet's rich mineral resources, and try to save lives by accurately predicting earthquakes, volcanic eruptions, and tsunamis.

Once your appetite has been whetted, the final chapter of *Handy Geology* provides a solid foundation for further research, as well as helpful advice on how to begin enjoying geology as a hobby or even a career. A helpful glossary will tell you what, for example, an alluvial fan or a talus slope is.

As scientists and writers, we have used our experience to write this book. Our backgrounds and love of nature have taken us to many of the geologic rarities of the world. We have traveled extensively across the Earth, doing fieldwork, conducting interviews with scientists, and managing many "rock hunts" of our own. We have crossed Antarctica's Transantarctic Mountains, hiked the European Alps, traveled the oceans, and witnessed the upheaval caused by the shift of tectonic plates in New Zealand and other places. We have visited almost every geologic "hot site" in America, hiking into the Grand Canyon, walking the rim of Meteor Crater, and watching an active Cascade Range volcano. We have also explored plenty of dormant sites and experienced several California earthquakes, just to name a few of our adventures. Not that we have shunned our own backyard—we live just about where the edge of the last glacial ice sheet existed over 10,000 years ago, right near the famous New York Finger Lakes, where deep troughs were carved by glacial advances and retreats over millions of years.

In part, this book represents a distillation of what we discovered during our travels. And it answers hundreds of basic questions on that most interesting of topics—planet Earth. So sit back, grab your rock hammer and hand lens, and enjoy the scenery.

Acknowledgments

The authors thank the United States Geological Survey, National Science Foundation, American Geophysical Union, numerous university geology departments, and all the sundry geologists who gather data and spend long hours in the field to help us better understand our planet.

We also thank Kevin Hile for his patience, project management, skillful editing, and line art design; Christa Gainor for her project oversight; and Marty Connors for the green light. Thank you also to proofreader Dana Barnes, indexer Larry Baker, permissions editor Christopher Scanlon, image digitizer Robert Huffman, cover and interior designer Mary Claire Krzewinski, and typesetter Jake Di Vita of the Graphix Group. And, of course, our special thanks as always to our agent, Agnes Birnbaum, for her help and friendship.

DETAILS OF GEOLOGY

STUDYING THE EARTH

What is **geology**?

Simply put, geology is the study of the Earth. Like so many scientific words, "geology" is constructed from root words dating back to ancient times: *geo* comes from the Greek word meaning "the Earth"; it is a prefix used in other related fields, too, such as geography, geodesy, and geophysics. The *ology* suffix comes from the Greek *logos*, meaning "discussion," and is roughly translated as "the study of."

How **old** is the science of **geology**?

Although the science of geology as we know it today is a relatively young field, insightful observations of Earth processes were made as far back as the ancient Greeks. Some of these early ideas were handed down through the ages. For example, Herodotus (c. 484–425? B.C.E.) had rather modern insights about the formation of the Nile River delta and the important role sediment (deposited by flooding) played in producing the fertile Nile Valley. The Greek historian also applied a primitive form of a principle known as uniformitarianism, the idea that existing processes are sufficient to explain all geological changes that have occurred over time. (See below for more information about uniformitarianism.)

But many other "geological" observations by the ancient Greeks seem fanciful today. For example, Aristotle (384–322 B.C.E.), the famous philosopher and tutor to Alexander the Great, believed that the heat from volcanic eruptions was produced by underground fires. He also believed that air moving through caverns became heated by friction, causing these fires.

1

What are some **subdivisions of geology**?

Geology is a vast field, stretching from paleontology to mineralogy. It is easy to see why, since there are so many features and processes taking place on the Earth and beyond. The following lists some important subdivisions of geology:

Economic geology—the study of how rocks are used, mined, bought, and sold, such as in the search for metals. In other words, economic geologists explore our natural resources and their development.

Environmental geology—the study of the environmental effects produced by changes in geology, such as the determination river flow and its connection to flooding, and conversely, how the geology is affected by environmental problems, such as pollution and urban development.

Geochemistry—the study of the chemical composition of rocks and minerals; geochemists use this information to determine more about the internal structure of materials.

Geomorphology—the study of landform development, such as how a river forms and develops over time.

Geophysics—the physics of the Earth, including such fields as seismology (including interpretation of the Earth's interior), and the effects of the Earth's magnetic and electric fields.

Glacial geology—how ice sheets and glaciers affect each other and the geology of an area.

Hydrology—how water, such as groundwater flow in a karst terrain or how pollution moves underground, affects the geology of an area.

Limnogeology—the study of ancient and modern lakes.

Marine geology—the study of the geology of the ocean floor and/or coastline, especially with regard to how they change over time.

Paleontology—the study of ancient life in the form of fossils, including specializations in invertebrates, vertebrates, plants, and dinosaurs.

Petroleum geology—the study of how petroleum products are formed, found, and extracted.

Planetology—the study of the planets and satellites of our solar system, especially with regard to their formation and how they compare to the Earth.

Volcanology—the study of volcanoes and volcanic phenomena.

What is the **law of superposition**?

There are no real laws in geology. After all, most of the processes, events, and sundry items attached to geology don't work in all situations at all times. But many geologists think there are some truths that work most of the time, including the law of superposition. This idea states that the rock layers on the bottom are the oldest, while those on the top were formed more recently. Of course, this is not always the case, especially when mountain formation folds rock layers over, resulting in older layers lying on top of younger layers.

What does the term **cross-cutting relationships** mean in geology?

The term cross-cutting relationships means that the rock layer being cut by, for example, an intrusive igneous chunk of magma, will be older than the magma doing the cutting. This is usually the case in practical applications of geology, especially when one is discussing faults, sills, and dikes.

HISTORY AND GEOLOGY

What is **uniformitarianism**?

Uniformitarianism (or the uniformitarian principle) is a doctrine that states that current geologic processes and natural laws are sufficient to account for all geologic changes over time. This is exemplified in the saying, "the present is the key to the past." The concept was first formally introduced around 1788 by Scottish geologist James Hutton (1726–1797). Later, in 1830, British geologist Charles Lyell (1779–1875) originated the term "uniformitarianism" to describe this doctrine.

What were **neptunism, catastrophism, and plutonism**?

By the end of the 18th century—often called the "heroic age" of geology—there were three major divisions in geological circles. Most of the debates at the time had to do with how surface features developed on Earth, and all three divisions had support by major geologists of the day. The following describes the primary camps. (Note: these three beliefs are often capitalized):

Neptunism—Neptunism was a popular theory in the late 19th century and was especially favored by such famous geologists of the time as German scientist Abraham Werner. It held that the world was once all ocean, and that the rocks of the Earth's crust, including basalt and granite, seen today were precipitated out of the ocean.

Catastrophism—This doctrine maintained that most geologic formations were created by sudden, violent, catastrophic events, such as earthquakes, floods, asteroids strikes, and volcanoes. Today, most geologists believe that Earth's features are formed by slow natural processes that, at times, are punctuated by certain catastrophic events.

3

Although German scientist Alfred Wegener is credited with the idea of continental drift, there were others who proposed the same idea even earlier. One was American geologist Richard Owen (1810–1897), who suggested that the continents drifted apart over time. The essay that explained his theories, *Key to the Geology of the Globe: An Essay,* was largely forgotten for decades, until Wegener published his treatise on the subject in 1912. (For more about Alfred Wegener, see "The Earth's Layers.")

Plutonism—Scottish geologist James Hutton was an avid follower of plutonism, which also included the volcanists. This idea recognized that rock could be created from both the oceans and subterranean processes, such as shallow ocean deposits and volcanoes; in addition, it held that landforms could form from these processes, such as heat from molten magma forcing mountains to uplift. The theory also stated that the Earth formed by solidification of a molten mass. Hutton proposed these ideas in his book *Theory of the Earth*.

FAMOUS GEOLOGISTS

Who was **Georgius Agricola**?

Georgius Agricola (Georg Bauer, 1494–1555) was a German scientist who is also thought of by many as the "father of mineralogy." Agricola was originally a philologist (a person who studies ancient texts and languages); later, he worked in the greatest mining region of Europe of that time, near Joachimsthal, Germany. As a result of this experience, he produced seven books on geology that set the stage for the development of modern geology two centuries later. His main contributions were compiling all that was known about mining and smelting during his time and suggesting ways to classify minerals based on their observable properties, such as hardness and color.

What did **Nicolaus Steno** contribute to geology?

Nicolaus Steno (Niels Stensen; 1638–1686) was a Danish geologist and anatomist. In 1669, he determined that the so-called "tonguestones" sold on the Mediterranean island of Malta as good luck charms were actually fossilized sharks' teeth. He also developed what is called Steno's law, or the principle of superposition. This theory says that in any given rock layer, the bottom rocks are formed first and are the oldest.

Steno proposed two other principles: that rock layers are initially formed horizontally (often called the law of original horizonality), and that every rock outcrop in which only the edges are exposed can be explained by some process, such as erosion or earthquakes (often called the law of concealed stratification). All of these principles are generally held to be true today. Some people consider Steno to be the "father of modern geology," a title also given to several other early geologists, including James Hutton.

Charles Lyell (1779–1875) was at the forefront of the study of the composition, structure, and processes that have formed the Earth's geology. *Library of Congress.*

Who was **Abraham Gottlob Werner**?

Abraham Gottlob Werner (1750–1817) was a German geologist and mineralogist who first classified minerals systematically based on their external characteristics. He also was a great believer in neptunism.

Who started the **Geological Survey in Britain**?

Sir Henry Thomas de la Beche (1796–1855) was an English geologist who established the Geological Survey in Britain, officially recognized in 1835, when de la Beche was appointed director. He was also internationally known for his geological studies of the United Kingdom, France, Switzerland, and Jamaica.

What was **James Hutton's contribution** to geology?

James Hutton (1726–1797) was a Scottish natural philosopher, but his contribution to geology was even more important: He is considered by some to be the "father of modern geology." Hutton was also an avid follower of plutonism and the author of *Theory of the Earth,* a book that emphasized several fundamentals of geology, including that the Earth was older than 6,000 years, that subterranean heat creating metamorphic material is just as important a process as rock forming from sediments laid down underwater, and that the exact same agents that are operating today created the landforms of the past, a principle also known as uniformitarianism.

What famous geologist **expanded** upon **James Hutton's theories**?

British geologist Charles Lyell (1779–1875) further developed James Hutton's theories in his book *The Principles of Geology*. He was one of the first to describe geology as the study of the Earth's composition, history, structure, and processes. Lyell believed that geological features were eroded, shaped, and formed at a constant rate over time, and he was the first to coin the term "uniformitarianism."

Who published the first simple **geological map** of the **United States**?

The first simple geological map of the United States was actually published by Scottish-born merchant William Maclure (1763–1840). He became an American citizen in 1796 and later wrote *Observations on the Geology of the US* (1809), in which the map was presented.

Who was the **founder of English geology**?

William Smith (1769–1839) was a surveyor employed in canal building who eventually became known as the founder of English geology. While working on the canals, he kept copious notes of his observations. He eventually published *A Delineation of the Strata of England and Wales* in 1815, the first true geologic map of its kind and the one that made him famous. Along with Hutton and Steno, Smith is also sometimes called the "father of modern geology."

Who was **John Playfair**?

Scottish geologist John Playfair (1748–1819) went against most of his contemporaries in proposing that river valleys were actually carved by streams, an idea that is readily accepted today. Many of his peers believed that valleys formed during cataclysmic upheavals of the land, with the rivers flowing through much later.

What did **James Hall** contribute to geology?

James Hall (1811–1898) was an American geologist and one of the first to establish an acceptable theory of mountain building. Hall also was the first to compose a record of all fossils found below the Upper Carboniferous Period rock layers throughout the United States, and he was known as the chief American invertebrate paleontologist of his time.

Who was **James Hall**?

Scottish geologist Sir James Hall (1761–1832) was one of the first to establish experimental research as an aid in geological investigations. For example, one of his experiments demonstrated how lava forms different kinds of rocks as it cools. Hall was also friends with James Hutton and John Playfair; his rock studies helped confirm many of Hutton's views regarding intrusive rock formations.

Louis Agassiz (1807–1873) was the first geologist to propose that the Earth has undergone several ice ages. *Library of Congress.*

Who was **Louis Agassiz**?

Jean Louis Rodolphe Agassiz (1807–1873), best known simply as Louis Agassiz, was the Swiss-born geologist and paleontologist who introduced the concept of the ice ages, periods of time when glaciers and ice sheets covered much of the Northern Hemisphere. Agassiz announced this astonishing idea in a famous speech to the Swiss Society of Natural Sciences in 1837. "Ice ages" was a term he adopted from Karl Schimper (1803–1867), who coined the phrase the year before. Agassiz later moved to the United States, where he was a dominant force in the fields of geology and paleontology until his death. Interestingly, he was one of many scientists who rejected his contemporary Charles Darwin's theory of natural selection.

What was **James Dwight Dana's contribution** to geology?

James Dwight Dana (1813–1895) was an American mineralogist who compiled a book that is still one of the most respected works in mineralogy: *Dana's Manual of Mineralogy*. First published in 1862, the book was only one of Dana's to become a standard reference book in geology. It covers most of the known minerals and metals on Earth and includes their chemical formulas, characteristics, uses, and sundry other useful information.

Who was **Clarence Edward Dutton**?

Clarence Edward Dutton (1841–1912) was a American geologist who, among other geologic ventures, pioneered the theory of isostasy, which describes how land can rise up as a result of events such as retreating glacial ice sheets. He also worked on a Tertiary history of the Grand Canyon, Arizona, detailing the rock layers of that period in the region in his classic book, *Tertiary History of the Grand Canyon District* (1882).

Why is **William Gilbert** so well known?

William Gilbert (1544–1603), the English physicist and physician to Elizabeth I and James I, made most of his contributions in the field of magnetism. His experimental work dealt with lodestones (rocks with magnetic properties). He also proposed that the Earth was like a giant magnet with a north and south pole.

Who was **Grove Karl Gilbert**?

Grove Karl Gilbert (1843–1918)—no known relation to William Gilbert—was an American geologist and geomorphologist who laid the foundations for much of the 20th-century's advances in geology. His monographs, including *The Transportation of Debris by Running Water* (1914), greatly contributed to theories of river development. He was also known for making other contributions, such as to theories about glaciation and the formation of lunar craters, as well as to the philosophy of science.

Why is **Thomas Chrowder Chamberlin** remembered?

Thomas Chrowder Chamberlin (1843–1928) was an American geologist whose primary interest was in glacial geology, and he was a proponent of the idea of multiple glaciations. He also established the origin of loess (wind-blown deposits of silt); discovered fossils in Greenland, suggesting that the landmass had experienced an earlier, warmer climate; and developed theories on the Earth's origin (Chamberlin proposed the planetesimal origin of our planet, going against the more commonly accepted nebula-gas-cloud theory), formation, and growth.

Who was **Matthew Fontaine Maury**?

Matthew Fontaine Maury (1806–1873) was an American oceanographer who wrote the first text on modern oceanography, *Physical Geography of the Sea and Its Meteorology* (1855). He also provided guides for ocean currents and trade winds—compiled from ships' logs—which helped cut sailing time for ships on many routes.

John Wesley Powell, seen in this late 1800s photo, made a name for himself leading a now-famous 99-day expedition to explore the Colorado River in 1869. *AP/Wide World Photos*.

Who was **Vasily Vasilievich Dokuchaev**?

Vasily Vasilievich Dokuchaev (1846–1903; also seen as Vasily Vasilyevich Dokuchaev) was a Russian geographer who is considered by many scientists to be the founder of modern soil science. He reasoned that soils form by the interaction of climate, vegetation, parent material, and topography over a certain amount of time. He also suggested that soils form zones, but the idea of zonal soils would not be fully developed until later.

Why was **John Wesley Powell** important to geology?

John Wesley Powell (1834–1902) was an American geologist, soldier, and administrator who, despite losing an arm in the Civil War, led the first successful expedition into the Grand Canyon. He described the Colorado canyons and led the United States Geological Survey for a time.

What **father and son** were known for their **geological works**?

Father and son Albrecht and Walther Penck were German geologists who contributed a great deal to European geology. Albrecht Penck (1858–1945) was a pioneer in his studies on the glaciations of the Alps, showing evidence of four advances and retreats of the ice sheets. Walther (1888–1923) did not live as long, but he advanced theories of landscape development based on the balance between rivers cutting into the strata and uplift.

Who was **Robert Elmer Horton**?

Robert Elmer Horton (1875–1945) was an American engineer, hydrologist, and geomorphologist who was one of the first to develop a qualitative way to describe landforms. He also studied the phenomenon of overland flow of rainwater runoff, a process that was named after him (Horton overland flow).

Who is considered the **greatest petrologist** of the 20th century?

Many scientists consider the greatest petrologist of the 20th century to be Norman Levi Bowen (1887–1956). He developed the idea of phase diagrams of common rock forming minerals—the Bowen's reaction series is named after him—thus providing scientists with information about how to interpret certain rock formations.

Who **greatly advanced** the field of **geomorphology**?

William Morris Davis (1859–1934), an American geologist, geographer, and meteorologist, was an authority on landforms. His articles on the subject advanced the field of geomorphology more than anyone in his time. His most influential concept was the "cycle of erosion" theories, an indication that Davis was greatly influenced by Charles Darwin's organic evolution theory. In a 1883 paper Davis stated that "it seems most probable, that the many pre-existent streams in each river basin concentrated their water in a single channel of overflow, and that this one channel survives—a fine example of natural selection."

Because of his studies and theories, he is often called the "founder of geomorphology" and the "father of geography" (some scientists consider both subjects to be very similar, if not the same). He also founded the National Geographic Society.

Who was **Beno Gutenberg**?

Beno Gutenberg (1889–1960) was the foremost observational seismologist of the 20th century. He analyzed seismic records, contributing important discoveries of the structure of our solid Earth and its atmosphere. He also discovered the precise location of the Earth's core and identified its elastic properties. Besides a plethora of other major contributions to seismology, Gutenberg discovered the layer between the mantle and outer core, a division that is now named after him. (For more information about the Earth's layers, see "The Earth's Layers.")

What contribution did **Preston Cloud** make to geology?

Preston Ercelle Cloud Jr. (1912–1991) was a biogeologist, paleontologist, and humanist who left a diverse legacy that cuts across many scientific and other disciplines. As an historical geologist, he contributed to our understanding of the atmosphere's evolution, oceans, and Earth's crust; he also added to the understanding of the evolution of life. In addition, he was concerned about how humans would continue to evolve in this environment, noting problems with population increases and related activities (for example, pollution) that could greatly affect our planet.

MEASURING THE EARTH

VIEWS OF THE EARTH

Why does the Earth have **seasons**?

Contrary to popular belief, the Earth's seasons are not caused by the varying distance from the sun. They occur because of the tilt of our planet's axis (23.5 degrees) in relation to the plane of the sun. As the Earth orbits the sun, our planet's tilt causes the intensity of sunlight to change over parts of the Earth at different times. The following describes the seasons in terms of the Northern Hemisphere:

Summer—When the geographic North Pole is tilted toward the sun, it is summer in the Northern Hemisphere. During this time, the North Pole receives 24 hours of daylight, while the South Pole experiences total darkness. On the Northern Hemisphere's first day of summer—or the summer solstice—the sun's direct rays are shining at 23.5 degrees north latitude, also known as the Tropic of Cancer. This occurs on June 21 or 22, a day that contains the most daylight hours.

Fall or Autumn—The first day of autumn in the Northern Hemisphere occurs on September 22 or 23 and is called the autumnal equinox. On this day, the sun's direct rays are directly overhead at the equator and the hours of day and night are equal. (The North Pole experiences "sunset" on this day, while the South Pole experiences "sunrise.")

Winter—December 21 or 22 is the first day of winter for the Northern Hemisphere and is called the winter solstice. During this time, the South Pole receives 24 hours of daylight, while the North Pole experiences total darkness. On the Northern Hemisphere's first day of winter, the sun's direct rays are directed at the Tropic of Capricorn, or 23.5 degrees south latitude. This day contains the shortest number of daylight hours in the Northern Hemisphere.

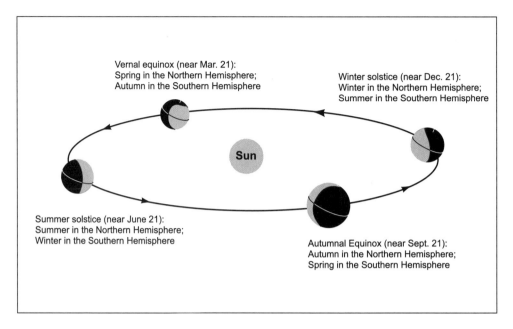

Vernal equinox (near Mar. 21):
Spring in the Northern Hemisphere;
Autumn in the Southern Hemisphere

Winter solstice (near Dec. 21):
Winter in the Northern Hemisphere;
Summer in the Southern Hemisphere

Sun

Summer solstice (near June 21):
Summer in the Northern Hemisphere;
Winter in the Southern Hemisphere

Autumnal Equinox (near Sept. 21):
Autumn in the Northern Hemisphere;
Spring in the Southern Hemisphere

Because the Earth is tilted on its axis at a 23.5° angle, we experience various climate changes according to the location of the Earth in relation to the sun and our latitudinal location on the planet's surface.

Spring—Similar to autumn, the Northern Hemisphere's spring—or vernal equinox—occurs when the sun's direct rays are once again at the equator. Spring occurs on March 21 or 22, and day and night are again of equal length. (The North Pole experiences "sunrise" on this day, while the South Pole experiences "sunset.")

What are **perihelion and aphelion** when talking about the Earth and sun?

The Earth does not spin around the sun in a perfect circle, but follows an elliptical or oval orbit. Because of this, the Earth's orbit has a point at which it is closest to—and another point at which it is farthest from—the sun. When the Earth is at its closest point to the sun, it is called perihelion; the point farthest away from the sun is called aphelion. Perihelion occurs when the Earth is about 91 million miles (147.5 million kilometers) from the sun, usually around January 3 of each year. Aphelion occurs when the Earth is about 94 million miles (152.6 million kilometers) away from the sun, which happens usually around July 4 of each year.

What are the **Van Allen radiation belts**?

No one knew the Van Allen radiation belts existed until the first American satellite, *Explorer I,* discovered the intense inner radiation zone in early 1958 with the use of onboard

If the Earth is closest to the sun in January, why is it so cold?

This is definitely a question that would be asked by someone in the Northern Hemisphere, since January in the Southern Hemisphere is definitely warm! From a global perspective, when the Earth is farther away from the sun in its orbit, the average temperature does increase by about 4°Fahrenheit (2.3°Celsius), even though the sunlight falling on Earth at aphelion is about 7 percent less intense than at perihelion.

So why is it warmer when we are farther away from our star? The main reason is the uneven distribution of the continents and oceans around the globe. The Northern Hemisphere contains more land, while the Southern Hemisphere has more ocean. During July (at aphelion), the northern half of our planet tilts toward the sun, heating up the land, which warms up easier than the oceans. During January, it's harder for the sun to heat the oceans, resulting in cooler average global temperatures, even though the Earth is closer to the sun.

But there is another cause for warm temperatures in the north when the Earth is at aphelion: the duration of summers in the two hemispheres. According to Kepler's second law, planets move more slowly at aphelion than they do at perihelion. Thus, the Northern Hemisphere's summer is 2 to 3 days longer than the Southern Hemisphere's summer, giving the sun more time to bake the northern continents.

Geiger counters. The space probe *Pioneer 3* also detected the outer radiation belt in late 1958. Although it did fail in its mission to reach the Moon, its objectives were changed to measure the Van Allen outer radiation belt during its 38 hour and 6 minute flight.

Named after James Van Allen of the University of Iowa, who first interpreted the data, the two doughnut-shaped rings of ionized gas (plasma) circle the Earth's equator. They form as rapidly moving charged particles from the solar wind become trapped by the Earth's magnetic field. The outer belt stretches from 11,806 miles (19,000 kilometers) to 25,476 miles (41,000 kilometers) in altitude; the inner belt lies between 4,722 miles (7,600 kilometers) and 8,078 miles (13,000 kilometers) in altitude. Similar radiation belts have been found around other planets in our solar system.

The inner and outer belts are not the only radiation zones encircling the Earth, however. In 1990, the Combined Release and Radiation Effects Satellite (CRRES) discovered a third radiation belt between the inner and outer Van Allen belts. In addition, scientists determined that other radiation belts can periodically form between the inner and outer belts. For example, around May 8, 1998, a series of large solar disturbances caused such a new radiation belt to form, but it eventual disappeared as the solar activity subsided.

THE EARTH IN NUMBERS

How **fast** does the **Earth orbit** around the sun?

The Earth travels around the sun at a velocity of about 18.6 miles (29.8 kilometers) per second. This number is based on the distance divided by the time it takes to get around the sun: The circumference of the Earth's orbit is 584,116,000 miles (940 million kilometers) and the travel time is 365.25 days (or 8,766 hours), so 584,116,000 miles divided by 8,766 hours equals 66,634 miles (107,215 kilometers) per hour, or 18.6 miles (30 kilometers) per second. In other words, every second you read this book, the Earth's orbit is carrying you 18.6 miles (30 kilometers) through space.

Where is the **Earth situated** in relationship to the other planets?

The Earth is the third planet from the sun; it is also the fifth largest planet, and the densest major body in the solar system. It is the only planet whose name ("Earth") does not come from Greek or Roman mythology, but rather from an Old English and Germanic name. (For more information about the Earth and the solar system, see "Geology and the Solar System.")

How **fast** does the **Earth rotate** on its axis?

If you stand at the Earth's equator, you are moving approximately 25,000 miles (40,075 kilometers) in each 24-hour day. To calculate this speed, divide the distance around the Earth by the length of a day—25,000/24—which equals just over 1,000 miles (1,670 kilometers) per hour. To figure out how fast you are moving in your hometown, find out your latitude; then multiply the cosine of your latitude by 1,000 miles per hour. For example, Los Angeles is at about 34 degrees north latitude, so $1{,}000 \times \cos(34) = 829$ mph; in terms of kilometers, the calculation would be $1{,}670 \times \cos(34) = 1{,}384$ kph. New York, to use another example, is at about 41 degrees north latitude, so the speed of someone in New York would be $1{,}000 \times \cos(41) = 755$ mph (or 1,260 kph). Thus, one moves faster as the latitude decreases because spots closer to the equator have to travel farther on our globe to complete a rotation.

To be even more precise, scientists also use the Earth's rotation at the equator based on the sidereal period, or the time relative to the stars, not the sun. The Earth's sidereal rotation period is about 23 hours, 56 minutes, and 0.409053 seconds. This is not equal to our 24-hour day, because by the time the Earth has rotated once, it has also moved in its orbit around the sun. To make up the balance, it has to rotate about 4 minutes more to equal 24 hours so that the sun is back in the same spot in the sky as the day before.

At the poles, there is no contest. Because the poles are located just about at the axis of the planet, you would move very little, covering perhaps inches per hour.

Has the Earth always rotated at today's speed?

No, the Earth has rotated at various speeds over the billions of years of its existence. Scientists estimate that 900 million years ago, when single-celled organisms were the only life forms in the oceans, the length of a day on Earth was 18 hours, and the year was 481.

The Earth is continuing to slow its rotational speed as the years pass. This is because of the gravitational interaction between our planet and the Moon, which slows the Earth's rotation by about 0.0015 second per century. This also causes the Moon to move farther from the Earth by a few centimeters each year. Some scientists estimate that in several billion years, the lunar month will go from its current 27.3 days to 47 days.

Directly on the axis, you would spin in a complete circle (360 degrees) every 24 hours—that's 15 degrees per hour—and you would wobble a little.

What is the exact **circumference of the Earth**?

The Earth's circumference is 24,901.55 miles (40,075.16 kilometers) at the equator. Because the Earth bulges slightly at the equator, if one measures the circumference around the poles instead of around the equator, then the result is only 24,859.82 miles (40,008 kilometers).

What are some **major statistics** about the **Earth**?

The following lists some major statistics about our planet. (Note: as with most large measurements of the Earth, these numbers represent the closest approximations):

Diameter—7,926 miles (12,753 kilometers)

Mass—6.5×10^{21} tons (5.972×10^{24} kilograms)

Total Earth volume—259.8 billion cubic miles ($1,083.16 \times 10^{9}$ cubic kilometers)

Density—0.34 pounds per cubic foot (5.515 kilograms per cubic meter)

Surface gravity—32.1 feet (9.78 meters) per second squared

Equatorial escape velocity—6.95 miles (11.18 kilometers) per second

What **elements** are found in the **Earth's atmosphere**?

The Earth's atmosphere is composed of about 77 percent nitrogen, 21 percent oxygen, and traces of argon, carbon dioxide, water, and other compounds and elements. It is interesting that the Earth maintains free oxygen, as it is a very reactive gas. Under most circumstances, it combines readily with other elements. But our atmosphere's oxygen is pro-

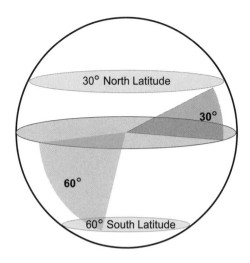

Latitude is determined by measuring the angle between the plane formed by the Equator and a line radiating from the center of the Earth out to the surface.

duced because of biological processes. Without life on Earth, there would be no free oxygen in our atmosphere.

When the Earth formed, it is believed that the atmosphere contained a much larger amount of carbon dioxide—perhaps as much as 80 percent—but this diminished to about 20 to 30 percent over the next 2.5 billion years. Since that time, the gas has been incorporated into carbonate rocks, and to a lesser extent, dissolved into the oceans and consumed by living organisms, especially plants. Today, the movement of the continental plates, the exchange of gas between the atmosphere and the ocean's surface, and biological processes (such as plant respiration) help continue the complex carbon dioxide flow, keeping the amount of carbon dioxide in balance.

Why is the **carbon dioxide** in our atmosphere so **important**?

We should all be thankful for the carbon dioxide in our atmosphere. This gas is important in maintaining the Earth's surface temperature by creating a greenhouse effect. Carbon dioxide (and smaller amounts of other greenhouse gases, such as methane) form a kind of "greenhouse glass" barrier, keeping temperatures higher than if there were no greenhouse gases. Carbon dioxide absorbs some of the heat, keeping it near the ground and raising the average surface temperature. In fact, without the greenhouse effect, the oceans would freeze, and life as we know it would not exist.

Scientists are debating whether the carbon dioxide that humans add to the atmosphere, as a result of industry, transportation, and sundry other processes, can eventually cause catastrophic changes in the Earth's surface temperatures. Many scientists believe we are already seeing the effects of the more than 25 percent rise in carbon dioxide that humans have added to the atmosphere since the beginning of the industrial age around 1750. Since that time, the planet's overall surface temperatures have warmed by at least 1°C. But no one knows how much extra carbon dioxide—even gas additions from natural causes—is truly detrimental.

Why are **longitude and latitude** important to geologists?

In order to describe any location on Earth, most scientists, including geologists, use two numbers: latitude and longitude. This coordinate system is not only used by scientists but many other professions as well, including sailors and military personnel.

Why is Greenwich, England, called the Prime Meridian?

The reason why Greenwich, England, is designated the Prime Meridian has its roots in history: The imaginary line passes through the old Royal Astronomical Observatory, which was chosen by astronomers of the day as zero longitude. The observatory is now a public museum located at the eastern edge of London. It is a great tourist spot, where people go to see a brass band stretching across the museum's yard that marks the "prime meridian." Here it is possible to straddle the line with one foot in the Earth's Eastern Hemisphere and the other in the Western Hemisphere.

This system is used by geologists to pinpoint the location of a certain rock or rock type, keep a record of where the outcrop was identified, and return to the discovery later.

If you look at a globe of the Earth, the latitude lines encircle the Earth parallel to the equator, and they differ in length depending on their location. The longest line is at the equator (latitude 0 degrees); the shortest lines—which are actually pinpoints—are at the poles (90 degrees north at the North Pole; 90 degrees [or –90 degrees] south at the South Pole). In the Northern Hemisphere, latitude degrees increase as you move north away from the equator; in the Southern Hemisphere, latitude degrees increase as you move south away from the equator.

Longitude lines or meridians (once called "meridian line" and eventually shortened to "meridian") are those that extend from pole to pole, slicing the Earth like segments of an orange. Each meridian crosses the equator; in the Western Hemisphere, longitude increases as you move west from Greenwich, England (0 degrees); in the Eastern Hemisphere, longitude increases as you move east from Greenwich, England. All points on the same line of longitude experience noon (and any other hour) at the same time. (Note: Longitude lines are not to be confused with time zones, most of which follow a more erratic demarcation.)

Where did the abbreviations **a.m.** and **p.m.** come from?

The abbreviations "a.m." and "p.m."—or how we differentiate between morning and afternoon/evening hours—originally came from the use of the meridians that make up longitude. The term "meridian" is derived from the Latin *meri,* which is a variation of *medius,* meaning "middle," and *diem* meaning "day." Meridian once meant "noon" long ago, and the time of day before noon was called "ante meridiem," and after noon was called "post meridiem." These terms were then abbreviated "a.m." and "p.m." (the sun at noon is said to be "passing meridian.")

19

Do you capitalize a.m. and p.m.? There is no real agreement on this issue. When accompanied by numbers, usually both are understood to mean the times before noon and after noon, or before midnight and after midnight, whether or not they are capitalized.

What are some **statistics** on the Earth's **geologic features**?

The Earth's geologic statistics include many interesting figures regarding the planet's longest, highest, deepest, and largest features:

Longest river—The Nile River is the longest river in the world, measuring 4,100 miles (6,598 kilometers).

Largest freshwater lake—Lake Superior is the largest freshwater lake in the world, measuring 31,700 square miles (82,103 square kilometers).

Highest point on land—Mount Everest is the highest point on the Earth's land surface, measuring 29,035 feet (8,850 meters) in height above sea level.

Highest waterfall—Angel Falls, Venezuela, is the highest waterfall in the world, measuring 3,212 feet (979 meters) in height; also, it has the world's greatest uninterrupted drop at 2,648 feet (807 meters).

Deepest depression known in the ocean—The deepest depression known in the ocean is the Mariana Trench, Pacific Ocean, measuring 36,201 feet (11,033 meters) in depth. (This is also sometimes seen as Marianas Trench, but the plural form is usually indicative of the Marianas Archipelago along the subduction zone that created the trench.)

Deepest depression on land—The deepest depression on land is the Dead Sea shoreline, measuring about 1,312 feet (400 meters) below sea level.

Largest island—Greenland is the largest island in the world, measuring about 840,000 square miles (2,175,590 square kilometers).

Largest desert—The largest desert is the Sahara in Africa, measuring about 3,500,000 square miles (9,065,000 square kilometers).

What are the **highest and lowest temperatures** ever reported in the world?

The highest temperature ever recorded on Earth was 136.4°F (58°C) in Azizia, Libya, in 1922. The coldest temperature ever recorded on the Earth was at Vostok Station, Antarctica, in 1983, where it dropped to –128.6°F (–89.22°C).

How much does the Earth's **temperature vary**?

The Earth's surface temperatures vary greatly—from the coldest ice sheets at the North and South Poles to the searing heat of the Sahara Desert. For example, in terms of annual average temperature, the hottest place in the world is Dallol, Ethiopia, with a mean temperature of 94°F (34.4°C). In eastern Siberia, the town of Oimyakon on the Indigirka River (population 4,000) is the coldest permanently inhabited place on Earth. Some of the world's coldest temperatures are recorded here every year, averaging about –60°F (–51°C), and sometimes falling to –90°F (–68°C).

The interior of the Earth also varies in temperature depending on depth. For example, at an underground depth of about 50 feet (15 meters), the temperature remains a constant 55°F (11°C), as is experienced in many caves around the world. One good example of this is the 74-foot- (23 meter-) deep cellar at the Paris Observatory in France, where recorded temperatures since 1718 have not varied by more than 1°C. And one of the hottest places is inside the Earth: the inner solid core reaches about 11,690°F (6,477°C).

What are some important **temperature statistics** relating to geology, and how do they compare to some other interesting average temperatures?

Temperatures of various features on Earth vary greatly. The following lists some of these. All temperatures are approximate:

| | Temperature | |
Description	°F	°C
A lightning bolt	53,540	29,727
Solid inner core of the Earth	11,690	6,750
An incandescent light bulb	4,940	2,727
Fresh lava from a volcano	2,240	1,227
Boiling point for water	212	100
Average temperature of water at the vent of Old Faithful	204	95.6
Steam from Old Faithful	350	177
Average temperature of hot springs in Iceland	167	75
Human body's average temperature	98.6	37
Average water temperature in the Mariana Trench	34–39	1–4
Average temperature at South Pole in winter	–108	–78
Average temperature at North Pole in winter	–30	–34
Average temperature at the top of Mt. Everest	–100	–73
Average temperature of the universe	–454.8	–270.4

What is the **Earth's average density**?

The Earth's average density is about 5.52 grams per cubic centimeter—the units most scientists (and everyone else in the world) use to measure the density. The density of water is about 1 gram per cubic centimeter; the density of rocks at the Earth's

Water covers more than 70 percent of the Earth's surface, and 97 percent of the Earth's water is found in oceans with an average depth of about 12,460 feet (3,798 meters). *AP/Wide World Photos.*

surface (usually in the form of granite) is about 2.7 grams per cubic centimeter; the upper mantle's density (mostly considered to be composed of peridotite) is about 3.4 grams per cubic centimeter. Many scientists believe the density of rocks at the Earth's core is very high, which is why they believe the center of our planet is composed of dense iron.

Who was the **first person** to determine the Earth's **average density**?

In 1798, English physicist Henry Cavendish (1731–1810) calculated the Earth's density, basing his idea on specific gravity, or a ratio of the Earth's density to that of water. Although his data was good, he still made an error in calculating the Earth's density to be 5.48. This may have been due to an omission error of one number. Though Cavendish was close, the accepted value today is 5.518.

Contrary to popular belief, the famous "Cavendish experiment" (or "Cavendish apparatus") to determine the Earth's density was actually designed by the geologist Rev. John Michell, who is also known as the "father of seismology" to some. Five years after Michell's death, Cavendish rebuilt the apparatus to calculate the Earth's density, making it smaller than the original. Using two fixed lead spheres hanging from a metal rod, he measured the twisting in a silver wire that supported the spheres when exposed to masses of metal. This allowed him to find the attraction force on a mass the size of the Earth—and eventually to determine the planet's density.

Why do oceans look blue from space?

The oceans appear blue from space for the same reason the sky appears blue: The ocean waters absorb the other colors of the light spectrum while reflecting light in the blue part of the spectrum. When ocean waters carry little organic matter, the color is a deep blue. If there are marine plants present, more blue light will be absorbed and more green light reflected, thus changing the ocean's color. For example, during hurricanes in the Atlantic Ocean, the blue waters mix with the churned up yellow plant pigments to form a green color; this is just like combining blue and yellow paints to form the color green.

Scientists have used this observation to determine marine plant abundance and productivity in the oceans, as well as to gather data about harmful algal blooms (large growths of algae that are often toxic and harmful to other organisms, especially along coastal waters) and other types of pollution. One such study is provided by NASA's SeaWIFS (an acronym for sea-viewing wide field-of-view sensor), a satellite that provides data on global ocean bio-optical properties, or the changes in ocean colors caused by organisms. Along with other information, such as the overall productivity of the oceans, the satellite documents and studies algal overabundance by detecting the various tell-tale hues of harmful blooms or other biological activity.

What is the **average ocean temperature**?

Although ocean temperatures vary depending on depth and latitude, scientists estimate that 87 percent of the world's overall ocean water averages about 40°F (4.4°C) or less. But there are many variations. For example, water temperatures in the shallow Persian Gulf can reach as high as 104°F (40°C). The water temperature at the equator averages about 75 to 85°F (23.9 to 29.4°C). The coldest oceans are the Arctic and Antarctic, both polar oceans containing a permanent covering of ice that feeds cold water into the global ocean circulation. Temperatures there range from 32 to 40°F (0 to 4.4°C).

How much **water** is in the Earth's **oceans**?

Oceans contain about 97.2 percent of the total water on the planet. The rest is held in the ice caps, glaciers, the atmosphere (as clouds, snow, and rain), seas, rivers, and lakes. (For more information about the distribution of water on our planet, see "Geology and Water.")

Another way of looking at the Earth's ocean water is by volume. Combining all the oceans and seas in the world, the total estimated volume of water on Earth is about 330 million cubic miles (1,400 millon cubic kilometers), but this number can fluctuate with the growing or melting of polar and glacial ice. It is also interesting to note that the Earth's oceans are estimated to weigh about 1.45×10^{18} tons, representing a mere 0.022 percent of the Earth's total planetary weight.

What is the **average depth** of the Earth's **oceans**?

The average depth of oceans overall is about 12,460 feet (3,798 meters) below sea level. This is about five times the average elevation on land, which has an average height of about 2,757 (840 meters) above sea level. Some examples of average individual ocean depths are as follow:

Pacific—13,740 feet (4,188 meters)*

Indian—12,704 feet (3,872 meters)

Atlantic—12,254 feet (3,735 meters)

Arctic—3,407 feet (1,038 meters)

Who **first determined** the **Earth's circumference**?

Eratosthenes, a Hellenistic geographer (273–193 B.C.E.) in Alexandria, was the first to calculate the Earth's circumference around 225 B.C.E.; he was only off by a half of a percent. Eratosthenes knew that the Earth was a sphere, and that the sun was at least 20 times farther away than the Moon (actually, it is about 400 times farther away). Because of this, he thought that the rays of the sun should be parallel to the Earth's surface when they reached our planet.

At local noon on the summer solstice in Syene, the sun would be directly overhead (at zenith). Eratosthenes calculated that, while in Alexandria, an observer would measure the sun to be 7 degrees south of the zenith. Knowing the distance between the two cities to be 4,900 stadia (1 stadium equals 0.099 miles or 0.16 kilometers), Eratosthenes calculated the Earth's circumference by reasoning the following: Divide 360 degrees (the degrees equal to the circumference of the Earth) by 7 degrees (the angular distance he found between the cities); then multiply that times the number of stadia (4,900), which equals a circumference of 252,000 stadia. This is equal to about 25,054 miles (40,320 kilometers).

His results were not perfect, but he came very close to today's measurement of the Earth's circumference at 24,873 miles (40,030 kilometers). Eratosthenes' genius did not end there. He was also the first to draw a world map with latitude- and longitude-like lines—the original way to figure out where you are on the planet.

---

* The Pacific also contains most of the ocean trenches in the world, the deepest spots known on the Earth's surface; the greatest is the Mariana Trench, measuring 36,201 feet (11,033 meters) deep. If Mt. Everest, the tallest mountain on land at 29,035 feet (8,850 meters), was dropped into this trench, it would still be covered by more than a mile (about 1.5 kilometers) of ocean water.

KEEPING TRACK OF TIME

What were some **early ideas** about the **Earth's age**?

In 1644, John Lightfoot (1602–1675) calculated the age of the Earth using the Bible. Based on this, he determined that our planet began at 9 a.m. on September 17, 3928 B.C.E. Not long after, another attempt was made by James Ussher (1580–1655), Archbishop of Armagh, Ireland. He determined that the Earth began on October 23, 4004 B.C.E., based on the genealogies and ages of people he found in the Bible. It wasn't until the latter part of the 18th century that someone took a different approach to calculating the Earth's age. French naturalist Jean-Baptiste Lamarck (1744–1829) believed the Earth was much older, although he couldn't give a convincing reason how or why this would be true.

By the 19th century, many scientists knew fossils found in rock layers could not have come from animals that lived a mere 6,000 years ago. But the only method used to date rocks at the time was relative dating, in which the age of a rock was estimated based on how deep it was beneath the surface. The verification that some rocks were hundreds of millions of years old didn't come until the turn of the 20th century, when French physicist Antoine Becquerel (1852–1908) discovered radioactivity. Scientists could then use radioactive techniques to determine the decay rate of certain isotopes within rocks (absolute dating), allowing more precise measurements of overall rock ages and determining a more accurate age of the Earth. (For more information about relative and absolute dating, see "Fossils in the Rocks.")

How **old** is the **Earth**?

Thanks to meteorites from space, rocks brought back by the Apollo astronauts from the Moon, and sundry other long-distance readings (mostly from satellites) taken of planetary bodies throughout the solar system, scientists have been able to calculate the age of the Earth. They believe the planets, including the Earth, formed between 4.54 to 4.58 billion years ago. In general, most scientists say that the Earth formed somewhere in between—about 4.55 to 4.56 billion years ago. (For more information about the Earth's age, see "The Earth in Space.")

The reason for the reliance on other space bodies to determine the Earth's age is simple: The movement of the lithospheric plates around our planet has recycled and destroyed the Earth's oldest rocks. If there are any primordial rocks left on Earth, they have yet to be discovered. Therefore, scientists must use other means to infer the age of our planet, including the absolute dating of planetary rocks that probably formed at the same time as the Earth.

What are some of the **oldest rocks** so far discovered on Earth?

Scientists have found rocks exceeding 3.5 billion years of age on all the Earth's continents. But the oldest rocks uncovered so far are the Acasta Gneisses in north-

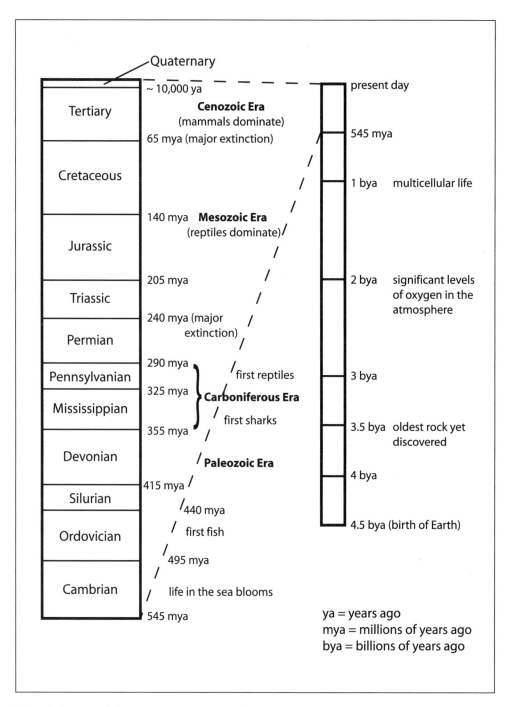

The geologic time scale (times are approximations only).

western Canada near Great Slave Lake, which has been dated at about 4.03 billion years old. Others that are not as old include the Isua Supracrustal rocks in West Greenland (3.7 to 3.8 billion years old), rocks from the Minnesota River Valley and northern Michigan (3.5 to 3.7 billion years old), rocks in Swaziland (3.4 to 3.5 billion years old), and rocks from western Australia (3.4 to 3.6 billion years old). These ancient rocks are mostly from lava flows and shallow water sedimentary processes. This seems to indicate that they were not from the original crust, but formed afterward.

The oldest materials found on Earth to date are tiny, single zircon crystals uncovered in younger sedimentary layers of rock. These crystals, found in western Australia, have been dated at 4.3 billion years old, but the source of the crystals has not yet been discovered.

How is **geologic time** divided?

The geologic time scale is primarily divided into geochronological units. The following lists the longest to the shortest divisions:

Eon—On most geologic time scales, eons are the longest time intervals represented.

Era—One of the longest eras was the Precambrian Era, meaning "before old life," a time that shows little fossil evidence. Thus, when talking about eras, most scientists refer to the eras of the Phanerozoic, a time when life became more abundant on the Earth. There are three divisions: the Paleozoic, Mesozoic, and Cenozoic Eras.

Period —Each era is broken down into periods. Most of the division names are from the geographic localities in which the period was first identified, Latin roots for a deposit's characteristic, or even ancient peoples who lived nearby.

Epoch—Some periods are further broken down into epochs. For example, the Tertiary Period (Cenozoic Era) is broken down into the Paleocene, Eocene, Oligocene, Miocene, and Pliocene Epochs.

Age—This unit (although not used in most general geologic time scales) is even shorter than the epoch, with most epochs divided into ages that usually number in the thousands of years only. The term "age" is not to be confused with the nicknames for certain geologic times. For example, the "Age of the Dinosaurs" is often the nickname for the Paleozoic Era.

Chron—Chron is an even shorter unit of time than age and is not used as frequently as the larger units on the geologic time scale. It is usually used to describe a local division of rock in an area.

What was the **Precambrian**?

Most sources list the Precambrian as an era; others just refer to it as "Precambrian Time." In other words, like many other parts of the geologic time scale, there are disagreements among charts.

Most charts do agree that the Precambrian is broken down into several other divisions. In some countries, it is divided into the following: Hadean (after Hades, with no rock record so far discovered); Archean (meaning ancient—it contains little evidence of life, and the Earth conditions were very dissimilar to today's planet); and Proterozoic (meaning early life—a time when multicellular organisms started to appear as fossils and conditions on Earth were becoming more similar to today's). Other charts divide the Precambrian into the Priscoan (oldest), Archean, and Proterozoic, while still other scales mention merely the Archean and Proterozoic.

But there is one thing everyone seems to agree upon so far: The Precambrian included about 80 percent of Earth's history, lasting from about 4.56 billion years ago to about 545 million years ago. During this time, the most significant Earth events occurred, including the formation of the Earth, the beginnings of life, the first movement of tectonic plates, the formation of eukaryotic cells, and the enrichment of the atmosphere with oxygen. Just before the Precambrian ended, multicellular organisms evolved, including those that eventually produced the first plants and animals.

How are **divisions** of geologic time **determined**?

The first geologic time scales were based on the natural breaks within the rock layers, which were thought to be evidence of worldwide mountain building events. Scientists soon discovered that not all mountain building events affected the entire Earth but were instead usually limited to a single continent (or a part of a continent) during a specific interval of time.

Today, the majority of divisions are most often delineated by a major geologic or life form event. For example, at the end of the Permian Period (at the beginning of the Triassic Period), a major catastrophic event occurred on Earth, wiping out close to 90 percent of all species on land and in water. This also is the demarcation between the Paleozoic and Mesozoic Eras. The smaller divisions (most often epochs) are usually divided by more "minor" changes in life forms or events. For example, the division between the Pleistocene and Holocene Epochs (Quaternary Period) signaled the end of the ice ages about 10,000 years ago.

How are the **divisions** on the geologic time scale **named**?

Most charts refer to four major eras, basing the names on fossils associated with the rock strata. The name Precambrian means "before Cambrian"; the Paleozoic refers to "ancient (or old) life," the Mesozoic to "middle life," and the Cenozoic to "modern (or recent) life." The names of most eons and eras end in "zoic" because the time intervals are usually based on animal life. For example, "paleo" means ancient and "zoic" means life—thus, paleozoic means "ancient life."

Smaller divisions on the geologic time scale are named after places in which the rocks were found or after ancient peoples endemic to the area. For example, the Cambrian Period is named after *Cambria,* the Roman name for Wales, because the sandstones and shales from this period were first described in North Wales; the Ordovician Period is from *Ordovices,* a Celtic people who lived in northwestern Wales where the rocks were first studied; the Devonian Period was first studied in Devonshire, England.

Are all geologic **time scales** around the world given the **same names**?

No, not all geologic time scales are the same. In fact, that is what makes determining the age or name of a division so difficult from country to country, especially when naming the periods, epochs, and ages. The confusion stems from naming time periods for local geographical, geological and/or paleontological reasons.

Besides geochronologically, what **other ways** do geologists **distinguish rock layers**?

Scientists not only label rock layers for geochonological reasons (according to time units), but also based on the layers' chronostratigraphy (types of rocks formed during each time unit) and lithostratigraphy (specific rock units):

> *Chronostratographic units*—The breakdown of chronostratographic units is as follows: *Erathem* represent rocks of a particular era; *system* represents rocks of a certain period; *series* represents rocks of a particular epoch; *stage* represents rocks of a certain age; and, finally, *zone* or *chronozone* represents rocks of a particular chron. To illustrate, fossils from the Cambrian Period are found in rocks of the Cambrian System. In addition, because geochronological units are time units, geologists often divide them into early, middle, and late (and combinations of these). However, chronostratigraphic units repre-

sent a geological column of a given time period and are divided into lower, middle, and upper. For example, during the Early Cambrian, rocks of the Lower Cambrian System were deposited.

Lithostratigraphic units—Lithostratigraphic units are divided into supergroup, group, formation, and member. The formation is a rock unit distinct enough that its extent can be marked on a map with a scale of 1:25,000 (or one inch on the map is equal to 25,000 inches on the ground). Formations are usually named after local towns, ancient peoples, or representative fossils within the rock layer, and seem to be a way to best communicate certain rock layers between scientists. For example, some of the most famous formations in the world include the Morrison Formation in Colorado, which contains abundant dinosaur fossils, and the Stephen Formation in Canada, which includes the famous Burgess shale fossils.

OUT IN THE FIELD

What is a **topographic map**?

A topographic map (also called a "topo" map) is a field map that represents a scale model of part of the Earth's surface. Using special symbols and lines, it shows the three-dimensional shapes of the surface using two dimensions. Topo maps are used a great deal by geologists in the field, primarily to gather information about features, map certain interesting rock areas, and to generally get around rougher terrain.

What are **contour lines** on topographic maps?

The most important features on topographic maps are contour lines: the brown lines of differing widths that represent points of equal elevation. These lines symbolize the shape of the Earth's surface, with each line representing the land as if it were sliced by a horizontal plane at a particular elevation above sea level. Thicker contour lines called index contours—usually shown as every fifth contour line—make it easier for the user to determine elevations. Contour lines that are very close together represent steep slopes, while widely spaced contours—or no contours at all—represent relatively level ground.

The contour lines also represent a distance called the contour interval, or the difference in elevation represented by adjacent contour lines. Each map has a different contour interval listed on the map's legend. For example, a relatively flat area may have a contour interval of 10 feet (3 meters) or less, meaning that the difference between each contour line will be 10 feet (3 meters) up or down in elevation; a mountainous area may have a contour interval of 100 feet (30 meters) or more.

What are some of the **major rules for contour lines** on a map?

There are several major rules contour lines must follow. In particular, contour lines must not cross (except in the rare case of an overhanging cliff); spacing represents either a very steep slope (lines close together) or wide plains (lines wide apart); a hill is represented by a series of closed contour lines "stacked" on one another like a lopsided bull's eye (depressions are the same, but contain hatch marks within the closed contour lines—or the downhill side); contour lines must never diverge; and when contour lines cross stream or river valleys, they must form Vs that point upstream.

What are **bathymetric contours**?

Bathymetric contours are similar to regular contours, except they depict the elevations, shape, and slope of marine features offshore (usually the bottom floors of bays, seas, and oceans). These contours are in black or blue, and they are usually written in meters at various intervals, depending on the map scale. They should not be confused with maps that depict depth curves, which usually represent water depths along coastlines and inland bodies of water. The contours of these maps are usually show in blue, with the data coming from hydrographic charts and depth soundings.

How do you **determine the scale** of a **topographic map**?

A topographic map's scale—no matter which scale is used—represents the horizontal distances on the map (not elevation distances, which are shown by contour lines). Similar to a street or highway map, the scale can vary widely, depending on the map.

But the topographic map's scale differs in a major way, by allowing the easy interpretation of each map's distances. Topographic maps are notorious for using different scales, depending on how much detail is desired. Each scale comes with a map ratio. For example, a map with a scale of 1:25,000 means one inch on the map is equal to

25,000 inches on the ground. And because both numbers use the same units, it can also be interpreted as any unit measure. For example, the same map could also be interpreted as 1 centimeter equals 25,000 centimeters on the ground. For those who prefer to measure in miles and kilometers, most topographic maps also offer a graphic scale in the legend.

What is a **topographic profile**?

A topographic profile is used to draw a cross-section (or "side view") of a part of the Earth's surface. You can make a topo profile by selecting any area you want to profile on a topographic map; draw a line from point A to point B; lay down a strip of paper along the profile line; then mark the paper where each contour line, hill crest, and valley low intersects while at the same time labeling each mark with the corresponding elevation.

Next, take a piece of grid-lined paper. On the vertical axis, draw in elevation units from the lowest to highest elevations from the profile line; place the strip of paper that marked the contour intervals along the horizontal axis; and project each topographic tickmark upward to its proper elevation. Finally, connect all the points of elevation, and this will reveal the profile of that area from the topographic map.

What is a **geologic map**?

A geologic map is actually a form of topographic map, but in this case it shows the type of sediment or rock outcrops exposed at the Earth's surface, along with the contour lines. The information on these maps can range from the rock type and age to the orientation of rock layers and major (and sometimes minor) geologic features.

Who uses these maps? Most geologists involved in almost every phase of field geology use geologic maps. For example, petrologists use these maps to determine the location of possible economic resources, such as metal ores, water, or oil. Geomorphologists use such maps to detect potential hazards in various areas, such as areas prone to earthquakes, flooding, or landslides. Occasionally, geologic profiles are also provided on these maps to help scientists understand, for example, the rock underlying an area.

How do geologists use **strike and dip** while in the field?

Strike and dip are not baseball terms; rather, they are used by geologists in the field to determine how rock layers and/or outcrops lean (or don't lean) in certain directions. Both are very useful to geologists as they map rock outcrops and geologic features.

Dip is the angle at which a layer or rock is inclined from the horizontal. It is usually measured with a clinometer. This instrument contains a straightedge that is lined up against the dip of the rock; a weight is used to measure the angle. Strike is the opposite—a line that a dipping rock layer makes with the horizontal (one way of visu-

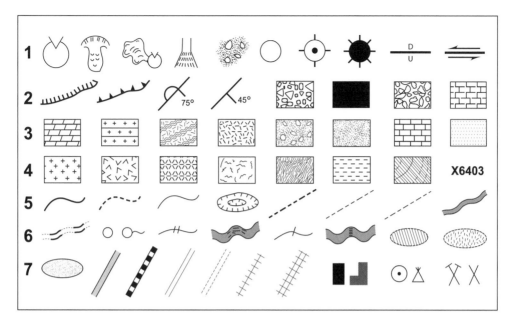

Some commonly used topographical symbols (from left to right): **Row 1:** volcano; landslide; lava flow; alluvial fan; sand dunes; water well; dry hole; oil and gas well; fault (down and up sides); fault (lateral displacement). **Row 2:** normal fault; thrust fault; overturned bed; dip and strike; breccia; coal; conglomerate; chalk. **Row 3:** dolomite; marble; gneiss; granite; gravel; sand; limestone; sandstone. **Row 4:** massive igneous rock; massive igneous rock (alternative symbol); mica; quartz; schist; shale; slate; spot elevation. **Row 5;** index contour; supplementary contour; intermediate contour; depression contours; national boundary; state boundary; county or parish boundary; perennial streams. **Row 6:** intermittent streams; water well and spring; small rapids; big rapids; small falls; big falls; intermittent lakes; glacier. **Row 7:** dry lake bed; primary highway; secondary highway; light-duty road; unimproved road; single-track railroad; multiple-track railroad; buildings; landmark and windmill; quarry and prospect.

alizing it is to think how a waterline would form if the rock layer dipped into a lake). Geologists often use a compass to measure strike.

What are some **common map symbols** found on topographic maps?

There are numerous map symbols found on topographic maps—far too many to mention here. The above illustration provides examples of several common topographical symbols.

THE EARTH IN SPACE

THE EARTH AND THE UNIVERSE

Where is the **Earth in relation to the universe**?

The Earth is merely a small speck in the universe. Starting from big to small, the Earth is thought to be some 15 to 20 billion light-years from the theoretical "edge" of the universe, in Orion's (or the local) spiral arm of the Milky Way Galaxy, and third in orbit from our solar system's sun. (The sun is about 24,000 light-years from the galactic center; a light-year is 5.88 trillion miles [9.46 trillion kilometers].) As a part of the Milky Way Galaxy, the Earth is accelerating outward toward the outer regions of the universe (or toward its border if it is truly finite); the Earth and the other members of the solar system are orbiting the galaxy at about 140 miles (225 kilometers) per hour.

The Earth takes one standard year to travel around the sun traveling faster at aphelion (the farthest distance in the Earth's orbit from the sun) than at perihelion (when the Earth is closest to the sun). As it travels, the Earth and other planets oscillate back and forth with the sun across the galactic plane. Because of the overall motions of the galaxy and solar system, the Earth's orbit can be thought of as a perpetual spiral, never quite visiting the same place twice in space.

What is the **big bang theory** of the universe?

When it comes to the formation of the universe, theories abound. The most well-known is the big bang (often capitalized as the Big Bang). This theory claims that about 10 to 20 billion years ago, a big bang, or explosion, occurred, creating the universe. (Not all scientists believe the universe started with a bang, but just started; astronomer Fred Hoyle despairingly gave the theory its nickname.) The universe before the explosion was confined to a singularity—essentially a spot compressed into the confines of an atomic nucleus.

Simply put, the big bang created matter, energy, space, and time. Quantum theory suggests that just after the explosion, at 10^{-43} second, the four forces of nature—strong nuclear, weak nuclear, electromagnetic, and gravity—were combined as a single "super force." Elementary particles called quarks then began to bond in trios, forming photons, positrons, and neutrinos—all created along with their antiparticles. At this stage, there were also tiny amounts of protons and neutrons, about one for every billion photons, neutrinos, or electrons that existed in the universe. As they reached out into the expanding universe, certain particles and antiparticles collided, annihilating each other. The remaining particles are what constitute our universe today.

Is **Earth** in any **danger as it travels** through space?

Dangers do exist as Earth travels through space. But in most cases, chances of encountering one of these "cosmic pests" is very low. For example, as we travel through interstellar space, close encounters with thick dust clouds and other stars could cause damaging changes (due to gravity) in the orbits of the solar system planets, including the Earth.

Dangers also lurk closer to Earth. The better-known ones include asteroids and comets, both of which have crashed into the Earth's surface and oceans in the past, and could in the future. (For more on asteroids and comets, see below.) Also dangerous are coronal mass ejections (CMEs)—eruptions from the sun's surface that often send energetic particles toward the Earth. If intense enough, the particles heat up and expand our atmosphere, affecting satellites in orbit and electronics on land; some are even powerful enough to disrupt communications and cause power failures. In fact, the more powerful CMEs have also been known to disrupt magnetic compasses. (This is not the same effect as the magnetic pole reversals that have occurred throughout Earth's history.)

FORMING THE SOLAR SYSTEM

How did the **solar system form**?

Once upon a time, about 15 billion years ago or so, there was a collection of dust, gases, and sundry space chunks along a spiral arm of the Milky Way Galaxy. This tendril of starstuff—through the effect of something big such as a supernova explosion—gradually condensed, spun, and formed one main star and its surrounding planetary bodies. This included our world, the Earth.

Although there have been numerous solar system formation theories, the solar nebula theory is the best known. A more detailed step-by-step description follows:

1. The solar nebula cloud was somehow disturbed and collapsed under its own gravity.

2. As the cloud collapsed, it compressed and heated up at its center—enough for dust to vaporize—in a process that takes close to 100,000 years, a relatively

How old is the universe?

No one truly knows the definite age of the universe, but some scientists believe data from a NASA satellite may give a close answer. In 2003, scientists analyzed the data from the Wilkinson Microwave Anisotropy Probe (WMAP) to capture a new cosmic portrait—an image that revealed the afterglow of the big bang in the form of background cosmic microwaves. From this data, they determined the universe is about 13.7 billion years old (the start of the big bang), give or take 1 percent. They also discovered that the first stars only started forming about 200 million years after the big bang—much earlier than previously thought.

short period of time, geologically speaking. This compression led to a protostar (the precursor to our own sun). The gas within the nebula added mass to the beginning star—that is, to the thick disk of gas orbiting the center (called an "accretion disk").

3. At this point, the gases far from the protostar began to cool. Scientists believe the gas compressed under its own gravity, cooling enough to condense into tiny particles of dust. By about 4.55 to 4.56 billion years ago, metals condensed; rock formed a bit later, about 4.4 to 4.55 billion years ago.

4. Dust particles began to accrete, which means they collided to form larger objects the size of boulders or small asteroids. When large enough, the gravitational pull of the larger accretion objects pulled even more dust and debris together. The size of these early objects depended on the distance from the protostar—smaller ones were closer to the center, larger ones in the outer solar system. This process created what are called planetesimals, which took a few hundred to about 20 million years to form, depending on their distance from the protostar.

5. About 1 million years after the nebula cooled, the protosun released a strong solar wind, which cleared away all of the gas left in the early nebula. The inner solar system planetesimals and smaller bodies lost most of their gases, becoming rocky or icy. The larger planetesimals of the outer solar system retained most gases due to their immense gravity. As these bodies continued to orbit the protosun, they also continued to collide with each other, becoming more massive. Some 10 to 100 million years later, the "winners"—including the Earth—remained in the form of about ten or so planets orbiting the new sun.

How was our solar system's **solar nebula disturbed**?

There are many theories as to why the initial solar nebula contracted and eventually formed the sun and planets. The simplest theory states that an overly dense part of the dust and gas cloud started to collapse under the force of its own gravity. Still other theories point to the nebula collapsing after being perturbed by a star passing close to

What would happen if Earth traveled near a gamma ray burst?

Gamma ray bursts are thought to be the most brilliant and powerful explosions in the universe since the big bang. If one exploded only a couple of hundred light-years from our planet, scientists theorize it would have disastrous effects. At first, the sky would be dazzlingly bright as the explosion heated up the atmosphere for several days. Then the gamma rays would arrive, stopping about 15 miles (24 kilometers) above the surface and creating smog as the rays destroyed our ozone layer. The smog would darken the Earth, creating a bone-chilling cold. Since gamma rays from such a burst come complete with cosmic rays, the various types of accompanying radiation would damage the tissues of all organisms on the planet, including, of course, humans.

the nebula, by the shockwaves from a nearby supernova (exploding star), or by a massive dust cloud passing the nebula.

Do all **solar systems form the way ours did**?

The theory of how our solar system formed is very general. In fact, there is controversy over the true origin of the solar system since the discovery of dozens of extrasolar planets. Nearly all of the known extrasolar planets—over 100—around other stars travel in highly eccentric orbits. Even those with super-Jupiter masses are found orbiting improbably close to their host stars. These findings have already forced scientists to invent new theories, including the migration theory, which states that large planets form at greater distances than small ones, with gravity eventually pulling the planets toward the center of the protostar's disk.

Discrepancies may be due to a bias about our own solar system (maybe it is different from other solar systems); observations (strange solar systems may be easier to detect from Earth than those similar to our own system); or flaws in the solar nebula theory. Additional studies of extrasolar planets may one day point to—or verify—the true story of our system.

But our solar system may not be a cosmic oddball after all: In 2002, astronomers found a planet orbiting the star 55 Cancri—a star similar to our sun in age and size. The best part: The newly discovered planet resembled Jupiter's mass, and most importantly, had an orbit similar to Jupiter's orbit.

Where is the **Earth located** within our solar system?

The Earth is the third planet from the sun, orbiting between the planets Venus and Mars. The Earth's path follows the plane of the solar system (called the ecliptic) and takes about 365.26 days to travel around the sun.

Are planets forming elsewhere in the universe today?

Yes, scientists realize that our solar system is not alone in the universe—and we finally have evidence. These extrasolar planetary discoveries have sprung from close to 20 years of observations with ground-based telescopes of the nearest 1,200 sun-like stars. The astronomers cannot actually see the planets through their telescopes; they infer the presence of the planets by observing the tiny wobbling motions of the star caused by the planets' gravitational tuggings.

THE EARTH'S EARLY CRUST

How **long ago** did the **Earth's crust finally solidify**?

When asked the age of the Earth, most scientists answer about 4.55 to 4.56 billion years old. This is usually defined as the time that the Earth and other bodies of the inner solar system formed.

But the Earth's crust took much longer to form. (For more about the Earth's modern crust, see "The Earth's Layers.") About 4.55 to 4.6 billion years ago, the planets were quickly sweeping up the "leftover" material in their orbits. This lasted about 1 billion years, during a time called the Hadean Period. The hot underworld is an apt name for this period, as the Earth's surface was periodically heated, broiled, and punched by incoming space material. Large asteroids would vaporize on contact, becoming part of the surface as they impacted the Earth, creating an explosion of searing plasma that would reach around the globe. There were also volcanic eruptions, as molten magma easily ripped through the nonexistent crust. Because of all this activity, the Earth's crust could not cool down. Based on the oldest rocks found on Earth, it is estimated that the crust finally cooled sufficiently to be called a "crust" about 3.8 billion years ago—about the time bombardment of the inner solar system quieted down.

Who **calculated Earth's age** first?

Many scientists tried to calculate the age of the Earth during the 18th and 19th centuries. For example, in 1779, French naturalist Comte de Georges Louis Leclerc Buffon (1707–1788) suggested that the Earth was about 75,000 years old. This was contrary to the commonly held belief that was based on the Bible, which concluded that our planet was only 6,000 years old. Lord Kelvin (British physicist William Thomson [1824–1907]) proposed that the Earth was 100 million years old based on his theories about our planet's temperature. Both theories proved to be incorrect.

Physicist and chemist Bertram Boltwood (1870–1927) developed a new radioactive dating technique by 1907. His method determined that a certain mineral was

This image, known as the "Blue Marble," was taken by the *Apollo 17* mission, which launched on December 7, 1972. Astronauts had the first chance to get the perfect shot of Earth when hours after lift-off, the spacecraft aligned with the Earth and sun, allowing the crew to photograph Earth in full light for the first time. *AP/Wide World Photos.*

about 4.1 billion years old. Although the mineral was later found to be only about 265 million years old, his methods led the way to better radioactive dating techniques. Today, based on radioactive dating analysis of Earth (and Moon rocks as comparisons), scientists believe our planet is about 4.55 to 4.56 billion years old.

What was the **late heavy bombardment**?

The late heavy bombardment occurred around 3.8 to 4 billion years ago when huge chunks of rock crashed into the Earth, Moon, Venus, Mercury, and Mars. The telltale signs of this massive bombardment are seen everywhere on the Moon and Mercury. Venus shows little evidence of the bombardment today because its thick atmosphere eroded any trace of the event. On Earth, erosion by wind, water, and ice—and the movement of the continents over time—has erased any evidence of the bombardment.

Have there been **more recent asteroids or comet impacts on Earth** since the late heavy bombardment?

Yes, traces of impacts exist on every continent, except Antarctica. Currently, scientists have identified over 150 impact craters on Earth. The majority of known impact craters are found on the surface, but fewer than a dozen have also been found beneath the surface and on the ocean floor. Because of erosion by wind and water and deposition of sediment, we will never truly know the number of impacts that have occurred on Earth. (For more information on impact craters, see "Geology and the Solar System.")

THE EARTH AND THE SUN

Is our **sun an average star**?

Yes, the sun is considered an average star. It has a much lower temperature than, for example, Rigel, a blue-white star, and yet a much higher temperature than Antares, a reddish-orange star. It is one of 100 billion stars in the Milky Way Galaxy. It comprises 99 percent of all our solar system's mass and has a diameter of 864,950 miles (1,392,000 kilometers).

> ## How thick is the Earth's crust?
>
> The Earth's crust varies in thickness. It is between 3 to 6.8 miles (5 to 11 kilometers) thick under the oceans, and about 12 to 40 miles (19 to 64 kilometers) thick under the continents. (For more information on the Earth's structure, including the crust, see "Through the Earth's Layers.")

How **fast** is the **sun moving through space**?

Since the sun (and its solar system) is moving through three dimensional space, it is traveling in several directions at once. For example, the sun is moving toward the constellation of Hercules (or toward the region of Lambda Herculis) at 12 miles (20 kilometers) per second—or 45,000 miles (72,000 kilometers) per hour. At the same time, the sun is orbiting around the galactic center at about 140 miles (225 kilometers) per second.

The sun is also moving upwards, out of the plane of the Milky Way Galaxy, at a speed of 4.4 miles (7 kilometers) per second. Currently, our sun is moving outward from about 50 light-years above the galactic plane, but scientists believe the gravitational pull of the Milky Way stars will slow the sun's escape in about 14 million years. Eventually, the sun will be pulled back into the galactic plane and then drift below it, oscillating back and forth across the plane until its demise.

What is the **sun's distance** from the **center of the Milky Way Galaxy**?

The sun is about 26,000 light-years (a light-year is about 5.88 trillion miles [9.46 trillion kilometers]) from the center of the Milky Way. This places it approximately at the inner edge of the Orion arm of the galaxy. The sun (and the rest of the solar system) takes about 225 million years to revolve around the galactic center, a period sometimes referred to as a cosmic or galactic year. For each orbit, the sun travels 150,000 light-years; it has completed about 20 orbits since the solar system formed.

How does the **sun release energy**?

Simply put, matter is compressed at the sun's core by gravity. The immense pressure triggers a nuclear fusion reaction in which protons of hydrogen nuclei combine into a stable helium nucleus and energy is released. Helium is less massive than the initial combination of the hydrogen, and the extra mass is released in the form of gamma ray energy.

The lifetime of the sun is thought to be about 10 billion years. Right now, the sun is about 4.5 billion years old. For all that time it has converted an average of 4 million tons of matter into energy every second. Scientists believe the sun will eventually run out of fuel, but not for another 5 to 6 billion years.

How much time does it take for the sun's light to reach Earth?

Though the sun is millions of times closer to us than any star in the universe, it is still an average of 93 million miles (149.7 million kilometers) from the Earth. Because light travels at a finite speed of about 186,411 miles (300,000 kilometers) per second, it takes almost 10 minutes for the light from the sun to reach us. In contrast, it takes just over 4 years for the light from the closest star (Proxima Centauri) to reach us, and it takes the most distant starlight billions of years to reaches us. When we look at distant stars, we see them as they were when the light left those stars, so we are actually seeing them as they appeared long ago.

How does the **sun affect the Earth**?

Even though the Earth intercepts only 0.002 percent of the sun's energy output, our planet's life is still sustained by the heat and light from our central star. Even that small output is enough to stimulate air currents around the equatorial regions, creating our changing weather systems. Some people think that the sun is also responsible for the changing seasons, but it is actually the tilt of the Earth's axis as it revolves around the sun that creates spring, summer, fall, and winter.

There are other solar affects, too. For example, the Earth's magnetic field is shaped by solar wind particles. These high-energy particles stream from the sun at more than a million miles (1.6 million kilometers) per hour. When they reach Earth, the planet's magnetic field deflects the particles, creating a protective cocoon called the magnetosphere. On Earth, especially at the North and South Poles, an influx of these particles results in shimmering curtains of light called the aurora borealis (northern lights) and aurora australis (southern lights). Occasionally, the sun also sends out huge geomagnetic storms called coronal mass ejections (CMEs) that can wreak havoc with low-Earth-orbit satellites, power lines and grids, and electromagnetic communications of all kinds.

Some scientists also believe the fluctuations in the sun's energy output over time might influence the Earth's climate, though this idea is still a matter of debate. Some even suggest the Earth's many cold (ice ages) and warm periods are the result of these changes in solar energy.

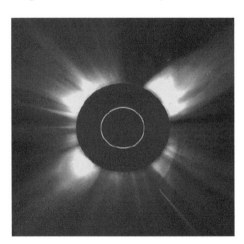

NASA's Solar and Heliospheric Observatory (SOHO) spacecraft, orbiting one million miles from Earth, captured this view of a comet (streak at lower right) as it plunged into the sun on October 23, 2001. *AP/Wide World Photos.*

A photographer prepares to take a picture of a partial solar eclipse observable from the eastern German town of Lebus in the early morning. Viewed from Germany, the sun was covered by the moon up to 84 percent. *AP/Wide World Photos.*

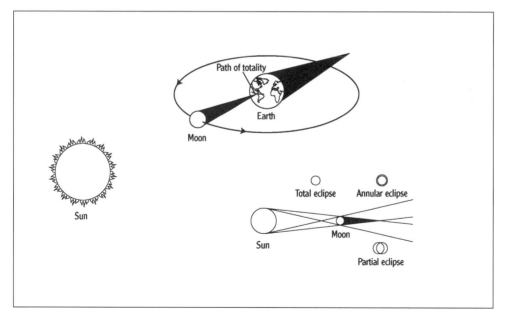

How a solar eclipse occurs.

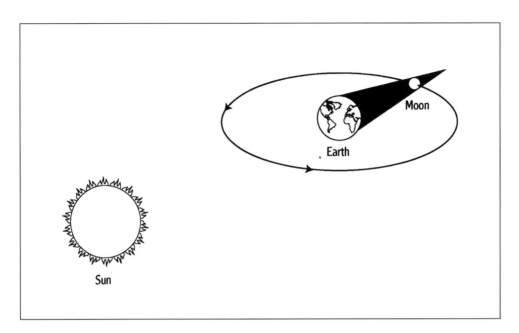

How a lunar eclipse occurs.

What causes **eclipses of the sun and Moon** as seen from Earth?

Because the Moon, sun, and Earth are in relatively the same plane (the solar system's ecliptic), the shadows and light of these space bodies often interact, creating eclipses. Eclipses do not occur each month, but only when the Moon's orbit intersects the shadow being cast by the Earth (lunar eclipse) or the Moon lines up directly with the sun at a point in its orbit when it is relatively close to the Earth (solar eclipse).

Here is a brief overview of the various eclipses as seen from Earth:

Total solar eclipse—This occurs when the Moon is in the new moon phase and comes between the Earth and the sun. The Moon's darkest shadow (umbra) blocks out the sun along a specific path on Earth called the "path of totality." This circular shadow is never more than 167 miles (269 kilometers) in diameter.

Annual solar eclipse—This occurs when the Moon is too far from the Earth to completely block the view of the sun. In this case, the sun's light is visible around the edges of the Moon, forming a bright ring.

Partial solar eclipse—Partial solar eclipses occur on the lighter, outer regions of the shadow (penumbra). In a partial eclipse only a part of the sun is covered by the Moon.

Total lunar eclipse—Total lunar eclipses occur when the Moon passes through the umbral shadow of the Earth at the full moon stage. At the peak of totality, when the entire Moon is covered by the Earth's umbral shadow, the Moon does not completely disappear. Instead, it will often be a deep coppery red, as some light is bent as it passes through the Earth's atmosphere (also called earthshine).

Partial lunar eclipse—This occurs when the Moon passes through the lighter penumbral shadow of the Earth; the Moon merely looks dimmer during this type of eclipse.

What **total solar eclipse feature** is caused by the **geology of the Moon**?

During a total solar eclipse, when the Moon comes in between the sun and the Earth, a bright ring resembling a string of beads is often seen just as the eclipse reaches totality. This is called "Bailey's beads," formed as the last vestiges of sunlight shine in between the rugged craters and mountains of the Moon's edge.

MOTHER EARTH, DAUGHTER MOON

What causes the appearance of the **"Man in the Moon"** we see on the **Moon's surface**?

People looking at the Moon over the centuries have said its face resembles everything from a man's face to a rabbit. The reason for such whimsical images are the light and dark regions caused by the satellite's geographical features. The large, dark areas of our Moon, which Italian astronomer Galileo Galilei called *mare,* or "seas" (maria is the singular), and light-colored highlands are made of chemically and mineralogically different rocks.

Mare are actually dark volcanic lavas (basalts) that poured out into the huge basins from volcanic eruptions about 3.5 billion years ago. The dark, relatively lightly cratered mare cover about 16 percent of the lunar surface. They are concentrated on the nearside of the Moon (the side facing the Earth), mostly within impact basins formed long ago. The relatively bright, heavily cratered highlands are called *terrae.* The craters and basins in the highlands were formed by meteorite impacts; and they are thought to be about 4 billion years old. (Scientists believe older parts of the Moon are more heavily cratered; therefore, the highlands are older than mare.)

The light and dark regions on the moon's surface, partially due to light and shadow, are made of chemically and mineralogically different rocks, also contributing to the contrast visible from Earth. *AP/Wide World Photos.*

45

Is the **Moon's composition** similar to Earth's?

Yes, the Moon's composition is similar to the Earth's, but only with regard to its mantle. (See "The Earth's Layers" for a definition of the Earth's mantle). Thanks to the Apollo manned missions to the Moon, scientists have analyzed about 882 pounds (400 kilograms) of lunar rock and soil. (Depending on the reference, the numbers range from 840 pounds (381 kilograms) to 842 pounds (382 kilograms.) The following table lists the general composition of lunar rocks:

Chemical Composition	Percent
Silicon dioxide	43
Iron oxide	16
Aluminum oxide	13
Calcium oxide	12
Magnesium oxide	8
Titanium dioxide	7
other elements	1

Does the **Moon** have an **atmosphere**?

No, the Moon has no atmosphere—at least, not an "atmosphere" similar to the one on Earth. Although the space just above the lunar surface is not a total vacuum, the lunar atmosphere is only about 10^{-15} as dense as the Earth's. Data from the *Apollo 17* manned lunar mission in the late 1970s identified helium and argon atoms at the Moon's surface, while Earth-based observations added sodium and potassium ions to the list by 1988.

For many planets with thick atmospheres, the exosphere (the tenuous part of the atmosphere beyond the ionosphere that blends into space) is miles thick. For example, the Earth's exosphere starts at about 298 miles (480 kilometers) above the surface; for the Moon, the exosphere begins directly above the lunar surface.

Though the Moon does not have a "true" atmosphere like our Earth, there is evidence of something just as interesting: Data from two spacecraft that visited the Moon—the *Clementine* and *Lunar Prospector*—suggest that there may be water ice in some deep craters at the Moon's south and north poles. This ice was detected in the permanently shaded areas of craters.

Are the **Moon's layers** similar to the Earth's?

No, the Moon's interior layers differ greatly from the Earth's interior. (See "The Earth's Layers" for more on the Earth's interior). Unlike the Earth's thin crust, moving mantle, semi-solid outer core, and solid inner core, the Moon is almost completely solid. Only a small amount of material near the Moon's core might still be hot enough to be molten rock. In fact, the Moon probably has a thick metallic core, a solid mantle of dense volcanic rocks, and a much thicker crust than the Earth that is made up of a layer of basalt and anorthositic gabbro on top (because the Moon's rocks are rich in a particular variety of plagioclase feldspar called anorthite, lunar rocks are often referred to as anorthositic). Capping the entire Moon's surface crust is a thin layer of regolith, a "soil" composed of shattered bedrock broken up by the action of impacting space bodies and the sun's radiation.

Is a **moonquake** similar to an **earthquake**?

Yes, a moonquake is similar to the minor tremors felt on Earth, though they are not as powerful as major earthquakes. Scientists know about moonquakes thanks to the *Apollo* manned missions to the Moon, where astronauts left behind quake monitors to keep track of any tremors. Astronomers on Earth can also observe regolith ("soil") that has slid down the Moon's slopes and is evidence of landslides there.

Each year, there are an average of about 3,000 moonquakes, many of them reverberating for more than an hour. Most of them are so small that you would hardly notice them if you were standing on the Moon. Moonquakes are detected more often when the Moon is closest to the Earth and our planet's pull on the Moon's crust is the strongest. Other smaller quakes might be caused by occasional meteoroid impacts on the Moon's surface.

What are some **early theories** on how our **Moon formed**?

Because the Moon can be readily seen by humans, its origin has been speculated on for centuries. Past theories include the capture theory, in which the Moon was captured by the Earth's gravity as it swung by our planet, and a theory that says the Moon formed in the same way as the Earth through the accretion of smaller space bodies.

Scientists also have speculated that the Moon actually came from the Earth. This theory was proposed long ago, especially by those who believed the "gaping hole" of the Pacific Ocean was created by the ripping away of material that then formed the Moon. More than a century ago, George H. Darwin proposed a mechanism to describe how this happened. He believed the Earth's rapid spin caused material to be thrown out into space, creating the Moon.

What is the **modern theory** of the **Moon origin**?

More recently, scientists proposed another explanation for the Moon's origin. In this theory, a huge, Mars-sized body struck the Earth during its early formation, ripping away part of our planet's mantle. (Rocks brought back by the *Apollo* astronauts show that they have characteristics similar to the Earth's mantle, which might mean something big was responsible for excavating rock deep within our planet.) The huge impact threw out material, forming a disk that orbited the Earth. Our planet's

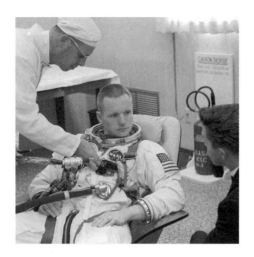

Astronaut Neil Armstrong, shown here during a suiting up exercise March 9, 1966, at Cape Kennedy, Florida, prepares for the *Gemini 8* flight. *AP/Wide World Photos.*

gravity held the material in orbit, with momentum eventually causing the molten material to spin and congeal to form the Moon.

Why are the **Earth and Moon** called a **double planetary system**?

Scientists often call the Earth-Moon pair a double planet. The Moon is close to the Earth at an average of about 238,857 miles (384,403 kilometers); it is also much smaller, with a diameter of 2,160 miles (3,476 kilometers). The Earth's diameter is about 7,926 miles (12,756.3 kilometers). But the reason for the double planet status is the Moon's rotation and revolution around the Earth: both are the same, at 27 days, 7 hours, and 43 minutes. This creates a synchronous rotation, allowing the Earth's gravity to keep one lunar hemisphere permanently turned toward Earth, or what we call the Moon's nearside.

The Earth-Moon system is not the only so-called double planet. The planet Pluto and its moon, Charon—the smaller moon orbiting the larger planet—are in synchronous orbit, too. Thus, there appear to be two "double planets" in our solar system.

What causes the **Moon's phases** as seen from the Earth?

The Moon's phases are caused by the movements of the Moon around the Earth. The Moon, as with the Earth, reflects the visible light from the sun. Thus, an observer on Earth can only see that portion of the Moon lit by the sun. As the Moon orbits Earth, different parts are illuminated, causing the Moon to pass through various phases. In order, the phases are: full moon (we see the entire face of the Moon); new gibbous moon; first quarter (we see the face of a half Moon); new crescent; new moon (we cannot see the Moon at all); old crescent; last quarter (we see the other half of the Moon opposite the first quarter); old gibbous; and then back to full moon.

Although the Moon revolves around the Earth in 27.32 Earth days, it still takes 29.5 Earth days to go through the entire phase cycle. This is because as the Moon is revolving around the Earth, the Earth is revolving around the sun about 1 degree per day. When the Moon completes one of its revolutions, the sun is no longer in the "same spot" in the sky, but in 27.32 Earth days it has moved 27 degrees. Moving at 13 degrees per day, the Moon takes about 2 days to "catch up" with the sun and get back to the "original" phase in the cycle.

Does the Earth have another moon?

Some scientists claim the Earth has another "moon," albeit it is much smaller than our main natural satellite. Discovered in 1986, this second moon is called Cruithne, also known as asteroid 3753, or 1986 TO. It takes an irregular horseshoe path around our planet as it is tossed about by the gravity of the Earth and Moon. At its closest approach to Earth, the 3-mile-wide (5-kilometer-wide) Cruithne was about 40 Moon lengths from our planet. This is equal to 0.1 Astronomical Units, in which 1 Astronomical Unit is equal to 93 million miles (149.7 million kilometers), or the average distance between the Earth and the sun. Currently, however, it does not come closer than 0.3 Astronomical Units.

In reality, Cruithne is not a moon because Earth and Cruithne are not gravitationally bound. Luna, the other name for our Moon, is a true moon because it is gravitationally bound to Earth. But Cruithne is locked into a 1 to 1 resonance with Earth, which may be why some scientists call this space body a moon. In other words, Cruinthe and Earth both orbit the sun at the same speed. In comparison, Neptune and Pluto have a 3 to 2 resonance, with Neptune orbiting the sun 3 times in the time it takes Pluto to orbit twice.

Does the **Moon** have any **effect on the Earth**?

Yes, the Moon does have an effect on the Earth. In particular, the Moon is responsible for our planet's ocean tides. (For more information about the Earth's tides and the Moon, see "Geology and the Oceans.")

Who are the **astronauts** who have **landed on the Moon**?

On July 20, 1969, Neil Armstrong became the first human to step onto the surface of the Moon; he was followed down the ladder of the lunar lander by Edwin "Buzz" Aldrin. Other astronauts followed: Pete Conrad and Alan Bean on *Apollo 12;* Edgar Mitchell and Alan Shepard on *Apollo 14* (the *Apollo 13* had a major explosion and had to turn back, so its astronauts never landed on the Moon); David Scott and James Irwin on *Apollo 15;* Charles Duke and John Young on *Apollo 16;* and the last two men to walk on the Moon, Eugene Cernan and Harrison Schmitt (the first geologist on the Moon) on the *Apollo 17*. Humans have not walked on the Moon since *Apollo 17*'s lunar lander left the surface of the Moon on December 14, 1972.

What did the moonwalkers see? Plenty of lunar geology, including tons of broken rock, boulders, dusty soil (the Moon's regolith), and impact craters. Since there is no atmosphere to obscure their view by diffracting light, they also saw a deep carpet of stars. The astronauts also experienced gravitational differences, as the moon's gravity is only one-sixth that of the Earth's. Thus, an astronaut who weighed 180 pounds on Earth weighed a mere 30 pounds on the Moon.

Are there any **more Earth moons**?

Astronomers are still searching for more natural satellites around the Earth, but it takes time to search the sky. They can also be fooled by space objects. For example, in September 2002, an amateur astronomer discovered what looked like the Earth's third natural satellite. The mysterious object, designated J002E3, was found to have a 50-day orbit around the Earth. Scientists took more observations of the strange satellite and determined the truth: By measuring the reflected light from the object, researchers discovered not a typical asteroid surface but colors consistent with the spectral properties of an object covered with white titanium oxide (TiO) paint. Scientists now believe the object was a discarded rocket booster. Perhaps it came from the *Apollo Saturn S-IVB*'s upper stages, all of which were painted with TiO paint.

THE EARTH AND SMALL SPACE DEBRIS

Does **Earth** come in **contact with meteoroids, meteors, and meteorites**?

Yes, the Earth does come in contact with these smaller space debris, as do all the planets. Meteoroids are particles of space rock that range from the size of a grain of sand to the size of a human fist. They usually enter our atmosphere by chance as they come close to our planet. As they fall, they incinerate at an altitude of 50 to 60 miles (80 to 97 kilometers) above the Earth's surface. The bright, fiery streak of light in the nighttime sky is called a meteor (or the more romantic term, "falling star"). A "meteor shower" happens when the Earth's orbital path crosses through a comet's debris tail, increasing the number of particles that enter our atmosphere and incinerate, literally creating a "shower" of meteors.

Most burned-up meteors result in meteoric dust. About 100 tons of dust fall onto the Earth every day. (That is close to 4,000 tons a year.) But if the space debris is any larger, it might make it through our thick atmosphere and strike the ground. These fallen space rocks are referred to as *meteorites*. Most meteorite pieces are thought to come from the asteroid belt; a small number are thought to come from the Moon or Mars.

How do scientists know **certain meteorites came from other planets**?

By knowing the general composition of a planet or asteroid—such as Mars, the second largest asteroid known as Vesta, or even Venus—scientists can speculate where other meteorites found on Earth originated. For example, while it is true that no human has ever been to Mars, much less collected rocks there, scientists speculate that several meteorites did come from Mars. They base this speculation on data from the *Viking Lander* spacecraft, which landed on the red planet in 1976. The lander took readings of the air, so scientists know the composition of the Martian atmosphere. Because

Leonid meteors are seen streaking across the sky over snow-capped Mt. Fuji, Japan's highest mountain, in this 7-minute exposure photo. *AP/Wide World Photos.*

some meteorites found on Earth contain traces of these gas, scientists believe the space rocks somehow made their way from the red planet to our planet.

What **types of meteorites** fall to the Earth?

The following is a list of the major types of meteorites that fall to Earth:

Chondrites—Chondrites represent the largest category of meteorites found on Earth. They are stony and contain chondrules, spheres of material around 0.03 inches (1 millimeter) in diameter that are melted together. Chondrites, which might represent the oldest material in the solar system, are broken into several groups: enstatite chondrites (very rare), ordinary chondrites, and carbonaceous chondrites (rare meteorites that contain carbon).

Achondrites—Achondrites are stony meteorites without chondrules. They are broken into several groups, including the howardites, eucrites, and diogenites (or HED meteorites that may have originated on the second largest asteroid, Vesta); the shergottites, nakhlites, and chassignites (or SNC meteorites that may have originated from Mars); and the lunar achondrites (thought to have come from the Moon).

Stony-Irons—Stony iron meteorites are a mix of iron-nickel alloy and non-metal materials called silicates. They are grouped into the pallasites and mesosiderites.

51

An enhanced photo of Asteroid Mathilde taken by the Near Earth Asteroid Rendezvous (NEAR) spacecraft on a 25-minute flyby. The asteroid is so battered that it is almost "all crater" with one pit big enough to swallow the District of Columbia and some of its suburbs. *AP/Wide World Photos.*

Irons—Iron meteorites are a iron-nickel alloy. The category is divided into several groups, including the hexahedrites and the octahedrites. A separate category of iron meteorites is broken down by its chemical classification.

What is the **largest meteorite** to land on the Earth?

To date, the largest meteorite is the Hoba, found in the vicinity of Grootfontein, Namibia, Africa. Discovered by J. Brits in 1920, the meteorite—thought to be between 200 and 400 million years old—fell to Earth about 80,000 years ago. It weighs 60 tons and measures about 10 by 9.3 feet (2.95 by 2.84 meters), with a thickness that varies between 30 and 48 inches (75 to 122 centimeters). The meteorite is 82.4 percent iron, 16.4 percent nickel, 0.76 percent cobalt, plus other trace elements.

Has anyone ever been **hit by a meteorite**?

Though it is difficult to verify in many instances, a number of people have claimed they were hit by a meteorite. But the odds of this happening are about a billion to one. As one scientist said, you would have a better chance winning the lottery. The following lists a few instances of such close encounters:

1. On November 30, 1954, housewife Ann Hodges of Sylacauga, Alabama was taking a nap when a 3-pound (1.4 kilogram) meteorite crashed through the roof, bounced off a radio, and hit her in the hip, leaving a bruise the size of a football.

2. As Jose Martin and his wife, Vicenta Cors, rode past Gatafe, Spain, on June 21, 1994, a 3-pound (1.4-kilogram) meteorite smashed their car's windshield on the driver's side. It ricocheted off the dashboard and bent the steering wheel, breaking the little finger on Jose's right hand. It then flew between the couple's heads and landed on the back seat.

3. In late August 2002, a young girl from Northallerton, England, was getting into the family car in mid-morning when she claimed a stone that looked like lava fell on her foot from the sky. (As of press time, this incident has not been confirmed.)

In two other instances, meteorites seemed to prefer one town: In 1971, Wethersfield, Connecticut, a meteorite weighing about 0.6 pound (272 grams) crashed through the roof of a house. The next morning, it was discovered sticking out of the ceiling by the residents. Then in 1982, just across town, another meteorite slammed through the roof of a different house. The space rock fell into the living room, bounced into the dining room, and landed under the table, much to the surprise of the homeowners, who were watching television.

Another fall was just as sensational. On October 9, 1992, the Peekskill meteorite broke up as it fell through the atmosphere, an event recorded by at least 16 independent videographers. One of the pieces announced its arrival by hitting an old parked Chevrolet automobile in suburban Peekskill, New York. No one was in the car at the time, but the owner heard the noise and arrived to find a warm, 26-pound (12-kilogram) meteorite under the vehicle. It was not a complete loss, however; the owner was able to sell the rather run-down automobile as a collector's item for tens of thousands of dollars.

Does the Earth hold any **evidence of interplanetary dust**?

Yes, interplanetary dust particles, which measure only about 0.00039 inches (0.001 centimeter) in diameter, have been discovered on Earth. This dust is not easy to analyze: The particles rain down on every part of our planet but are difficult to discern from the much more common ash, soot, and terrestrial dust created by humans and natural processes. In order to understand space dust more, scientists have collected the particles from a higher layer of the atmosphere (the stratosphere) using high-altitude weather balloons and aircraft.

Scientists still debate the origins of this dust. Though the most likely sources were thought to be asteroids or comets, observations made by spacecraft within the past few years now strongly support the idea that these dust particles are actually "space dust." Their chemical compositions are strongly suggestive of extraterrestrial origin, and they have a high abundance of helium, a major component of the solar wind.

RIGHT PLACE FOR LIFE

How did the **position of Earth** in the solar system aid in the **development of life** on our planet?

It is thought that the formation of most life on Earth depended on our proximity to the sun. Scientists believe that being in the "habitable zone"—an area that stretches from just outside Venus's orbit to Mars's orbit—was the perfect place to encourage life on Earth.

But there was another "helper": Astrophysicists believe Jupiter's location and gravitational force acted as a protective shield that (presumably) allowed life to take root and flourish on early Earth. This is because the gas giant deflected the heaviest asteroids and comets away from the inner solar system, putting a relative halt to bombardments, particularly in the "habitable zone" most suitable for life.

What were the **first living cells** on Earth?

The first living cells on Earth may have appeared between 3.5 and 2.8 billion years ago in the form of prokaryotes. (Scientists have yet to pinpoint the actual dates, as one-celled organisms are almost impossible to detect in ancient rock.) These most-primitive organisms still exist today and include bacteria and related organisms; they have cells that lack a nucleus and other membrane-bound internal structures. Eukaryotes cells, or cells with nuclei and membrane-bound internal structures (organelles), developed about 1.5 billion years ago. Eventually, these cells developed into multicellular organisms. (For more information on Earth's early life, see "Fossils in the Rocks.")

What **other types of life** can exist **without the sun's energy**?

There are several forms of life on Earth that can live without light—or even heat—from the sun. And if they can survive in such conditions on Earth, some scientists speculate life might have formed elsewhere in the solar system under similar conditions. Here are some Earth examples of sunless environments that are known to include life or that scientists believe might harbor life:

> *Caves*—Once thought to be lifeless, caves are now known to teem with life. Three kinds of cave animals are recognized: Those that come in from above ground (cave guests or trogloxenes), including bats and raccoons; those that can live in the cave or out (cave lovers or troglophiles), including cave salamanders and harvestmen ("daddy longlegs"); and organisms that spend their entire lives in the cave's total darkness (cave dwellers or troglobites), including the blind millipede and blind salamander. Most troglobites are sightless and without pigment. As for plants, mosses and ferns at the entrance zone give way to very little plant life, except for some mold and fungi. Bacteria usu-

What is the "universal ancestor" or "common ancestor" on Earth?

The "universal ancestor" or "common ancestor" is the first instance of life on Earth. It therefore had to survive in the Earth's early, harsh environment. This organism would have been anaerobic (needing no oxygen) because the Earth's atmosphere was mostly carbon dioxide at the time; it would have been hyperthermophilic (tolerant of high temperatures) and halophilic (tolerant of saline conditions); and it would have been a chemolithoautotroph, which means it could obtain its energy and carbon from inorganic sources, since there was no other true life yet on the planet.

Chemolithoautotrophic organisms are not from someone's overactive imagination. In the past several years, scientists have discovered such organisms in what is called "extreme environments," from hot sulfur springs on land to deep ocean hydrothermal vents. In fact, these environments mimic conditions on the early Earth: high temperatures, high sulfur and salt contents, and anaerobic surroundings completely independent from oxygen and sunlight.

ally enter caves via the water and are food for larger aquatic cavern animals such as flatworms and amphipods. They, in turn, are eaten by larger animals such as salamanders and crawfish.

Geothermal vents—Life in the deep oceans was discovered in the late-1970s around geothermal vents, places far from the sun where hot magma squeezes through cracks in the ocean floor. These organisms range from huge clams to long tube worms and live off bacteria found around the vents. In turn, the bacteria live off of the hot, hydrogen sulfide-rich waters around the vents. This is a form of chemosysnthesis in which organisms convert chemicals into food.

Meteorites—So far scientists have found carbon in meteorites, but no real signs of life (also see below about possible life in Martian meteorites). If life did exist in meteorites, in order for us to discover it the organism would have to survive the heat of reentry and the contamination from our own planet. This might be why it is so difficult to discover such life in these rocks.

Hot springs—In these hot spots, microbes exist without the benefit of the sun. Scientists know this because the microbes leave several distinctive signs of their presence, even if the microbes themselves can't be recognized. For example, microbial biofilm (or slime) is a good biomarker that is created by nearly all microbes. Not only does it resist weathering, but it is readily preserved by silica.

Under-ice lakes—Some of the most astounding possibilities for life in strange places are in the under-ice lakes of Antarctica. More than 70 lakes lie thousands

Voyager 1, an unmanned craft, and its sister, *Voyager 2,* carry gold records containing messages and pictures of Earth while the other side of the record is a diagram with "instructions" on how to read the messages. The voyager crafts have provided images of the farthest reaches of the Milky Way Galaxy. *AP/Wide World Photos.*

of feet under the continental ice sheet, including under the South Pole and Russia's Vostok station. Lake Vostok is one of the largest of these subglacial lakes, comparable in size and depth to Lake Ontario. Two separate investigations of ice drilled at Lake Vostok reveal some evidence that Vostok's waters may contain microbial life—bacteria that can survive thousands of feet below the ice sheet.

Does **extraterrestrial life** exist in our **solar system**?

To date, there has been no direct proof of life on other bodies in the solar system. but scientists are still looking and speculating about the possibilities. For example, scientists recently found strong evidence that beneath the surface of Jupiter's icy moon Europa might lie an ocean of liquid water, an essential ingredient for all living organisms as we understand them. Although the surface temperature of Europa is about –260°F (–170°C), some scientists believe this moon might have plenty of biological fuels useful for the basic chemical reactions necessary for life.

How? One theory surmises that charged particles from Jupiter constantly rain down on Europa, producing organic and oxidant molecules, such as formaldehyde, that are sufficient to fuel a substantial biosphere. Even though the ice is thought to be about 50 to 100 miles (80 to 170 kilometers) thick, cracks could allow the molecules to seep through to microbes below. In fact, on Earth there are certain bacteria—the tiny *Hyphomicrobium*—that survive on formaldehyde as their only carbon source. That Europa might be teeming with these tiny organisms is only theory, but if proven true it would show we are not alone in the universe.

Was there **evidence of life** in a **Martian meteorite**?

In 1994, scientists looking for meteorites near Allan Hills, Antarctica, found a space rock that has "rocked" the astronomical world: ALH 84001, which is thought to be a Martian meteorite. By 1996, several researchers had examined the rock and discovered what they believe to be evidence that life existed in the meteorite. Taking into consideration the discovery that more water exists on the red planet than previously thought and the rock evidence, the scientists believed they had found evidence of extraterres-

Could aliens theoretically discover that Earth exists?

If alien beings do exist, and they meet up with four craft sent by humans—*Pioneer 10, Pioneer 11, Voyager 1,* and *Voyager 2*—there is a chance they will know of our existence. All of these spacecraft are carrying plaques giving other space travelers an idea of who we are and our location in the solar system. But it might take a long time.

The *Pioneer* craft carry 6-by-9-inch (152-by-229-millimeter) gold-anodized, aluminum plates complete with pictures of a male and female human and where Earth is located. Where are these craft now? *Pioneer 11* lost touch with Earth-based antenna in 1995, but scientists know it is heading toward the constellation of Aquila (The Eagle), northwest of the constellation of Sagittarius. If it makes it, the craft will pass near one of the constellation's stars in about 4 million years. *Pioneer 10* is heading generally for the red star Aldebaran, a star 68 light-years away that forms the eye of Taurus (The Bull). At its current speed and trajectory, it will take *Pioneer* over 2 million years to reach Aldebaran.

If space beings miss the *Pioneer* craft, there are always *Voyager 1* and *2*. These two craft carry gold records containing messages and pictures of Earth; the other side of the record has a diagram with instructions on how to read the messages. Right now, it seems as if *Voyager 1* wins the "farthest out" race: On February 17, 1998, *Pioneer 10* was the most distant human-made object. But by then, *Voyager 1*'s distance was equal to *Pioneer 10*'s—at 69.419 Astronomical Units (one Astronomical Unit is the average distance between the sun and Earth—or about 93 million miles [149.7 million kilometers]). *Voyager 1*'s distance will continue to exceed *Pioneer 10*'s at an approximate rate of 1.016 AUs per year. Thus, *Voyager 1* has surpassed the elder craft, and it might be the first found by a space traveler sometime in the very, very distant future.

trial life. Note, however, that the scientists did not claim that they had found Martian bacteria—only the evidence of dead, fossil bacteria and chemical traces that might have come from bacteria. In other words, only the smallest of possible organisms, not multicelled creatures, plants, animals, or even little green men.

By 2002, another study of a 4.5 billion-year-old Martian meteorite showed new evidence confirming that 25 percent of certain material in the meteorite may have been produced by ancient bacteria on Mars. Like the first claim, scientists are currently trying to duplicate such discoveries in other Martian meteorites. And it might be a long time before we are certain that the meteorites hold (or do not hold) evidence of primitive life from the red planet.

So far, no one has proved beyond a shadow of a doubt that the shapes and chemicals in ALH 84001 or other meteorites are due to the presence of fossilized bacteria.

But no one has disproved it, either. This debate brings up a major question we may have to face in the near future: Just how *do* you recognize ancient life in rock when you see it?

Will scientists continue to **search for life on other planets** in the future?

Yes, because humans have a burning desire to determine whether they are alone in the universe, the search for life on other planets will continue. As of this writing, there are several spacecraft searching—indirectly and directly—for evidence of life in our solar system:

Mars Global Surveyor—This craft measures the geology and topography of Mars. It will also provide data that might determine if Mars once had an ocean and how the red planet's climate has changed over time. Both factors are connected to finding out if Mars also had life at one time in its past.

Mars Odyssey—This craft is orbiting Mars, studying the chemical composition of the red planet to find a possible indication of previous life on Mars.

Cassini—This craft will orbit the planet Saturn in 2004. The objectives of the mission are threefold: conduct detailed studies of Saturn's atmosphere, rings, and magnetosphere; conduct close-up studies of Saturn's satellites; and characterize Titan's (Saturn's largest moon) atmosphere and surface. Scientists believe that Titan's surface may be dotted with lakes of liquid hydrocarbons formed from photochemical processes in the upper atmosphere. In other words, a place that could possibly contain life.

THE EARTH'S LAYERS

THE INSIDE STORY

What are the **major layers** of the Earth?

The Earth is generally divided into four major layers: the crust, mantle, inner core, and outer core. The following defines each division. (Note: numbers representing the thickness and depth of these layers differ depending on the reference; thus, the numbers here should be taken as approximations):

Crust—The Earth's crust is the outermost layer and is the most familiar, since people live on the outer skin of the crust. It is rigid, brittle, and thin compared to the mantle, inner core, and outer core. Because of its varying characteristics, this outer layer is divided into the continental and oceanic crusts.

Mantle—In general, the Earth's mantle lies beneath the crust and above the outer core, averaging about 1,802 miles (2,900 kilometers) thick and representing 68.3 percent of the Earth's mass. A transition zone divides this layer into the upper and lower mantles.

Outer core—The liquid outer core is a layer between 1,793 and 3,762 miles (2,885 and 5,155 kilometers) deep in the Earth's interior. It is thought to move by convection (the transfer of heat through the circulating motion of particles—in this case, the material that makes up the outer core), with the movement possibly contributing to the Earth's magnetic field. The outer core represents about 29.3 percent of the Earth's total mass.

Inner core—The inner core is thought to be roughly the size of the Earth's Moon. It lies at a depth of 3,762 to 3,958 miles (5,150 to 6,370 kilometers) beneath the Earth's surface and generates heat close to temperatures on the sun's surface. It represents about 1.7 percent of the Earth's mass and is thought to be composed of

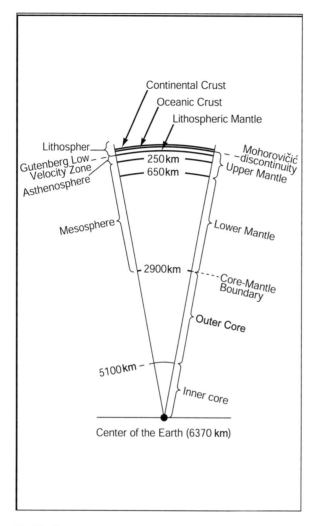

The Earth's cross section, illustrating the cores, mantle, crust, and other divisions.

a solid iron-nickel alloy suspended within the molten outer core.

What **percent** of the Earth's volume **constitutes the crust, mantle, and core**?

Although the core and mantle are about equal in thickness, the core actually forms only 15 percent of the Earth's volume, while the mantle comprises 84 percent. The crust makes up the remaining 1 percent.

Do geologists **subdivide the Earth** in any other way?

Yes, geologists have another way of looking at the Earth's interior layers. The following list refers to this view:

Lithosphere—The lithosphere (*lithos* is Greek for "stone") averages about 50 miles (80 kilometers) thick and is composed of both the crust and part of the upper mantle. Overall, it is more rigid than deep, yet more molten mantle and cool enough to be tough and elastic. It is thinner under the oceans and in volcanically active continental regions, such as the Cascades in the western United States. The lithosphere is physically broken up into the brittle, moving plates containing the world's continents and oceans. These lithospheric plates appear to "float" and move around on the more ductile asthenosphere. (For more on plate tectonics, see below).

Asthenosphere—A relatively narrow, moving zone in the upper mantle, the asthenosphere (*asthenes* is Greek for "weak") is generally located between 45 to 155 miles (72 to 250 kilometers) beneath the Earth's surface. It is composed of a hot, semi-solid material that is soft and flowing after being subjected to high temperatures and pressures; the material is thought to be chemically similar to the mantle. The asthenosphere boundary is closer to the surface—within a few miles—under oceans and near mid-ocean ridges than it is beneath landmasses. The upper section of the asthenosphere is thought to be the area in which the lithospheric plates move, "carrying" the continental and oceanic plates across our planet. The existence of the asthenosphere was theorized as early as 1926, but it was not confirmed until scientists studied seismic waves from the Chilean earthquake of May 22, 1960.

What is the difference between **compositional and mechanical layering** of the Earth?

When scientists talk about the Earth's crust (oceanic or continental), mantle, and cores, they are discussing layers with distinct chemical compositions; thus, it is referred to as compositional layering. The lithosphere and asthenosphere differ in terms of their mechanical properties (for example, the lithosphere moves as a rigid shell while the asthenosphere behaves like a thick, viscous fluid) rather than their composition, so this is why the term mechanical layering applies.

Who gave the first **scientific explanation** of the **Earth's interior**?

Empedocles, a philosopher who lived during the 400s B.C.E., was one of the first to formulate a scientific description of the Earth's interior. He believed the inside of the Earth was composed of a hot liquid. In fact, Empedocles was close to the truth. Modern scientists realize that the Earth's interior does not hold mythical beings but megatons of rock and molten matter.

What is **isostatic rebound**?

Scientists know the lithosphere can "bend," usually thanks to a heavy load on the crust. For example, a large ice cap, glacial lake, or mountain range can bend the lithosphere into the asthenosphere, causing the lower layer to flow out of the way. The load will then sink until it reaches its buoyancy point; then it is supported by the asthenosphere. If anything changes the load, the lithosphere will rise back up over geologic time—usually measured in thousands of years—in a process called isostatic rebound. One recent example is the Northern Hemisphere regions that sat under the miles-thick glacial ice sheets during the last ice ages. Many, such as the Adirondack Mountains in northern New York, are now experiencing isostatic rebound, often resulting in small earthquakes as the land rebounds.

Why is it so **important to study** the Earth's **internal structure**?

The Earth's internal structure is important because of its many influences on our planet's past, present, and future geology. For example, the mantle and crust are directly involved in plate tectonics, or how our planet's landmasses and ocean floor move around the planet. Convection in the Earth's mantle generates and recycles the planet's crust at the plate boundaries over time. And the inner and outer cores appear to be directly related to the generation of the Earth's magnetic field.

CRUST, MANTLE, AND CORES

How do the **oceanic and continental crusts** differ?

The Earth's crust varies in thickness and composition beneath the oceans and continents. The oceanic crust measures from 3 to 6 miles (5 to 10 kilometers), averaging about 4 miles (7 kilometers) in thickness; the continental crust measures between 16 to 62 miles (25 to 100 kilometers), averaging about 19 miles (30 kilometers) thick. The thickest continental crust regions exist mostly under large mountain ranges such as the Sierra Nevada, Alps, and Himalayas, where it can be as thick as 62 miles (100 kilometers).

The oceanic and continental crusts also differ in composition and density. The oceanic crust is composed of dark, iron-rich rock similar to basalt. It is high in silica and magnesium (sometimes referred to as SIMA); it is often distinguished from the next layer (the mantle) by having more silica. The continental crust's composition is more complex. In general, continental rocks are light-colored, with an average com-

12/08/2018

Item(s) Checked Out

TITLE	Encyclopedia of
BARCODE	33029042852426
DUE DATE	**12-29-18**

TITLE	The handy geology
BARCODE	33029054181789
DUE DATE	**12-29-18**

TITLE	Plates : restless earth /
BARCODE	33029048196711
DUE DATE	**12-29-18**

TITLE	Volcano & earthquake /
BARCODE	33029099052292
DUE DATE	**12-29-18**

TITLE	Plate tectonics / Alvin
BARCODE	33029047089370
DUE DATE	**12-29-18**

Terminal #   85

position between diorite (mostly hornblende and plagioclase feldspar with a little quartz) and granodiorite (the same composition as diorite, but with more quartz present). These are rocks high in silica and aluminum and are often referred to as SIAL. The continental crust's distinguishing compositional difference from oceanic crust is its higher amount of silica, with many regions having relatively high concentrations of quartz. This is mechanically crucial because—particularly in the presence of minor amounts of water—quartz-rich rock becomes relatively flexible at mid-crustal temperatures and pressures. Density differs, too: The oceanic crust has a density of 233 pounds per cubic foot (3,000 kilograms per cubic meter); the continental crust has a lower density of only 195 pounds per cubic foot (2,500 kilograms per cubic meter).

Beside the fact that both crusts lie on the mantle, there is one similarity between the two: Both have a temperature range of between 1,148 to 32°F (700 to 0°C). (For more information about the rocks mentioned above, see "Rock Families.")

What are the **major minerals** that make up the **Earth's crust**?

Although the oceanic and continental crusts differ in mineral composition, the overall Earth's crust is abundant in two major mineral groups: silica (around 12 percent) and orthoclase feldspar (just over 50 percent). (For more information about these and other minerals, see "All about Minerals.").

What is the **Mohorovičić discontinuity**?

The boundary between the crust and mantle is named in honor of the man who first proposed its existence, Croatian geologist Andrija Mohorovičić. In 1909, Mohorovičić analyzed data from a Croatian earthquake, calculating a jump in seismic wave velocity at a depth of about 34 miles (54 kilometers). This turned out to be the mantle boundary underneath the continental crust, a division now called the Moho or the Mohorovičić discontinuity.

Scientists believe the change is compositional, the rock type changing from the crust (with a seismic velocity of around 4 miles [6 kilometers] per second) to the denser mantle (with a seismic velocity around 5 miles [8 kilometers] per second). Like most interior layers, the Moho varies in depth. Beneath the continents, the Moho averages around 22 miles (35 kilometers) but overall it ranges from around 12 miles (20 kilometers) to 44 or 56 miles (70 or 90 kilometers) in depth. Beneath the oceans, the Moho averages about 4 miles (7 kilometers) below the ocean floor.

What are the **upper and lower mantles**?

The mantle is divided into the upper and lower mantle, represented by seismic and chemical changes in the layer. The upper mantle falls between 12 miles (20 kilometers) to 44 miles (70 kilometers) beneath the continental crust, and about 3 miles (5

63

kilometers) beneath the oceanic crust. It begins to transition to the lower mantle at about 255 miles (410 kilometers) deep. Past the transition zone, the lower mantle starts about 416 miles (670 kilometers) to about 1,793 miles (2,885 kilometers) deep.

Are there many **discontinuities** in the Earth's **mantle**?

As technology improves, scientists uncover more about our planet. This includes the discoveries of the many "boundaries" (discontinuities) within the Earth's mantle created by seismic or chemical changes. For example, the Hales discontinuity is found in the upper mantle at depths of about 37 to 56 miles (60 to 90 kilometers), a region in which seismic velocities change. Other seismic discontinuities include the Gutenberg and Lehmann discontinuities. (For more information about these boundaries, see below.)

Other discontinuities occur in the mantle at about 255, 323, and 416 miles deep (410, 520, and 670 kilometers). Each is either a chemical (416-mile [670-kilometer]) or seismic (255- and 323-mile [410- and 520-kilometer]) change, the entire range representing the gradual transition between the upper and lower mantles. At the lowest depth, and with accompanying high pressures, the crystalline structure of the material gets tighter, changing the rock type.

Have any **upper mantle rocks** been discovered?

It was once thought that rocks found deep underground, such as kimberlites and lamproites (ancient rocks that also yield most of the world's diamonds), would give scientists a good idea of the composition of mantle rock, but even the deepest mine in the world, a gold mine in South Africa that reaches about 2,225 feet (3,581 meters), does-n't come close to reaching the mantle. One might think, too, that molten rock brought up from the mantle—by way of volcanoes and hot spots—might yield some information. However, that is still not a good representative sampling because as vol-canic rocks rise to the surface they are contaminated by other minerals.

But many scientists believe we do not have to drill to get to the mantle. They have found older, deep crustal and mantle rock thrust up in what are called ophiolite belts. Such rocks are found in the Bay Islands of western Newfoundland and in Oman, an area that contains the world's largest fragment of ancient oceanic lithosphere exposed on the Earth's surface. (The so-called Oman United Arab Emirates ophiolite belt con-tains 2-billion-year-old rock and is part of a large ophiolite belt running from Spain to the Himalayas.) In fact, certain Oman ophiolite rocks called boninites are thought to represent melted material formed by partial fusion of the hot, shallow mantle.

What major minerals may make up the Earth's lower mantle?

The Earth's lower mantle appears to contain three important minerals: just less than 80 percent perovskite $(MgFe)SiO_3$; about 20 percent magnesio-wüstite—$(MgFe)O$; and a small trace of stishovite, SiO_2, a mineral synthesized by Russian mineralogist Sergei Stishov (1937–) in the late 1950s and first found at Meteor Crater, Arizona, in the 1960s. To some scientists this implies that perovskite is the most abundant mineral in the Earth, though this has not yet been proven.

What is the Gutenberg discontinuity?

In 1913, German geophysicist Beno Gutenberg (1889–1960) was the first to discover the approximate location of the mantle-outer core boundary, a transition zone now known as the Gutenberg discontinuity. This is where seismic waves slow down, indi-cating a zone between the semi-rigid inner mantle and the molten, iron-nickel outer core. (Gutenberg also published many papers on seismic waves with Charles Richter, who developed the Richter scale that measures earthquake intensity. For more infor-mation about earthquakes and seismic waves, see "Examining Earthquakes.")

Who first discovered the Earth's core?

In 1906, R. D. Oldham, the first to use seismic data to determine the interior of the Earth, postulated the existence of a fluid core. In 1915, Beno Gutenberg published a measurement of the core's radius.

Who discovered the solid inner core?

In 1936, Danish seismologist Inge Lehmann (1888–1993) presented a paper titled, "P'" (or P-prime, after the seismic waves), which announced the discovery of Earth's inner core. The division between the inner and outer core is now called the Lehmann discontinuity. (Lehmann later became an authority on the structure of the upper mantle.)

The size of this core within a core was not calculated until the early 1960s, when an underground nuclear test was conducted in Nevada. Because the precise location and time of the explosion was known, echoes from seismic waves bounced off the inner core provided an accurate means of determining its size. These data revealed a radius of about 756 miles (1,216 kilometers). The seismic P-waves passing though the inner core move faster than those going through the outer core—good evidence that the inner core is solid. The presence of high-density iron thought to make up the inner core also explains the high density of the Earth's interior, which is about 13.5 times that of water.

What is the **importance of the outer core** to the Earth's **magnetic field**?

Some scientists believe the molten, iron outer core has a powerful effect on the Earth: Acting as a geodynamo, it might create the magnetic field. Others point to the differences in the inner and outer core flow rates. This could create what is called a hydromagnetic dynamo that would explain the Earth's magnetic field.

Either way, such a planetary-scale dynamo would be similar to how an electric motor generates a magnetic field, one that acts like a giant bar magnet with a north and south pole. The basic physics of electromagnetism fits here: Iron, whether liquid or solid, conducts electricity; when you move a flowing electric current, you generate a magnetic field at a right angle to the electric current direction.

The molten outer core of our planet releases heat by convection, which then displaces the flowing electrical currents. This generates the magnetic field that is oriented around the axis of rotation of the Earth, mainly due to the rotational effects on the moving fluid. Thus, invisible geomagnetic lines stretch from one pole, curve far out into space, then go back to the opposite pole. The curved lines are further shaped by the electrically charged particles of the solar wind into a teardrop shape called the magnetosphere.

Paleomagnetic records indicate this magnetic field has been around for at least 3 billion years. Scientists know that without some mechanism—such as the core's interactions deep within the Earth—the field would only last about 20,000 years, mainly because the Earth's inner temperatures are too high to maintain any permanent magnetism. And without this magnetic field, all organisms—including humans—would be exposed to the extremely damaging effects of solar wind radiation.

Has the Earth's **magnetic field** ever **reversed polarity**?

Yes. Based on data from ancient (and not-so-ancient) rock, scientists know the Earth's north and south magnetic fields have reversed polarity many times. The switching from north to south (an individual reversal event) seems to take only a

What are "core-rigidity zones" and why are they important?

Scientists have long believed the outer core is liquid, but they recently discovered the existence of core-rigidity zones, or small patches of rigid material within the fluid outer core that seem to congregate at the core-mantle boundary. They believe these patches might influence many phenomena, such as the behavior of the Earth's magnetic field; the formation of volcanic hot spots, such as the Hawaiian Island chain; and even why the Earth wobbles on its axis as it rotates (called nutation).

Some scientists suggest that these chunks of material form as the Earth cools and heat flows out of the core, allowing the molten outer core to solidify into the inner core. This causes an increase in the lighter elements in the outer core. Because they are lighter than iron, they can "float to the top" of the outer core, collecting as solid material at the core-mantle boundary. Of course, the reality of how this works is probably much more complex.

couple thousand years to complete; once the reversal takes place, periods of stability seem to average about 200,000 years. No one really knows why the poles reverse, but theories range from the changes in lower mantle temperatures to the imbalance of landmasses on our world (most of the continental landmass is in the Northern Hemisphere).

Interestingly enough, the last magnetic reversal was 780,000 years ago, giving us our current northern and southern magnetic poles. Scientists believe our magnetic field is slowing weakening, so we might be heading for a long-overdue magnetic reversal. They also know that reversals tend to occur when there is a wide divergence between the magnetic poles and their geographic equivalent, as is currently the case. No one really knows when or if a reversal will occur, but it will probably not be during our lifetimes. Researchers know our ancestors did survive the last few magnetic reversals to continue our species. In fact, as one scientist mentioned, the only consequence of a reversal might just be the purchase of a new compass.

SEISMOGRAPHY AND EARTHQUAKES

How do scientists use **earthquake waves** to study the **Earth's interior**?

All knowledge about what happens in the mantle and cores of the Earth has been derived from circumstantial evidence—mostly the analysis of seismic data. Scientists use earthquake waves, or waves generated by the seismic shaking from earthquakes, to better understand the Earth's interior. (For more information about earthquakes and earthquake waves, see "Examining Earthquakes.") When a quake occurs, the Earth reverberates like a ringing bell, generating seismic waves with speeds that range between 1.9 to 9.4 miles (3 to 15 kilometers) per second.

The two major seimic waves are called Love and Rayleigh (also seen erroneously as Raleigh) waves. Both are surface waves that are more likely than P- or S-waves to make tall buildings sway. Love waves, named after the English mathematician A. E. H. Love (1863–1940), move in a side-to-side direction at about 2.7 miles (4.4 kilometers) per second, perpendicular to the earthquake direction and parallel to the surface. Rayleigh waves travel at about 2.3 miles (3.7 kilometers) per second and cause the most damage. Named after English physicist Lord John William Strutt Rayleigh (1842–1919), they are similar to waves caused by throwing a rock into a pond, with the waves moving outwards in all directions from the center.

Two other seismic waves—P- (primary) and S- (secondary or shear) waves—go all the way through to the Earth's interior, moving at different velocities through various geologic materials. Similar to sound waves, P-waves compress and dilate the matter they travel through (rock or liquid) and can move twice as fast as S-waves. On the other hand, S-waves move through rock but can't travel through liquids. In general, both waves slow down as they move through hotter materials and refract or reflect as they travel through layers with varying physical properties.

Why are these factors important? The differences in speeds and direction of the waves helped researchers determine the various liquid and solid layers—along with discontinuities—within the Earth's interior. By mapping the travel time and dividing the distance traveled by that time, researchers were able to determine the velocity of the seismic waves. (These changes in speed occur because parts of the mantle and core are made of different materials.) Once scientists were able to pinpoint these velocity changes, earthquake data became a way of "seeing" inside the Earth's interior.

Do **P- and S-waves** travel through the **crust**?

Yes, even though scientists use P- and S-waves mostly to study the Earth's mantle and cores, the waves also can be used to interpret information about the crust. In general, it is known that P-waves travel about 4 miles (7 kilometers) per second through oceanic crust (similar to their speed through the rocks basalt and gabbro), and through the continental crust at about 3.7 miles (6 kilometers) per second (similar to their speed through the rocks granite and gneiss).

Can **P- and S-waves be detected everywhere** they occur during an earthquake?

No, not all places on Earth can detect P- and S-waves during certain earthquake events. Some areas do not receive these seismic waves because of the way the core and mantle interact with the waves. For example, P-waves are bent at the core-mantle boundary, creating an arc called the P-wave shadow zone, which is about 105-142 degrees from the earthquake's epicenter. Because of this, P-waves are not received at seismic stations along the shadow zone. S-waves also have an S-wave shadow zone, which is a much larger arc of about 105 to 180 degrees. Scientists also believe the S-waves can't go through the Earth's outer core, which is why they believe the outer core must be liquid.

Was **seismic data** used to determine the **lithosphere and asthenosphere**?

Yes, earthquake waves were used to determine the boundaries of the lithosphere and asthenosphere in a way similar to how the crust, mantle, and core were discovered. At a depth of about 62 miles (100 kilometers), both P- and S-waves decrease in velocity. This boundary marks the base of the lithosphere and the top of the asthenosphere, called the low-velocity zone (LVZ).

Are there **problems with determining the Earth's interior** using earthquake waves?

Like any study in science, understanding the Earth's interior using earthquake waves is not perfect. For example, seismic waves do not travel in a straight line, but will bend, refract, and reflect if they run past, through, or bounce off an irregularity (usually a difference in rock density). Also, most seismic stations are located on the continents in the Northern Hemisphere, leaving huge gaps in data elsewhere around the world (although scientists are working to build up the network of seismic stations).

What is **seismic tomography**?

Seismic tomography is a relatively new technique that looks into the Earth's interior. This method is similar to medical ultrasound, in which doctors use high frequency sound through a mother's belly to map out the various densities within the womb, using the changing velocities of the ultrasound waves. The image, in the form of a tomograph, is actually a cross section of the mother and child, with the various density readings interpreted as the baby's bones, skin, and organs.

Seismic tomography, first invented by Adam Dziewonski of Harvard University in the 1970s, depends on a somewhat similar procedure. Instead of using ultrasound, this tomography looks at seismic waves generated by numerous earthquakes, with the travel time of the seismic waves compared to a reference model. By combining the data from many earthquakes, subtle changes in wave speed can be identified. (In general, if the waves move quickly, it usually indicates cooler or denser rock; slower waves indicate warmer or less-dense rock.) Scientists then use computers to construct three-dimensional density images of the Earth's interior, including upwelling and downwelling of molten rock in the mantle.

MOVING CONTINENTS AND PLATE TECTONICS

Who **initially proposed** the idea of **moving continents**?

The idea of the continents moving around our planet was mentioned as early as 1587 by the Flemish map maker (with German origins) Abraham Ortelius (1527–1598) in his work *Thesaurus geographicus*. In 1620, Francis Bacon (1561–1626) also mentioned the idea, noting the fit of the coastlines on both sides of the Atlantic Ocean. By the 1880s, many other scientists were mentioning the connection. For example, in 1885, Australian geologist Edward Seuss proposed that the southern continents had once been a huge landmass that he called Gondwanaland.

But it was German scientist Alfred Wegener (1880–1930) who first formally published the idea of continental displacement (or drift) in his 1915 book, *The Origins of Continents and Oceans*. He believed the continents were once joined together into one supercontinent, a place he named Pangaea (also spelled Pangea, meaning "all land") that was surrounded by a superocean called Panthalassa. He also suggested that the massive continent divided about 200 million years ago, with Laurasia moving to the north and Gondwana (or Gondwanaland) to the south. Wegener based his ideas of continental motion on numerous observations: The continental distribution of fossil ferns called *Glossopteris* (from studies by Seuss); the discovery of coal in Antarctica by Sir Ernest Henry Shackleton (1874–1922); similar glacial erosion seen in the tropical areas of India, South Africa, and Australia; the apparent fit of the South America and west African continental shorelines; and, although it may only be legend, by watching ice floes drifting on the sea.

Although Wegener is now considered "the man who started a revolution" in geology, his ideas were hotly debated by scientists of his time. Not only was he a meteorologist in a community of geologists, but he could offer no logical mechanism for the movement of the landmasses. It wasn't until the 1960s, long after his tragic death in Greenland (he died at the age of 50 while on a rescue mission), before Wegener was vindicated. By then, scientific measurements, observations, and technology had

A man looks at the rubble of a building destroyed in Bumerdes, Algeria, after an earthquake struck the region on May 21, 2003. It was estimated that more than 1,500 people died and more than 7,200 people were injured in this quake, which registered 6.7 on the open-ended Richter scale. *AP/Wide World Photos.*

advanced enough to prove that, indeed, the continents are moving around the planet on giant lithospheric plates. Wegener's theory of continental displacement was replaced by the new field of plate tectonics, which is the basis for modern geology.

What **physical evidence** shows that the **continents move**?

Scientists have gathered plenty of evidence that shows the continents move over time. For example, the shape of the continents and their fit was determined by Sir Edward Bullard in 1965. He did not site the usual continental shapes we see, but he measured the "real" edge of the continents: the continental slope, an area that shows a much better fit at the 6,560 foot (2,000 meter) depth contour than at the shorelines of continents.

Other scientists matched the continental geology on either side of an ocean. For example, the mountain belts of the Appalachians and the Caledonides are relatively similar geologically, as are the sedimentary basins of South Africa and Argentina.

Another way to prove that continents move over time includes paleontology, in which similarities or differences of fossils on certain continents indicate a match. For example, there are similar Mesozoic Era reptiles in North America and Europe, a time when scientists believe those two continents were joined together; similar Carboniferous and Permian flora and fauna are found in South America, Africa, Antarctica, Australia, and India. In contrast—no doubt after the continents were well separated— there is a wide diversity of organisms in the Cenozoic Era.

Can magnetism in rocks determine continental movements over geologic time?

Yes, paleomagnetism, or the record of a rock's magnetism based on the polarity of magnetic minerals within, can be used to determine continental movements throughout the geologic record. Taking paleomagnetic data from all over the world, scientists discovered that the geomagnetic poles (not the geographic poles) wandered over time. The data also revealed that the magnetic poles have never been more than 20 degrees from the geographic poles.

Taking this data one step further, researchers knew there can be only one north pole and one south pole at any time. They discovered that ancient rocks vary in magnetic orientation and direction. Scientists determined that the magnetic poles did not wobble in different orientations for each continent, but that the continents moved and the poles stayed in relatively the same locations. Thus, scientists were able to reconstruct the location of continents over the Earth's long history based on the polarity changes in rocks from all over the world.

What is **plate tectonics**?

The Earth's crust and lithosphere are broken into over a dozen thin, rigid shells, or plates, that move around the planet over the plastic aesthenosphere in the upper mantle. The interaction between these plates is called tectonics, from the Greek *tekon* for "builder"; plate tectonics describes the deformation of the Earth's surface as these plates collide, pass by, go over, or go under each other. In other words, plate tectonics describes how these plates move, but not why.

Overall, plate tectonics combines Wegener's theory of continental displacement (or drift) and Hess's discovery of seafloor spreading (see below). The theory has truly revolutionized the study of the Earth's crust and deep interior. It allows scientists to study and understand the formation of such features as mountains, volcanoes, ocean basins, mid-ocean ridges, and deep-sea trenches, and to understand earthquakes and volcano formation. It also gives clues as to how the continents and oceans looked in the geologic past, and even how the climate and life forms evolved.

Who contributed to **early work in plate tectonics**?

There were several key scientists who contributed to the study of plate tectonics as it became more favored in the late 1960s. One of the most popular scientists to discover evidence for plate tectonics was J. Tuzo Wilson (1908–1993). By 1965, he described the origin of the San Andreas fault, the large crack in the Earth's surface near San Francisco, California, as a transform fault (or strike-slip)—one of the major plate bound-

What was the shrinking Earth theory?

Before the idea of plate tectonics took over modern geologic thinking, some scientists believed in the shrinking Earth theory. They believed that the Earth started as a molten ball of rock; as it cooled, a skin of crust formed. As the rest of the molten ball cooled, the Earth shrunk, causing the crust to buckle, much like how an apple shrinks and wrinkles as it dries in the sun. The large wrinkles became the continents and ocean basins, while the small wrinkles turned into the long mountain belts.

But there were problems with this theory. For example, the shrinking Earth theory predicts that mountain ranges continually rise from the shrinking of the globe, but in reality mountains rise and are worn down over geologic time. The theory also has a difficult time accounting for the movement of the continents around the world. It can't explain the presence of fossils in high places, such as the Alps. The theory also predicts that volcanoes and mountain ranges occur randomly worldwide, an idea that we now know is false. (Mountain ranges are found in narrow belts, such as the Alps and Appalachian Mountains, and most volcanoes occur along plate boundaries or above areas known as hot spots.) And of course, all these events and features are better explained by plate tectonics.

aries. In 1968, Xavier LePichon (1937–) participated in the definition of the overall "plate tectonics" model and published the first model quantitatively describing the motion of six main plates at the Earth's surface; in 1973, he wrote the first textbook on the subject.

Other geologists have made major contributions to the development of the plate tectonics theory: William Jason Morgan published a landmark paper in 1968 explaining the many tectonic plates and their movements; he also recognized the importance of mid-plate volcanic hot spots that create island chains such as the Hawaiian Islands. Walter Pitman III was instrumental in interpreting the pattern of marine magnetic anomalies detected around mid-ocean ridges, an indicator of active seafloor spreading and evidence of plate tectonics. And Lynn R. Sykes used seismology to refine plate tectonics, and he noted the connection between transform faults at the mid-ocean ridges and plate motion. He also coauthored *Seismology and the New Global Tectonics* in 1968, which relates how existing seismic data could be explained in terms of plate tectonics.

What is the connection between **earthquakes and plate tectonics**?

Only the lithosphere has the strength and brittle behavior to fracture in an earthquake. And as lithospheric plate boundaries push, pull apart, or grind against each other, earthquakes occur. In 1969, scientists published the locations of all earthquakes

The city of Palm Springs nestles at the base of Mount San Jacinto in this computer-generated perspective. The San Andreas Fault passes through the middle of the sandy Indio Hills in the foreground. This 3-D perspective view was generated using topographic data from the Shuttle Radar Topography Mission (SRTM) and an enhanced color Landsat 5 satellite image. *AP/Wide World Photos.*

that occurred from 1961 to 1967. They discovered most earthquakes (and volcanoes, too, they later learned) occurred in narrow belts around the world. Thus, it is now known that areas with frequent earthquakes and volcanoes help define the plate boundaries. (For more information about earthquakes, see "Examining Earthquakes," and for volcanoes, see "Volcanic Eruptions".)

What did the **continents** look like **in the past**?

Because of the movement of the lithospheric plates, the continents' positions have changed over time. For example, some scientists believe that about 700 million years ago a huge continent called Rodinia formed around the equator; about 500 million years ago, the continent broke apart, forming Laurasia (today's North America and Eurasia) and Gondwana (or Gondwanaland; today's South America, Africa, Antarctica, Australia, and India). Then, about 250 million years ago, the continents were once again together in one massive supercontinent called Pangea (or Pangaea, translated as "all land"). Eventually, the huge continent began to break up, forming Laurasia and Gondwana again.

What are the **four types of lithospheric plate boundaries**?

There are just over a dozen major lithospheric plates, and all include continental and oceanic crust, plus part of the mantle. As each one moves continuously over the face

Can plate tectonics help in earthquake prediction?

In an indirect way, plate tectonics can help in earthquake prediction, but only to a point. Scientists know that earthquakes along plate boundaries are some of the strongest. Thus, they can predict the general areas where larger earthquakes will occur in the future. In fact, based on past earthquake data, scientists estimate that about 140 earthquakes of magnitude 6 or greater will occur on the lithospheric plate boundaries each year. Unfortunately, we do not have the technology or knowledge to pinpoint when or where such events will occur.

of the Earth, they interact along their boundaries. There are four major types of plate boundaries: divergent, convergent, transform, and plate boundary zones. The following lists and defines each boundary:

Divergent—Divergent (or constructional) plate boundaries are where the plates pull away from each other, creating new crust; seafloor spreading takes place at divergent boundaries (see below). Earthquakes that occur along these boundaries are usually shallow, and such boundaries are usually very young, geologically speaking. Overall, the spreading rates of divergent plate boundaries seems to range from fractions of an inch to just over 3 inches (1 to 8 centimeters) annually. For example, the Mid-Atlantic Ridge is a divergent plate boundary in which the Eurasian plate is pulling away from the North American Plate.

Convergent—Convergent (or destructional) boundaries are often thought of as trench boundaries. An example is the Marianas Trench in the Pacific Ocean, which is the world's deepest trench and is where the Pacific plate converges with the Philippine plate. These boundaries also mark subduction zones, an area where the oceanic plate subducts into the mantle. At convergent plate boundaries, plates are destroyed by subduction. For example, in addition to the Mariana Trench, the Andes Mountains of South America's west coast sit on a convergent boundary (the Nazca plate is pushing into and being subducted under the South American plate). Earthquakes that occur along these boundaries are shallow to deep, and the crust is much older than at divergent boundaries. There are three subtypes of convergent plate boundaries: ocean-ocean, in which a volcanic island arc forms above the downgoing slab, such as in the Mariana Trench; ocean-continent, in which a volcanic arc forms along the edge of a continent (continental crust is too buoyant to subduct), such as with the Andes of South America; and the continent-continent, in which a collision between the boundaries produces a continental crust up to twice as thick as normal, such as with the Himalayas, created by the collision of the Indian and Eurasian plates.

Transform—Transform boundaries are plates that slide by one another but do not destroy or create new material. For example, the San Andreas fault in California is a transform boundary, in which the North American plate slides by the Pacific plate.

Plate boundary zones—For the want of a better name, scientists consider those broad belts as undefined boundaries; the effects of plate interaction are unclear at plate boundary zones. For example, the Mediterranean-Alpine region between the African and Eurasian plates is not well defined, as it contains several smaller fragmented plates (called microplates) in between the major plates. Because of this, the geological structure of the area—and even the earthquake patterns—are very complex.

What is **seafloor spreading**?

Seafloor spreading is one of the processes that helps move the lithospheric plates around the world. The process is slow but continuous: Like a hot, bubbling stew on the stove, the even hotter asthenospheric mantle rises to the surface and spreads laterally, transporting oceans and continents as if they were on a slow conveyor belt. This area is usually called a mid-ocean ridge, such as the Mid-Atlantic Ridge system in the Atlantic Ocean.

The newly created lithosphere eventually cools as it gets farther from the spreading center. (This is why the oceanic lithosphere is youngest at the mid-ocean ridges and gets progressively older farther away.) As it cools, it becomes more dense. Because of this, it rides lower in the underlying asthenosphere, which is why the oceans are deepest away from the spreading centers and more shallow at the mid-ocean ridges. After thousands to millions of years, the cooled area reaches another plate boundary, either subducting, colliding, or rubbing past another plate. If part of the plate subducts, it will eventually be heated and recycled back into the mantle, rising again in millions of years at another or the same spreading center.

How was **seafloor spreading discovered**?

In the 1950s, scientists realized that as igneous rocks cool and solidify (crystallize), magnetic minerals align with the Earth's magnetic field like tiny compass needles, essentially locking the magnetic field into the rock. In other words, rocks with magnetic minerals act like fossils of the magnetic field, allowing scientists to "read" the rock and determine the magnetic field from the geologic past. This is called paleomagnetism (see above).

The idea was proposed by Harry Hess (1906–1969), a Princeton University geologist and U.S. Naval Reserve rear admiral, and independently by Robert Deitz, a scientist with the U.S. Coast and Geodetic Survey, both of whom published similar theories

If the ocean floor is always reforming, how old are the rocks?

The age of the ocean floor varies from the Jurassic Period (200 million years ago) to modern time at the major mid-ocean ridges. For example, the continually forming Mid-Atlantic Ridge between the South American and African plates is a range of some of the youngest "mountains" on the planet.

that became known as seafloor spreading. In 1962, Hess proposed the idea of seafloor spreading, but had no proof. As Hess formulated his hypothesis, Dietz independently proposed a similar model, which differed by noting the sliding surface was at the base of the lithosphere, not at the base of the crust.

Support for Hess's and Dietz's theories came only one year later: British geologists Frederick Vine and Drummond Matthews discovered the periodic magnetic reversals in the Earth's crust. Taking data from around mid-ocean ridges (seafloor spreading areas), Vine noted the magnetic fields of magnetic minerals showed reversed polarity. (The Earth's magnetic field has reversed its polarity around 170 times in the last 80 million years.) From the spreading center outward, there was a pattern of alternating magnetic polarity on the ocean floor—swaths of opposing polarity on each side of the ridge. As the spreading center continues to grow, new swaths develop, pushing away material on either side of the ridge. Thus, these strips of magnetism were used as evidence of lithospheric plate movement and of seafloor spreading.

How **fast** does the **seafloor spread**?

Today, the rates of seafloor spreading vary from about 1 inch (2.54 centimeters) per year in the mid-Atlantic ridge area to about 6 inches (15 centimeters) in the mid-Pacific Ocean. Scientists believe seafloor spreading rates have varied over time. For example, during the Cretaceous Period (between 146 to 65 million years ago) seafloor spreading was extremely rapid. Some researchers believe this quick movement of the lithospheric plates may have also contributed to the demise of the dinosaurs: As the continents changed places over time, so did the climate. In addition, more plate movements might have meant more volcanic activity, releasing dust, ash, and gases into the upper atmosphere and contributing to more climate variation. This change in climate and vegetation may have cause several species of dinosaurs to die out or become diseased, contributing to the dinosaurs' extinction.

How **fast** are the **lithospheric plates moving**?

The speed of the lithospheric plates around the world depends on which plate you are observing. For example, the fastest is the Australian plate, which is moving northward at about 6.5 inches (17 centimeters) per year. The Atlantic Ocean, with plates such as the Eurasian to the east and North American to the west, is moving about 0.5 to 1 inch (1 to 2.54 centimeters) per year on each side. This is a more typical number for the majority of the major plates. In fact, the Atlantic Ocean has opened more than 33 feet (10 meters) since Columbus sailed across in 1492.

Why is the **Mariana Trench** so famous?

The Mariana Trench is the deepest point on the Earth's crust, measuring at a depth of 35,840 feet (10,924 meters) on the Pacific Ocean floor, or about 8 miles (13 kilometers) deep. If you sank the Earth's highest surface mountain (Mt. Everest, which stands at 29,022 feet [8,848 meters] high) into the trench it would still be covered with more than 5,000 feet (around 2,000 meters) of water.

What **mountains** have been formed by **plate collisions**?

There are a number of beautiful mountain ranges built by plate collisions. Some of the more famous ones are the Rocky Mountains in North America, the Alps in Europe, the Pontic Mountains in Turkey, the Zagros Mountains in Iran, and the Himalayas in central Asia. All these mountain ranges were formed by plates slamming into one another, creating the uplift of the land.

What is the **Wadati-Benioff zone**?

The Wadati-Benioff zone is named in honor of seismologists Kiyoo Wadati and Hugo Benioff. At a convergent plate boundary, the downgoing slab (or subducting chunk of a boundary) is defined by a zone of earthquakes known as the Wadati-Benioff zone, an area that reaches to a depth of about 435 miles (700 kilometers) from the Earth's surface.

What are **rifts**?

There are numerous deep rifts on the Earth's surface. Many associated with powerful tensional tectonic forces that are constantly trying to rip or split apart parts of the

Earth's crust. Over thousands of years, the movements of the rifts' associated plates will open many of these cracks even wider, while moving continents into new positions. We will not see these changes in our lifetimes because it takes thousands to millions of years to notice such great movements. Right now, we can only measure plate motion a few inches per year, not miles.

What is the **African Rift System**?

The African Rift System (or Afro-Arabian Rift System) is a group of rifts or cracks in the Earth's crust. In general, this system has three "arms" of rifts, or what geologists call a triple junction structure:

Red Sea Rift—This rift separates Arabia from Egypt and Ethiopia. (It extends northward into Israel to the Dead Sea, Jordan River, and Sea of Galilee Rift Valley; from there it becomes less distinct as it continues north into Lebanon and Turkey.) It is up to 167 miles (270 kilometers) across and increases about 1 inch (2.5 centimeters) per year; it currently has a maximum depth of up to 4,921 feet (1.5 kilometers).

Gulf of Aden Rift—This separates southern Arabia from Somalia. Its width is similar to the Red Sea Rift at the western edge; as it reaches the Indian Ocean, the width increases to over 205 miles (330 kilometers).

East African Rift Valley—This cuts through Eastern Africa in a southwest direction and is about 2,423 miles (3,900 kilometers) in length; its width in Kenya varies from 25 miles (40 kilometers) in the south to about 62 miles (100 kilometers) toward the north. In Ethiopia, its width varies from about 19 to 81 miles (30 to 130 kilometers); at the junction with the Red Sea and Gulf of Aden Rifts, the width reaches about 186 miles (300 kilometers).

What **mechanism** makes the **lithospheric plates move**?

Even though the idea of moving lithospheric plates has been accepted by most scientists, no one can agree on the mechanism(s) that cause them to move over time. But there are plenty of theories. One of the most commonly acknowledged theories was proposed in 1928, when geologist Arthur Holmes suggested that, like a pot of bubbling water on the stove, convection in the mantle was the driving force behind continental drift. He also suggested that the crust was recycled by divergence and subduction, but he had no proof at that time.

Today, most scientists believe in a modified version of Holmes's idea: they theorize that a dozen or so rigid plates slide around on the partially molten asthenosphere, with the continents embedded in the plates. In addition, because basalt is more dense than granite, the oceanic crust "sags" lower as it "floats" on the asthenosphere. As for the driving forces behind plate motion, they are probably some combination of mid-ocean ridge pushing and plates being pulled deep into the mantle (subduction), all the result of a fluid mantle in motion.

Can the **middle of a lithospheric plate** experience **earthquakes and volcanoes**?

Yes, the middle of a lithospheric plate can experience plate boundary events in the form of earthquakes and volcanoes, but it does not happen very often. Called intraplate ocean tectonics, it involves something scientists often refer to as hot spots: rising molten rock from the mantle that breaches the mid-plate surface, forming volcanoes. If the plate is moving in one direction, the volcanoes line up in one direction in a youngest-to-oldest fashion. One of the most famous hot spot areas is the Hawaiian Islands chain, which is located in the "middle" of the Pacific lithospheric plate. (For more information about the Hawaiian Islands, see "Volcanic Eruptions.")

There are also unexplained movements in the middle of lithospheric plates. For example, the strongest earthquakes ever recorded in the United States (1811–1812) occurred along what scientists call the New Madrid fault line, which is an area near New Madrid, Missouri, located at the "middle" of the North American plate. Another occurred in 1886 at Charleston, South Carolina. Both these areas seem to be strange places for such earthquakes, and scientists still can't explain why. (For more information about these and other earthquakes, see "Explaining Earthquakes.")

ALL ABOUT MINERALS

MINERAL FORMS

What are **minerals**?

The Earth's crust—and for that matter, any rock, sand, soil, gravel, or mud you pick up—is a rich source of minerals. The following lists several ways a material qualifies as a mineral:

It is naturally occurring—Minerals are naturally occurring as a consequence of natural processes in or on the Earth. They are not made in a laboratory. Thus, a diamond that forms naturally within the Earth is a mineral; a synthetic diamond made in a laboratory is not a mineral.

It is made of inorganic matter—A mineral has never been "alive." Thus, coal or amber (hardened, ancient tree resin) are not minerals.

It is represented by a chemical formula or symbol—Minerals are either elements or compounds and can be written as a chemical symbol (for example, Au for the element gold) or formula (for example, SiO_2 for quartz).

It has a crystalline form—The atoms or molecules that make up a mineral are the same throughout the entire mineral. When they are joined in a certain order, they create the internal structure (a definite pattern within the mineral) called the crystalline form. If this pattern can be seen with the unaided eye—and even with a microscope in the case of microcrystals—the solid is called a crystal (for more information about crystals, see below).

What is **mineralogy**?

A subset of geology, mineralogy is the study of minerals in the Earth's crust. In general, mineralogy is broken down into the following sspecialties (although it is difficult to actually separate these divisions when analyzing minerals):

A freshly washed crystal found at the Colemans Crystal mine near Hot Springs, Arkansas. People from around the country come to the area to dig in the mines for the popular crystal. *AP/Wide World Photos.*

Crystallography—Crystallography is the study of the internal (for example, atomic structures) and external geometry of crystals. Crystallographers study the growth, shape, and geometric characteristics of crystals.

Physical mineralogy—Physical mineralogy is the study of the physical properties of minerals.

Chemical mineralogy—Chemical mineralogy is the study of the chemical properties and structures of minerals.

What is some of the **history** behind the **study of minerals**?

The Greeks were one of the first to write about minerals as early as 300 B.C.E. But, as with so many other scientific disciplines, after the fall of the Greek and Roman empires, there was a huge gap before minerals were studied again. It took until the mid-1500s—when the German physician Georgius Agricola (1494–1555) focused his writings on minerals and mining lore before mineralogy once again came under scientific scrutiny. Agricola is often thought of as the person who built the foundation for mineralogy as a science.

By 1669, Danish scientist Nicolas Steno (1638–1686) had demonstrated that angles between crystal faces of certain minerals were always similar, an observation known as Steno's law. In 1768, Carolus Linnaeus (1707–1778), also known as Carl von Linné, presented one of the first classifications of minerals based on their external forms. (Linnaeus was also responsible for one of the first comprehensive classifications of living organisms). Around the mid-19th century, crystallography rapidly developed, with German scientist Johann Friedrick Christian Hessel's (1796–1872) discovery that certain geometric conditions restrict the number of crystal classes to exactly 32, and only two-, three-, four-, and six-fold axes of rotation symmetry in minerals are possible (for more information about mineral axes, see below).

From that time on, technology improved our knowledge of minerals. For example, the invention of the polarizing microscope in 1870 and the discovery of X-rays in 1895, along with the first crystal structure determined by 1913, all increased our understanding of minerals' interiors. And more recently, the electron probe microanalyzer, the scanning electron microscope, and the transmission electron microscope have contributed to our understanding of the intricacies of minerals.

Who is the true **father of mineralogy**?

German physician Georgius Agricola (1494–1555) is most commonly thought of as the "father of mineralogy." Agricola's two greatest works were *De natura fossilium* (1546) and *De re metallica* (1556). The first book includes a classification of minerals (then called fossils) based on their geometric form; the second book is a compendium of then-current knowledge of mining and metal production.

Running a close second for the "father of mineralogy" was German geologist Abraham Gottlob Werner (1749–1817), who is considered by some to be the true founder of the field. He was the first to systematically classify minerals according to their obvious physical properties, such as color, crystal habit, cleavage, luster, streak, and hardness. Years later, better measuring techniques and technology helped scientists identify minerals by chemical composition, specific gravity, fusibility, and internal crystal structure, things that were not easily determined or even possible in Werner's time.

Why do modern scientists **study mineralogy**?

Modern scientists study mineralogy for many reasons. First, and most important, minerals are the building blocks of all rocks. Thus, knowledge of minerals can be tied to the understanding of Earth and other planets (and satellites) of the solar system. Second, many minerals are of great economic value, including in industry, health care, and even in the politics of various countries. And finally, almost all aspects of geology are essentially tied to mineralogy—from geophysics (a rock's properties are definitely associated with the minerals within the rock) to geochemistry (the properties of chemical reactions with minerals).

What are **polymorphs**?

Minerals that are closely related but look different are called polymorphs; or, put yet another way, polymorphs are two minerals with totally different properties that have the same chemical composition. This is because the same elements can be joined together in more than one way.

Two well-known examples are graphite (a soft, gray mineral found in pencil leads and also used as a lubricant) and diamond (the hardest mineral known to humans). The composition of both minerals is carbon (represented by the symbol C), but the atoms in these minerals are joined together in different ways. In the case of graphite, the carbon atoms are in widely spaced sheets, with weak bonds between the sheets that allow them to easily slip past one another (which is why graphite is such an excellent lubricant). Diamonds are formed much deeper within the Earth; heat and pressure press the carbon atoms closer together, creating hard diamonds.

How do **minerals form**?

Minerals form under specific conditions—especially the right temperatures and pressures—when certain elements are present. In most cases, mineral formation depends on the rock family—igneous, sedimentary, and metamorphic (for more information about these rock classifications, see "Rock Families").

In general, igneous minerals crystallize from molten rock (magma) at temperatures between 1,112 to 2,192°F (600 to 1,200°C) and at about 19 miles (30 kilometers) deep, such as quartz or biotite mica. Sedimentary minerals form through the evaporation of water (for example, halite or table salt); precipitation from water because of a change in chemical conditions (for example, chert and carbonates); or through the deposition of such hard parts as bones or shells of organisms (for example, aragonite). And finally, metamorphic minerals form as minerals within rocks recrystallize in response to changes in heat and pressure.

What is **Bowen's Reaction Series**?

Bowen's Reaction Series was created by N. L. Bowen and others at the Geophysical Laboratories in Washington, D.C., in the early 1900s. The researchers determined that as molten rock (magma) cools, common silicate minerals crystallize in a certain order

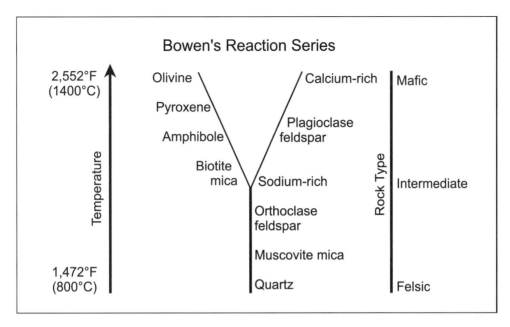

Bowen's Reaction Series

2,552°F (1400°C)

Temperature

Olivine

Pyroxene

Amphibole

Biotite mica

Calcium-rich

Plagioclase feldspar

Sodium-rich

Orthoclase feldspar

Muscovite mica

1,472°F (800°C)

Quartz

Rock Type

Mafic

Intermediate

Felsic

A diagram illustrating Bowen's Reaction Series indicates which types of form depending on conditions.

based on specific temperatures and pressures. This model is still generally accepted, but as with most things in nature, there are exceptions.

What are the most **common rock-forming minerals**?

Rock-forming minerals are just what the term implies: Minerals that form rocks (for more information about rocks, see "Rock Families"). Of the about 3,500 named minerals, only about 24 rock-forming minerals are abundant in Earth's crust. Some of the most common rock-forming minerals include olivine, quartz, mica, pyroxene, and amphibole (note: all but quartz are actually mineral groups).

What are some **other familiar rock-forming minerals**?

The following includes several more familiar rock-forming minerals:

Calcite—Calcite is commonly white to gray in color; its crystals are most often clear. This mineral can be scratched with a steel file (making it softer than quartz) and is one of the most common minerals found in sedimentary rocks. Calcite also reacts strongly to diluted hydrochloric acid, which is often used in the field to distinguish it from such similar minerals as dolomite.

Clays—Clay minerals are very fine grained and difficult to tell apart in the field; they range from white to gray, brown, red, dark green and black. Like potter's clay, their fine grains make them slippery and smooth when wet.

85

Talc—Talc is found in granular or foliated (layered) masses; it is slick to the feel and easily scratched with a fingernail. It is also known as soapstone (you may have seen carvings or even woodstoves made of soapstone) and can be white to green in color.

Magnetite—The mineral magnetite—black with a metallic luster—is common in igneous and metamorphic rocks; it is present in some sediments (only about 1 to 2 percent). Its crystalline form resembles two pyramids stuck together (a shape called an octahedra) and is characterized by its magnetic properties.

Pyrite—Pyrite, with its pale, brassy, metallic look, often forms cubic shapes and is also known as "fool's gold." It is the most common sulfide mineral (minerals that have sulfur as their primary component). It is found in igneous, sedimentary, and metamorphic rocks, but mostly in small amounts.

What are the **eight chemical elements** that comprise most of the **rock-forming minerals** on Earth?

The eight chemical elements that make up most of the rock forming minerals comprise 98.5 percent of the Earth's crust by weight (all other elements combined make up only 1.5 percent). The most abundant elements are oxygen and silica—the two elements that bond together as the basic building blocks of the most common mineral group called the silicates. The following are the eight elements that comprise most of the rock-forming minerals:

Name	Chemical Symbol	Percent
oxygen	O	46.6
silicon	Si	27.7
aluminum	Al	8.1
iron	Fe	5.0
calcium	Ca	3.6
sodium	Na	2.8
potassium	K	2.6
magnesium	Mg	2.1

How do geologists **identify minerals** based on **color**?

The color of a mineral is often important, and thanks to the human eyes' ability to see a variety of colors, this criteria is often used to identify certain minerals. In fact, ancient peoples realized the value of colorful minerals, using charcoal and iron oxides in cave paintings, most of which still retain their original intensity.

Still it is often not easy to tell a mineral just from color. For example, olivine is usually a distinctive olive green, but it also can be yellow; garnets occur in every color and shade from black to colorless; and "watermelon" tourmaline can occur in a mass of dark and light green and red. Still other minerals have a definitive color, such as malachite (green) and azurite (blue).

Why are **minerals certain colors**?

There are many properties of minerals that give them color. The following lists why minerals look a certain hue:

Idiochromatic—These minerals are essentially "self-colored" because of their composition. Elements responsible for these colors (usually transition metals) are part of the mineral's chemistry, and even a small amount of the element can create a deep color in a mineral. When the color becomes a predictable component of the mineral, it can be used for identification. For example, cinnabar is usually red, as iron is part of the mineral's composition; azurite's blue and malachite's green colors are from the copper part of the minerals' composition; and manganese gives rhodochrosite its pink color.

Allochromatic—These minerals exhibit different colors because of small amounts of impurities in their composition or defects in the mineral's structure. In this case, the color is a changeable and unpredictable property; thus, it is not as useful in identification. For example, different impurities cause fluorite to come in every color of the rainbow. Smoky quartz often appears almost black because growth imperfections interfere with light passing through the crystal. Still other minerals get their color from minute air bubbles that cause different colors to be seen.

Pseudochromatic—The color of a pseudochromatic mineral is false, with neither the mineral nor atomic properties responsible for the mineral's colors. Instead, the minerals contain layers (or films) that create color by light interference. The color may vary, but it is often a unique property of the mineral. For example, precious opal, moonstone, and labradorite all reflect in a characteristic way, but the colors are not true to the types of minerals.

Does the **same mineral** ever display **different colors**?

Yes, some minerals are a light show in themselves, actually turning a different color when viewed from different angles under the same light source. This is called pleochroism. When only two colors are seen when viewed at different angles, it is called dichroism (also defined as pleochroism with two color components); trichroism indicates three colors.

The effect is caused by the variable light absorption patterns along different axes of the crystal. It occurs in minerals that demonstrate refraction, in which a beam of light

Yes. Apparently irradiation—exposing a mineral to radiation—affects the color of certain minerals. In fact, many materials, both natural and human-made, can be irradiated to produce different colors. For example, diamonds can be irradiated to produce blue, yellow, and green colors. But be aware: Certain irradiated minerals won't change color later unless they are heated. Still others fade when exposed to light, or even fade in the dark.

Another interesting example of irradiation includes century-old glass bottles. If you expose such a bottle to the ultraviolet radiation present in strong sunlight (such as in the desert), then come back ten years later, the glass will have turned a purple color; this is called "desert amethyst glass." When heated in an oven, the purple color disappears.

is split in half, with the two halves traveling in different directions. Each half is subjected to different types or amounts of color absorption, depending on the path and vibration of the light. If you look at the mineral from various angles against light, the color or deepness of color changes. For example, biotite mica exhibits pleochroism.

How do geologists **identify minerals** based on **luster**?

A mineral's luster means how the surface reflects light. All common rock-forming minerals do not look like a metal (called nonmetallic luster); only metallic, native (only one element) minerals look metallic, such as gold. Some minerals look metallic in reflected light, and although they are not native, they are still considered to have a metallic luster. For example, pyrite ("fool's gold") and galena are not native minerals, but still have a metallic luster.

Most nonmetallic minerals have specific lusters, including vitreous (bright like glass); resinous (similar to amber, or ancient resin); greasy (like an oily coating); pearly (iridescence similar to a pearl); silky (fibrous sheen similar to silk); adamantine (brilliant reflection similar to a diamond); and splendent (sparkly). For example, the minerals olivine and garnet look vitreous. Quartz looks vitreous, but also can appear greasy, or splendent. Feldspars are often described as pearly, and micas can be splendent or silky, depending on the size of the specimen.

How do geologists **identify minerals** based on **cleavage**?

Cleavage is the flat surface a mineral forms when it is broken. This happens because in some minerals the bonds between the layers of atoms aligned in a certain direction are weaker than bonds between other layers. Because of this, when a mineral with cleavage breaks, it splits, leaving a clean, flat face parallel to the zone of weakness (called a cleavage plane). In certain minerals, the direction of weakness may only be in one direction, such as the mineral mica. Other minerals may have two to six cleavage planes. For example, halite (salt) and fluorite have four cleavage planes, while calcite splits along three cleavage planes, leaving a "diamond" shape called a rhombohedron.

How are **fractures** used to **identify minerals**?

The way a mineral fractures (a fracture leaves a rough, uneven surface at the break) can be used for identification. When bonds between atoms are approximately the same in all directions within a mineral, breakage occurs on irregular surfaces. This includes splintery or irregular fracturing (usually in reference to fibrous minerals), such as with serpentine; conchoidal fracturing (along smooth, curved surfaces resembling a shell, similar to thick pieces of broken glass), such as quartz and garnet; hackly fracturing (jagged, sharp edges), such as with certain metals like silver; or even earthy fracturing, similar to broken pieces of clay or chalk.

How do geologists **identify minerals** based on **streak**?

Geologists also use streak to identify minerals, especially in the field. A mineral's streak, or the color of a mineral crushed to a powder, is usually found by rubbing the mineral on a streak plate (usually a piece of white, unglazed porcelain). Most non-metallic minerals have a specific streak—even though the same mineral varies in color. Streak is most important when identifying metallic minerals, especially non-native metals. For example, the streak of gold is metallic yellow; for pyrite ("fool's gold") the streak is dull black.

How do geologists **identify minerals** based on **hardness**?

Scientists often use hardness to identify minerals in the field, rubbing the surface of the mineral with an object of known hardness (such as a penknife). In particular, a mineral's hardness is a measure of its ability to withstand abrasion by other substances, which is often an important identifying characteristic.

What is **Mohs' Scale** of hardness?

In the early 19th century, the Mohs' Scale of hardness (often seen as Mohs Hardness Scale, or Mohs Scale, or even erroneously as Moh's Scale) was developed by Friedrich Mohs (1773–1839), a German-born mineralogist. The arbitrary scale is a crude but practical method of comparing hardness or scratch resistance of minerals. In reality, the scale should more accurately be called a table, because the numbers given to the different minerals are not proportional to their actual scratch resistance.

The table is based on nothing more than the idea that a mineral with a lower number can be scratched by a mineral with a higher number. In fact, since Mohs'

Scale was developed other scientists have added and moved several numbers to increase the usefulness of the scale.

The following is the scale based on Mohs' work. The standard minerals on the scale range from softest (1) to hardest (10):

Mineral	Hardness
talc	1
gypsum	2
calcite	3
fluorite	4
apatite	5
orthoclase	6
quartz	7
topaz	8
corundum	9
diamond	10

What are the **hardest and softest minerals**?

Friedrich Mohs developed his scale over a century ago, and since that time no harder or softer minerals have been found to change the minerals' status. Thus, the hardest mineral known is no surprise: It is the diamond. This mineral is made completely of carbon and has a hardness of 10 on Mohs' Scale; on other scales, its hardness often reaches into the mid-teens. The softest mineral known is talc, a specimen usually associated with metamorphic rocks. Talc has a hardness of 1 on Mohs' Scale, and is composed of magnesium and smaller amounts of water, silica, and oxygen.

Can the **minerals** on Mohs' Scale of hardness be **compared** to any **common items**?

Yes. The following lists several common items and their comparison to Mohs' Scale of hardness:

Mohs' Scale	Comparable Substance
1.5 to 2.5	fingernail (scales differ on the actual hardness of a fingernail, perhaps because everyone's nails differ!)
4	"copper" coin (U.S. pennies before various alloys made them softer)
5	glass
5.5	average penknife blade
6.5	steel file

Are there any **other scales of hardness**?

Yes, there are several others, but they are less practical for the average person to use. These tests measure the depth or area of an indentation left by an object (called an indenter) of a specific shape. This indentation is achieved by applying a specific force on the mineral for a specific amount of time. The three most common test methods include the Brinell, Vickers, and Rockwell, all of which are divided into a range of scales based on the indenter's shape and the amount of pressure applied.

Are there **other ways** to **identify a mineral**?

Yes, there are other ways to identify a mineral without any special equipment. The following lists some of these methods:

Specific gravity—Specific gravity is actually the mineral's density with respect to the density of water. In the field, specific gravity—whether the rock is estimated to be light or heavy—is often used to help identify a mineral.

Magnetism—Another method (applicable to magnetic minerals only) is to use a magnet to determine iron-rich minerals, such as magnetite or pyrrhotite.

Effervescence—Effervescence (fizzing) is one way to identify calcium carbonate minerals. By applying a weak acid solution (most often dilute hydrochloric acid) you can often identify such calcium carbonates as limestone.

Fluorescence—Still other minerals are fluorescent—they glow in the dark after being exposed to an ultraviolet ("black") light source (for more information on fluorescence, see "Rock Families"). Certain minerals are termed phosphorescent if the fluorescence continues for a short time after the ultraviolet light is turned off.

What is the **blowpipe or fusibility test**?

The blowpipe or fusibility test is often used as a simple test—albeit a more archaic one—to identify minerals. It is based on the idea that every mineral has a definite melting point and thus, can be compared to other mineral melting points. As the mineral is exposed to a flame, the observer must recognize behaviors of the mineral: Was the sample easy, normal, or difficult to melt; and does it change color, bubble, or expand upon heating? This information is then compared against what is already known about certain minerals. Although it is more subjective than many other tests, it is still used to determine a mineral's identification.

Are **minerals valuable** to humans?

Many minerals are valuable to humans—too many to mention here. For thousands of years, humans have used minerals (and rocks) to build structures, make tools, even to paint. Today, minerals are sources of raw materials for metallurgical, chemical, and electronics industries. They are even used in the food industry as mineral supplements for humans and other living organisms—from cats to plants in your backyard. Minerals are also used in everything from metal in our lamps and light bulbs to the quartz in our watches. There are those, too, who don't find a monetary value in minerals, but an aesthetic one, enjoying the diversity and beauty offered by these often awe-inspiring natural materials.

Are any minerals of **economic importance**?

Although the numbers vary (depending on the source), there are about 100 minerals of some economic importance. Not only are the well-known metallic minerals important (such as lead, gold, silver, copper, aluminum, molybdenum, and cobalt), but so are energy minerals (such as coal, oil shale, and uranium) and industrial minerals (for construction and agriculture, including sand, gravel, clay, and limestone). Still other minerals are valued as personal adornments (such as diamond, emerald, sapphire, and ruby). In fact, some minerals that would otherwise be of no economic value are highly sought after as gemstones.

What are the **main mineral groups**?

There are several main mineral groups. The largest group is the silicate minerals, with a basic structural unit being the Si-O tetrahedron, composed of one silica and four oxygen (SiO_4) atoms. Common silicate minerals include non-ferromagnesian or felsic minerals (feldspars—both orthoclase and plagioclase; quartz; muscovite mica) and ferromagnesian or mafic minerals (containing iron and/or magnesium), including amphiboles, pyroxenes; biotite mica; and olivine). The other mineral groups include oxides; sulfides; sulfates (such as anhydrite and gypsum); native elements (such as gold); halides (such as halite or common table salt, which is sodium chloride); and carbonates (such as calcite and dolomite).

DETAILS OF CRYSTALLOGRAPHY

What are **crystals**?

Crystals are homogenous materials that have a definite, orderly structure (thanks to their internal atomic arrangement). A crystal's form is also bound by smooth, planar surfaces that exhibit some type of symmetry. Crystals form in several ways: as a fluid—usually as liquid rock—gradually becomes solid; from the deposition of dissolved matter; or as the direct condensation of a gas to a solid. Most solid matter is of crystalline structure; solids with no crystalline structure—such as glass—are called amorphous.

What is meant by **crystal lattice**?

Particles in a crystal take up positions with definite geometrical relationships to each other. Internally (and on the atomic level), these form a kind of scaffolding called a

Does anyone collect very small crystal specimens?

Yes, many people collect tiny crystal specimens—microcrystals that are much smaller versions of larger crystals—called micromounts or thumbnail crystals. Collectors gather these specimens not only to save space (hundreds can be stored in small cabinets) and because it is much less expensive than collecting larger minerals, but also for the joy of collecting, identifying, and showing off exceptional mineral specimens.

A micromount is a natural crystal specimen that often measures fractions of an inch (less than one millimeter in size). Micromounts are often packaged in a 1 inch (2.5 centimeter) box and require magnification; thus, many of the boxes come with a small magnifying lens built into the box's removable top. There is an art to mounting the specimen to make it look aesthetically pleasing; scientifically, the collector also has to know the accurate mineral name and where it was found.

Thumbnail specimens are most often less than 1.25 inches (3.1 centimeters) in size, which again saves space and is a less expensive alternative to collecting larger minerals. These thumbnail specimens—often with pieces of the associated matrix (material where the mineral was formed) attached—are usually mounted in small plastic boxes called "perky" boxes. Thumbnail minerals are collected for their crystal quality, for scientific study purposes, and to enjoy the sight of beautiful mineral specimens.

crystalline lattice, with the specific lattice positions determined by the chemical composition of the crystal.

What is meant by **crystal (or crystalline) forms**?

A crystalline form occurs when a mineral's atoms or molecules join in a certain pattern or internal structure—all defined by its chemistry and structural arrangement of its atoms. (If the resulting solid can be seen with the unaided eye, it is called a crystal.) The crystalline form of a mineral determines its cleavage (the way the crystal breaks) and many other properties. In fact, crystalline forms were known even in the 1660s, when Danish geologist Nicolaus Steno (1638–1686) published his first law of crystallography (or Steno's law): The angles between corresponding faces of all crystals of any specific substance are constant.

Depending on the type, crystalline forms are grouped into 32 geometric classes of symmetry—crystals that are symmetrical with relation to planes, axes, and centers of symmetry. From there, crystals are further subdivided into seven crystal systems on the basis of the relationship of their axes (imaginary straight lines passing through a crystal's center).

What are the **seven common crystal systems**?

The following lists the seven common crystal systems (often with two similar definitions for clarification):

Cubic (or isometric)—These crystals have three axes that intersect at right angles, and are all of equal length. Iron pyrite and halite crystals are from the cubic system.

Tetragonal—These crystals have two equal, horizontal axes at right angles and one axis (that is longer than the other two) perpendicular to their plane. In other words, all the axes are all at right angles to one another, and two are the same length. Chalcopyrite (or copper iron sulfide) has a tetragonal crystal.

Orthorhombic—These crystals have three unequal axes intersecting at right angles. Topaz (a fluorine and aluminum silicate) is a good example of a mineral in the orthorhombic crystal system.

Monoclinic—These crystals have three unequal axes—two that intersect at right angles, and one at an oblique angle to the plane of the other two. Augite (an iron and magnesium silicate) is a member of this crystal system.

Triclinic—These crystals have three unequal axes intersecting at oblique angles. Albite feldspar (a sodium and aluminum silicate) is an example of a mineral in the triclinic system.

Hexagonal—These crystals have three equal axes that intersect at 60° angles in a horizontal plane. They also have a fourth (longer or shorter) axis perpendicular to the planes of the other three. Or there are four axes, three of equal length at 120° to one another and at right angles to the fourth. Beryl (a gemstone known as a beryllium silicate) is in the hexagonal crystal system.

Trigonal (or rhombohedral)—These crystals have equal axes intersecting at oblique angles. This crystal system is often combined with the hexagonal system.

What is meant by **crystal or mineral habit**?

The mineral or crystal habit is the preferred shape of a mineral as it grows; it often has very little or no resemblance to the ideal shape of an individual crystal. The mineral (or crystal) habit is often used to aid in the identification of a single mineral, especially when the mineral is small. The following lists some of the more common single mineral habits:

Acicular—These minerals are usually needle-like in shape, such as crystals of naturalite.

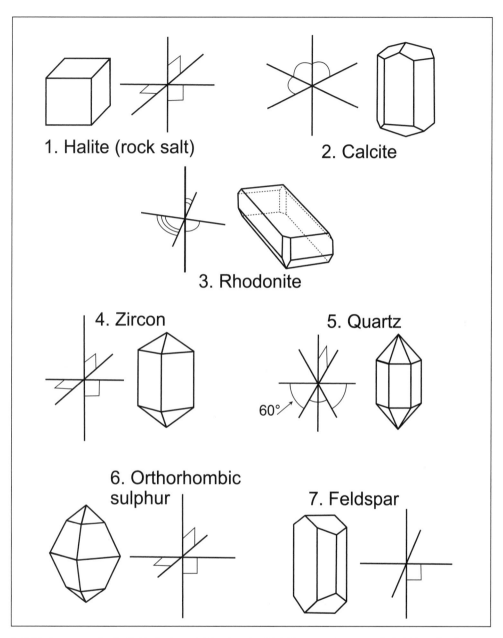

1. Halite (rock salt)

2. Calcite

3. Rhodonite

4. Zircon

5. Quartz

60°

6. Orthorhombic sulphur

7. Feldspar

1. Halite is an example of a Cubic (or isometric) system in which three equal axes are all at right angles. 2. Calcite is formed with a Trigonal (or hombohedral) crystal system in which the three axes are set obliquely at equal angles to each other. 3. Rhodonite is a crystal formed by a Triclinic system in which the three axes are unequal and set obliquely at unequal angles. 4. In a Tetragonal system minerals such as Zircon are formed by crystals with three axes, all at right angles, in which one axis is longer than the other two. 5. Quartz is an example of a Hexagonal system in which three axes at 60° angles to each other are positioned around a vertical axis that can be longer or shorter than the other three. 6. In an Orthorhombic system (such as Orthorhombic sulfur) the three axes are of unequal length and set at right angles to each other. 7. Finally, with a Monoclinic system, two of three axes are at right angles with the third set obliquely, such as in the example of Feldspar.

Bladed—These minerals are usually broad and flat. They are called bladed because they resemble a knife blade. Gypsum is an example of a bladed crystal.

Dendritic—These minerals have a dendritic (fingerlike or treelike) branching pattern. An example is the mineral copper.

Equant or equidimensional—These minerals are approximately the same diameter in ever direction. Garnets are often seen as equant crystals.

Fibrous—These minerals resemble long fibers that are often easy to pull off like threads. Serpentine is a good example of this mineral habit.

Prismatic—These minerals are elongated in one direction. Good examples are tourmaline and manganite crystals.

Striated—These minerals have evidence of striations (parallel grooves) on the crystal face. Pyrite crystals often exhibit striation.

Tabular—These minerals exhibit thick, flat plates of crystals. Good examples are orange, tabular plates of wulfenite.

Scientists don't use just a single crystalline habit to identify minerals. They also use aggregates of crystals, or a collection of crystals that form together in a "habit." For example, some crystals form outward in a wagon wheel-spoke type of pattern internally, while outwardly the crystal aggregate may be rounded and nodular. Such a collection of crystals resembles a bunch of grapes—called botryoidal—and includes the minerals hematite and psilomelene. Larger and more rounded shapes are called mamillated, and include chalcedony (a form of quartz) and goethite. Mineral aggregates that form flat sheets easily separated like pages of a book are said to be foliated—for example, the mineral muscovite mica.

What is the **softest crystal**?

The softest crystal is the same as the softest mineral—talc. While most talc is commonly found in granular or fibrous chunks, it also is found (though rarely) in crystalline form (it belongs to the monoclinic crystal system).

How is **talc used**?

Talc is used in many ways because of its softness and "perfect" cleavage in one direction. For example, it is used as a lubricant (where there is not too much pressure or stress), as so-called talcum powder, and for use in cosmetics. Some impurities in certain varieties of talc cause a more slippery, greasy feel to the stone, and is therefore called "soapstone." Often found in large masses, some soapstone is much "stronger" thanks to impurities, and it is used for woodstoves or in carving (although a statue of talc is much more fragile than marble).

A CLOSER LOOK AT A
FEW COMMON MINERALS

What is **quartz**?

Quartz (silicon dioxide, or SiO_2) is one of the most well-known and abundant minerals on Earth. Although it can eventually be broken down, it is very physically and chemically resistant to weathering. When quartz-bearing rocks weather, the quartz grains become concentrated in the soil, in the river, or on the beaches. White sands found along river banks and beaches are often composed of mostly quartz, with some white or pink feldspar.

Quartz can be crystalline or cryptocrystalline. The crystalline varieties have larger molecular structures, are more transparent (you can see light through the crystal), and are often used as a semiprecious gemstone (such as amethyst, citrine, rose quartz, and smoky quartz). The cryptocrystalline varieties have compact molecular structures, are usually opaque (light cannot penetrate) or translucent (light can penetrate, but it is not clear), and are often gemstones (such as agate, jasper, and onyx).

In its purest form, quartz is colorless (often called rock crystal or mountain crystal). The colors in other quartz crystals are caused by impurities within the molecular structure of the mineral. For example, when the impurities include tiny air bubbles, quartz becomes opaque (called milky quartz); if the impurities include other minerals, it can turn various colors (such as brown smoky quartz or violet amethyst).

What are some different **forms of silica**?

While quartz has a specific crystalline form (hexagonal), there are other forms of silica that are non-crystalline or a different crystalline form. Other forms of silica include opal, flint, and chert (non-crystalline). Chalcedony is also a different form of silica and is given other names depending on the color, such as brown (jasper), reddish brown (carnelian), and multicolored banding (agate). Four other forms of silica—cristobalite, tridymite, coesite, and stichovite—are actually polymorphs of quartz (same chemical composition but different crystal forms).

Why is **quartz** used in **electronics**

Besides being transparent, tough, and having a constant chemical composition, there are two other reasons why quartz is so versatile for use in electronics. First, quartz is one of several piezoelectric minerals. This means that when pressure is applied to the mineral, a positive electrical charge is created at one end of the crystal and a negative electric charge at the other end. Quartz is also strongly pyroelectric: Temperature changes create positive and negative charges within the crystal.

An inspector checks the quality and weight of individual pieces of quartz at a factory in Gus-Khrustalny, Russia. *AP/Wide World Photos.*

But natural quartz is not good for electronics because there are too many impurities and physical flaws in the mineral. Thus, scientists developed a way to make cultured quartz with a commercial process that manufactures pure, flawless, electronics-grade quartz crystals grown very carefully in highly controlled laboratory conditions. First, a seed crystal called a lascas—a small piece of carefully chosen, non-electronics grade quartz—is needed. From there, like an oyster pearl growing around a piece of sand, the manufactured crystal grows as a "perfect" crystal. Cultured quartz is always needed for the electronics industry, and about 200 metric tons are produced each year.

Where can someone **collect** good **quartz crystals**?

In many places, it is easy to collect quartz crystals. Quartz is a very common mineral, often found as veins cutting through various rocks all around the world. Spaces within rocks called vugs are often good places to uncover quartz. First, find the white tell-tale quartz vein, and then chip away, looking for vugs that may hold quartz crystals. (But *always* ask permission to collect on private land before you start looking for quartz, any other crystal, or rock.) You can also obtain good quartz crystals from rock shops or at rock, gem, and mineral shows (many of the quartz crystals you buy in the United States come from quarries near Hot Springs, Arkansas). And finally, many mines and quarries also let you search for quartz crystals for a fee.

How was mica once used in the past?

Mica breaks into thin, flat sheets similar to panes of glass. Thus, huge sheets of mica were once used for window panes, but its availability depended on a person's location. Because of its heat and electrical resistance, mica was also used long ago for oven windows in Old Russia, which was known as "muscovy." This mineral, often called Muscovy Glass (after the Latin term *vitrum Muscoviticum*), was formally renamed "muscovite" in 1850 by the famous American mineralogist James Dwight Dana (1813–1895).

What is **feldspar**?

Feldspars form one of the most abundant and important mineral groups. They are silicates of aluminum, with potassium, sodium, calcium, and (rarely) barium, crossing over two crystal systems: the monoclinic and triclinic systems (but all the crystals resemble each other closely in angles and mineral habit). Feldspars can be grouped into the potassic (potassium feldspars, including sanidine, orthoclase, and microcline) and plagioclase feldspars (sodium or calcium [the elements can substitute for one another within the feldspar], including albite, oligoclase, andesine, and anorthite).

Feldspars are useful for various purposes. For example, microcline feldspar is often used to manufacture porcelain by grinding it finely and then adding kaolin (or clay) and quartz. When the mixture is heated to a high temperature, the feldspar fuses, acting as a cement to hold the materials together. Polished feldspars are used for ornamental purposes, including amazon stone (a microcline feldspar) and labradorite (a plagioclase feldspar, also called moonstone or sunstone).

What is **mica**?

Mica is a group of silicate minerals (chemically all contain silica [SiO_4]), comprising over 30 members. All mica crystals are six-sided and rank about 2 to 3 on Mohs' Scale of hardness; they are heat-resistant and do not conduct electricity. Because their molecules combine to form distinct layers, mineralogists refer to micas as sheet silicates. The physical property of cleaving into thin layers is so distinctive that other minerals that break in a similar way are said to have micaceous cleavages.

What are the different **types of mica**?

There are a number of different mica minerals, including muscovite (also called white or clear mica, it contains aluminum), biotite (black and containing iron and magnesium), and lepidolite (purple, and containing lithium and aluminum). There are also several lesser-known micas, including glauconite, paragonite, phlogopite (brown, and containing iron and magnesium), and zinnwaldite micas. Both muscovite and phlogopite are the two micas most often used as a commodity.

What is the **concern** about the mineral **serpentine**?

Serpentine is a fibrous magnesium silicate with a hardness of 3 to 5 on Mohs' Scale of hardness. It forms in igneous rocks containing olivine and orthopyroxene, or in serpentines, rocks that have formed by the alteration of olivine-bearing rocks. And it also occurs mainly as the fibrous chrysotile, which is the most common and valued type of asbestos (also called white asbestos). In fact, of the five types of asbestos, chrysotile accounts for almost 90 to 95 percent used in United States buildings.

Long used for such items as fireproofing fabrics and in brake linings, asbestos no longer has a good reputation. According to the U.S. Environmental Protection Agency (EPA), asbestos can release microscopic bundles of fibers into the air, though usually only if it is disturbed. These particles can be inhaled, creating a multitude of long-term (but not short-term) health problems. The most common asbestos-related diseases are asbestosis, a lung disease first found in Naval shipyard workers, in which the lungs release an acid to dissolve the particles, which causes scar tissue buildup that accumulates in the lungs over many years; mesothelioma, a cancer of the outer lung and chest cavity; and lung cancer, which is often exacerbated by cigarette smoking.

Of course, serpentine as asbestos is not the only concern. Another mineral called vermiculite—a magnesium aluminum iron silicate—has been used for more than 80 years in industry, construction, and the horticultural markets. But because some vermiculite ores from certain areas sometimes contain asbestos, the EPA warns consumers to be cautious about using vermiculite, especially as a seed starter or amendment to potting soil for home gardeners, just in case the vermiculite is contaminated with asbestos. To be safe, use vermiculite in a well-ventilated area or outside, keep it damp to keep down the dust, or try other soil additives.

What is **salt**?

Salt ($NaCl$, sodium chloride, or halite) is a necessity to most living organisms on Earth. Humans and other animals need salt to live, and thus have always valued salt licks, springs, and marshes to satisfy the natural craving for salt. For humans through the centuries, it has been used to preserve meats, especially in hot climates.

Pure halite is clear and colorless, though it is often tinged by impurities. The mineral is soft, and crystals are cubic in shape. Salt is readily soluble in water, making it useful for cooking, food preservation, and chemical production.

GEMS GALORE

Why are certain minerals **classified as gemstones**?

There are numerous definitions of "gems" or "gemstones" (also seen as "gem stone"). Some define a gemstone as a special mineral that has some intrinsic or monetary value, especially if it is rare; some believe gems are objects of art, showing clarity, beauty, and durability; still others consider gems as stones that can be cut and polished to use for ornamentation. In general, the best definition for a gemstone seems to be all of the above.

Why do people **collect gemstones**?

Of all the reasons to collect rocks and minerals, one of the most popular is to obtain gemstones. The reason is almost as clear as many gems: Many of the crystals, especially if they are close to flawless, are beautiful. Because of this beauty—and for some crystals, their rarity—certain gems are collected both for their aesthetic and monetary value.

What **folklore** surrounds various **gemstones**?

Gems have been around since the beginning of human history. Certain cultures attributed various attributes ("powers") to owning such stones—from healing to courage. The following lists some of the attributes and folklore behind various gems (with many claims seemingly exaggerated):

Amethyst—This purple gem (a variety of quartz) is said to make the wearer gentle and amiable. Amethyst attributes include dreams, healing, peace, love, spiritual uplift, courage, protection against thieves, and happiness.

Garnet—Garnet attributes include healing, protection, and help to cure depression. In the past, friends would exchange garnet to symbolize affection—and to ensure they would meet again.

Jade—In ancient China and Egypt, jade was widely used as a talisman to attract good fortune, friendship, and loyalty. For many cultures, jade has various attributes, including protection for the kidney, heart, larynx, liver, spleen, thymus, and thyroid. The gem also helps give strength to the body and gives the wearer virtue, fidelity, humility, generosity, and balance.

101

This cushion-shaped 200-carat "ultimate blue" sapphire is worth more than a quarter of a million dollars. *AP/Wide World Photos.*

Opal—In France, opals are said to bring bad luck; in English-speaking countries, it brings good luck. Overall, opal helps to uncover buried emotions and symbolizes confidence, purity, and serenity. It also is thought by some cultures to cause tempers to flare and fits of passion, while others believe it helps the wearer to reach the highest spiritual levels and increase internal vision.

Turquoise—Turquoise has been found in ancient Egyptian tombs. Native Americans also placed the stones in tombs to guard the dead, while warriors tied the gem to their bow to ensure an accurate shot. Its attributes include the gaining of money, love, protection, healing, courage, friendship, and luck, and it eases mental tension.

Pearl—Pearls are actually organic gemstones, the natural ones forming in the shells of oysters on the ocean bed. Because pearls have been known for centuries, there is a great deal of folklore surrounding these gems: An ancient Chinese myth included pearls falling from the sky when dragons fought; the Greeks believed that wearing pearls promoted marital bliss. Overall, pearls are thought to give wisdom through experience, protect children, and to hold together engagements and love relationships. Even cultured pearls (those caused by human interference when a foreign tissue is inserted into a live oyster shell) have certain traditional influences: They are thought to offer the power of love, money, protection, and luck.

What are **birthstones**?

Birthstones are gems that humans have categorized by month. Depending on the month you were born, that stone becomes your birthstone. No one knows where the idea of the birthstone originated, but some researchers believe it is from the Breastplate of Aaron (the brother of Moses), a religious caftan arranged with twelve gemstones representing the twelve tribes of Israel. The following tables list the traditional and modern birthstones (note: there are even older gemstone lists, such as one from the Chinese that goes back over a thousand years):

Month	Modern*	Traditional**
January	Garnet	Garnet
February	Amethyst	Amethyst
March	Aquamarine	Bloodstone
April	Diamond	Diamond
May	Emerald	Emerald
June	Pearl, Moonstone	Amazonite
July	Ruby	Ruby
August	Peridot	Sardonyx
September	Sapphire	Sapphire
October	Opal, Tourmaline	Tourmaline
November	Yellow Topaz, Citrine	Citrine
December	Turquoise, Blue Topaz	Zircon, Lapis

* Modern birthstones were officially adopted in 1912 and sanctioned by the American National Association of Jewelers.

** Other "traditional birthstone" lists differ, depending on the culture of origin, but all such lists are thought to represent societal traditions.

Why are **emeralds famous** throughout history?

Emeralds (a green beryl) have evoked quite a few legends throughout history, it's a gemstone treasured by different worldwide cultures for close to 5,000 years. Egyptian mummies were often buried with emeralds on their necks; the gems were carved with the symbol for eternal youth. Cleopatra, the queen of Egypt, prized her emeralds more than any other gem she possessed. Her ancient emerald mines were lost for centuries, until they were rediscovered again in 1818 near the Red Sea. (No good quality emeralds have been found there since, as the mines were exhausted, they had been worked 2,000 years before Cleopatra, which explains why they were originally abandoned.) The Romans even got into the act: It is reported that the Emperor Nero peered through flat emerald crystals to watch the gladiators at the Coliseum.

But the Egyptians and Romans were not the only ones: The Mughals of India—who ruled from 1526 to 1858, and included the builder of the Taj Mahal, Shah Jehan— inscribed emeralds with sacred text, then wore them as talismans that were collectively called the Mughal emeralds. In 2001, a 217.80-carat Mughal emerald sold for a record 2.2 million dollars at auction; it was thought to have originally come from Colombia via Spanish traders. The Incas of South America had high-quality emeralds, too, probably mined from such places as Colombia. These gems were so valuable that in the 1500s the Spanish Conquistadors mined or took many of the more valuable stones. In fact, real treasures of emeralds might be at the bottom of the ocean in sunken Spanish ships.

What are **organic gemstones**?

Organic gemstones are just what the term implies: gemstones made from organic material. Some common organic gemstones are pearls (formed as an irritant enters the shell of an oyster; the animal responds by coating and recoating the irritant with layers of a hard material called aragonite); amber (ancient resin, usually from pine-like trees, that solidifies over time; dead insects are often found trapped in fossil amber); coral (a marine animal called a polyp excretes a calcium shell to protect itself; after it dies, the next generation of polyps builds on top of its predecessor); and jet (very compact black coal that can be made into beads and other jewelry).

What are some of the most **famous red spinel gemstones**?

When someone mentions famous gemstones, most people think of diamonds, emeralds, or sapphires. But there are other gemstones just as valuable, including one called red spinel, a gem often mistaken for a ruby. It is easy to see why this happens: Top quality red spinels and rubies have superb, pure-red colors, and they actually fluoresce, or glow, in natural light.

One of the most famous red spinels is the Black Prince's Ruby (it was misnamed "ruby" long ago). The stone weighs about 170 carats and is about the size of a hen's egg. It is also one of the world's most cherished jewels, as it is the mainstay piece at the center of the Crown Jewels of England.

Another famous red spinel is the Timur Ruby, also in the Crown Jewels of England, weighing in at 361 carats. It is inscribed with the names of six of its former owners. Still another red spinel is owned by the Russian government and is found at the Kremlin Museum in Moscow. This 414-carat gem probably belonged to the tsar.

The most dazzling collection of red spinels is found in the Crown Jewels of Iran. The majority were plundered from India when the Mughal Empire fell. The largest one weighs about 500 carats—the biggest to date on record—while most of the others weigh over 100 carats.

Australia's largest diamond, an uncut 104.73-carat gem, was mined in the Northern Territory and is larger than the Hope diamond and the Dresden diamond combined. *AP/Wide World Photos.*

Where are **diamonds found**?

Diamonds are usually found in pipes ("diatremes") in which certain Precambrian rocks exist. In general, these pipes are composed of an igneous rock called kimberlite—named after Kimberley, South Africa, where it was first discovered—an unusual rock made up of olivine, serpentine, mica, ilmenite, carbonates, and other minerals. Diamonds are most often found in rock that includes the kimberlite matrix, fragments of both or one of the rocks called eclogite (mostly garnet and pyroxene) and peridotite (of the variety harzburgite, mostly made of olivine and pyroxene), and chunks of sedimentary or local rock. Most kimberlite pipes are shaped like an upside-down ice cream cone and were former volcanoes.

But kimberlite is not the only place to find these gems. Diamonds from the world's most productive mine, the Argyle mine in western Australia, are found in lamproite, which is a close relative of kimberlite but richer in aluminum, potassium, silicon, and fluorine, and including less carbonate minerals. The shape of underground lamproite differs, too. Its upper extremes are wider where magma rose through the narrow pipes and came explosively into contact with cool, near-surface groundwater.

Diamonds also occur in secondary deposits, such as in sands and gravels that are a result of erosion of kimberlite or lamproite. Sources of these types of diamond deposits occur in India, Namibia, and Brazil.

Do some diamonds come from space?

Yes, scientists have discovered that some diamonds come from space, but don't start looking for them to rain from the sky. These space diamonds are microscopic and come from certain meteorites discovered on Earth. One in particular is the Murchison meteorite, a space rock that contains minute diamonds around 4.5 billion years old. Scientists would love to discover more such meteorites, as these diamonds contain the isotopic (radioactive) "fingerprints" of the exploding stars that originally tossed them into space.

Where are **major diamond deposits** found?

Diamonds were first found as secondary deposits—sands and gravels from erosion and deposition of diamond-bearing rock—in India over 2,000 years ago. (Diamonds were first discovered in that country around 600 B.C.E.) By 1730, Brazil also became known for its diamonds that were also found in secondary deposits. It was not until the 1870s that South Africa became known for diamonds from the now-famous kimberlite mines. Russia (1950), Botswana (1966), Australia (1979), and Canada (1991) all became diamond producers, too. Currently, over 20 countries produce more than 110 million carats of diamonds each year.

What are **black diamonds**?

Brazillian diamond hunters made an astounding discovery in the 1840s: rare black diamonds they called carbonados, the Portuguese word for "burned" or "carbonized." (Other carbonados have since been found in central Africa.) Unlike the single-crystal diamonds we are all familiar with, black diamonds are aggregates of individual crystals, giving the gemstone its dark color.

These coal-looking diamonds are over 3 billion years old, but no one knows how they originated. Some scientists believe they came from earthly organic carbon sources, similar to all diamonds found on Earth. But there is a definite problem with that idea: There were not enough organisms on Earth 3 billion years ago to form a carbon deposit, much less a diamond. In addition, chemical analysis of the carbon in the black diamonds shows they resemble surface carbons, unlike most diamonds forged deep in the Earth.

Because of the organic and crustal conundrum, other scientists believe these diamonds are from outer space, where a place in which carbon has been detected under many circumstances. Theories vary, but some believe the pressures from impacting space bodies (such as comets, asteroids, or other early planetary debris) could have carried the necessary carbon and created enough heat and pressure to form these diamonds.

What are the **largest diamonds** found to date?

The largest cut diamond in the world is the Cullinan I, also known as the Star of Africa, weighing in at around 520 carats with 74 facets. It was cut from the largest rough diamond ever found—the Cullinan—that weighed over 3,106 carats (over 1.25 pounds). In 1908, a diamond cutting firm cut the Cullinan I and 104 other diamonds from the original rough stone. Today, the Cullinan I sits in the Tower of London, set in the scepter of King Edward VII.

The second largest diamond ever found was the Excelsior, which in the rough originally weighed about 995 carats. It was found by an African mine worker as he loaded his truck. (His mine manager rewarded him with money, a horse, and a saddle.) The original stone was cut into ten pieces; those pieces were in turn cut into 21 gems weighing 1 to 70 carats. The third largest rough diamond ever found was the Star of Sierra Leone, weighing about 969 carats; the fourth largest, the Great Mughal (often seen as Mogul), weighing about 793 carats, was named after Shah Jehan, builder of the Taj Mahal.

The Hope Diamond, probably the world's most famous diamond, is a 45.52-carat blue diamond with an unusual history. It was discovered in India over 350 years ago and was sold to King Louis XIV of France in 1668. In Louis' hands the stone was called the French Blue and remained among the French crown jewels for more than 100 years. It was later stolen and reappeared in London, re-cut to its present size and shape. *AP/Wide World Photos.*

What are some **famous diamonds**?

Some of the most famous diamonds are not the biggest. For example, the Blue Hope (or Hope) diamond, weighing about 46 carats, was once owned by French king Louis

XIV. Officially called the "blue diamond of the crown," it was stolen during the French Revolution. It showed up again in 1830 and was bought by London banker Henry Philip Hope. Like the "curse of the mummy," there is also believed to be a curse on this stone because bad luck seemed to follow many family members of the diamond's owners (in reality, the reports of poverty, deaths, and suicides were probably due to the owners' and families' own doing, not a curse). The deep-blue, emerald-cut diamond became the property of the Smithsonian Institution in 1958.

Another famous diamond is the Koh-I-Noor ("Mountain of Light"), which weighs about 186 carats and was one of the large diamonds ever found. This gem has had a long and controversial history. It was first mentioned in records around 1304, when it was acquired from an Indian family that had allegedly owned the stone for generations. Another account places the origins at 1526, when it was mined near the Krishna River. Still a different report has the gem adorning the peacock throne of India's Shah Jehan in the 1600s. Its later history is more clear: After being stolen from India in the mid-1700s, it was taken to Iran, acquired by the Afghans, and finally lost to Indian rulers over a span of just under a century. When India came under British rule in 1849, the diamond's travels were over: It was recut down to a "mere" 108 carats to enhance its brilliance during the reign of Queen Victoria, and now forms part of the British Crown Jewels.

There are also diamonds with a mythical background. But, like all legends, few people know the real story. For example, the pear-shaped Idol's Eye diamond (weighing about 70 carats) gets its name from its legendary origin: The Sheik of Kahmir stole the diamond from an idol's eye to pay his daughter's ransom to the Sultan of Turkey. And like all long-term histories, there are diamonds with confounding backgrounds. One such diamond is the Sancy, the confusion being that another diamond had that same name.

What is the **difference** between **precious and semi-precious stones**?

Precious stones (or gems) are considered to be the most valuable gemstones and are the most rare and physically hardest. Diamond, emerald, and ruby are considered precious stones. Semi-precious stones are usually softer and less valuable, and thus not as rare as precious gems. For example, the mineral olivine has a semi-precious variety called peridot that can be cut and faceted like any other gemstone; jade, garnet, amethyst, citrine, rose quartz, tourmaline, and turquoise are all examples of semi-precious gems. There are two major exceptions to the rule: Because of their rarity, opals and pearls are considered to be precious, even though they are both physically soft.

Are **synthetic gemstones** as good as natural gemstones?

According to the Federal Trade Commission (the United States agency acting in the
interest of all consumers to prevent deceptive and unfair acts or practices in the gem-

stone trade) synthetic gemstones must be identical to the natural form in every way, especially with regard to composition, hardness, and optical qualities.

What are **natural, simulant (or simulated)**, and **synthetic** gemstones?

The following lists the differences between the various types of gemstones produced in various countries:

Natural—Natural gemstones are just what the word implies: They form naturally in the environment. They include most of the familiar gems (diamonds, emeralds) and organic gemstones (amber, coral, mother of pearl).

Synthetic—Synthetic, or laboratory-grown, gemstones are essentially the same in appearance—optically, chemically, and physically—as the natural materials they represent. In the United States, synthetic gemstones range from garnets and emeralds to rubies and sapphires. In many cases, it is very difficult to know the difference between synthetic or natural without being told.

Simulants—Simulants (or simulated gemstones) are also grown in the laboratory. They may look like the natural gemstone, but they differ optically, chemically, and physically. In the United States, one popular simulant is cubic zirconia, a stone that closely resembles a diamond to the untrained eye; if colored, they can resemble many other gems, such as emerald or topaz. Other gemstones such as coral, malachite, and turquoise can be used as simulants.

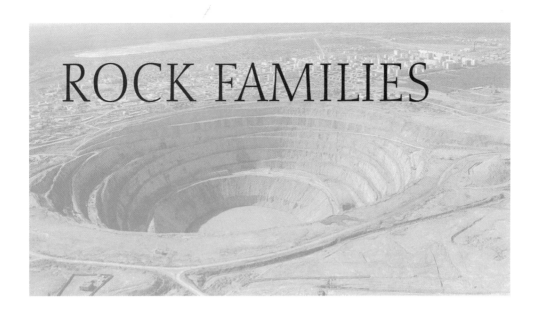

ROCK FAMILIES

ROCK DIVISIONS

What are **rocks**?

Simply put, rocks are naturally occurring aggregates (collections) of minerals. What makes rocks so different is their diversity—they can range from masses of minerals formed by volcanic action, from eroded sediment, or from great pressures and high temperatures. Rocks are usually composed of more than one mineral, although in rare cases they can be entirely composed of a single mineral, in which case they are called monomineralic rocks.

What is the **rock cycle**?

The rock cycle is the continuous process that changes rock types over time. For example, metamorphic rock melted into magma can cool into igneous rock, then, with heat and/or pressure, return to it's metamorphic state.

What are some of the **other natural cycles** that affect rocks?

There are many natural cycles on Earth, such as the nitrogen cycle that supplies nitrogen to the atmosphere, the water cycle, and the carbon cycle.

In terms of rock, the water cycle has the most effect. As rain falls, water erodes the land by weathering rock; in turn, it also deposits sediment elsewhere, building up certain landforms. Some water seeps into the groundwater or into river basins, helping to carve those landscapes, too. The surface water eventually evaporates to start the entire cycle again.

The carbon cycle involves mostly organic components, since carbon and carbon dioxide are the basis for all living organisms. But carbon also has a part in rock formation. It is stored in inorganic forms—fossil fuels like coal, oil, and natural gas and carbonate-based sedimentary rock like limestone—and in organic forms, such as organic matter and humic substances in soils. Carbon dioxide is also released by volcanoes as carbon-rich sediments, and sedimentary rocks are subducted and then partially melted beneath lithospheric plate boundary zones. (For more information about lithospheric plates, see "Examining Earthquakes.")

What are the **major rock families**?

The major rock families are igneous, sedimentary, and metamorphic. The following lists some details of these rock families:

Igneous rocks—These rocks form as magma (molten silicate material) from volcanic activity becomes solid. Igneous is from the Latin word *ignis,* meaning "fire."

Sedimentary rocks—These rocks form as sediment—erosional products of igneous, metamorphic, or other sedimentary rocks—crystallizes from a solution such as water.

Metamorphic rocks—Metamorphic comes from the Greek *meta,* meaning "change," and *morph,* meaning "form." These rocks form as existing rock layers—igneous, sedimentary, or metamorphic—are changed by extreme heat, pressures, and sometimes chemical means, all while the rock remains in a solid state. These rocks are most often found around mountain-building or volcanic regions.

What does the term **facies** mean when talking about rocks?

Facies is a word that refers to the characteristics of a rock that describe how it was formed. For example, a metamorphic facies is a group of rocks that were metamorphosed under similar temperature and pressure conditions, creating a characteristic metamorphic mineral group. Good examples are greenschist or blueschist, which form from the same temperature and pressure conditions regardless of the original parent rock. A sedimentary facies means rock from a particular depositional environment that distinguishes it from other facies within the same rock group.

Where does the **magma originate** that forms **igneous rocks**?

The magma that forms igneous rocks originates deep within the Earth's crust. Much of the magma comes from magma chambers, a rather loose term that means a place hot enough to melt the rock and with pressures low enough to cause minerals (actually their elements) to exist in a liquid state.

What are the **two major types of igneous rocks**?

In general, there are two types of igneous rock. *Plutonic* rocks slowly crystallize at depths far below the Earth's surface. Containing coarse crystals that can be seen with the naked eye, they include granite, granodiorite, gabbro, and diorite. *Volcanic* rocks form from erupting volcanoes with most of the associated magma quickly cooling. Containing fine crystals not obvious to the naked eye, types of volcanic rocks include rhyolite, andesite, and basalt.

Why is **texture** used to identify **igneous rocks**?

Texture is one main way to identify the multitude of igneous rocks and is based on the size of the individual mineral grains after magma turns to a solid. In general, mineral size is most often determined by how fast or slow the magma cooled. Grain sizes can vary greatly in igneous rocks, from plate-sized feldspar grains to dense, fine-grained rhyolites, materials that cooled so quickly that there are no visible mineral grains.

What are **intrusive and extrusive igneous rocks**?

In general, the faster the magma cools, the smaller the crystals within the resulting rock. Small-grained rocks generally form from volcanic eruptions and are often referred to as *extrusive* rocks; they are also called aphanitic. For example, basalt is a fine-grained aphanitic igneous rock.

If cooling occurred slowly—usually this happens deep inside the crust where rock layers form an insulation—the crystals are larger. These coarse-grained igneous rock are often called *intrusive* rocks. Gabbro and granite are examples of these coarse-grained types of igneous rocks, which are also referred to as phaneritic.

Of course, not everything is black and white in any science, including in the case of igneous rocks. Thus, there are many intermediate stages between these fine and coarse-grained igneous rocks.

What is **country rock**?

Among geologists, country rock is not a type of popular music; rather, it is any rock intruded by and surrounding an igneous intrusion. In other words, it is the rock that already exists in an area before molten rock forces its way through the rock layers.

What are **dikes and sills**?

Dikes (or dykes) are igneous rock masses that cut across cracks and fractures in country rock. They usually are offshoots of a magma mass, working their way into an area by following areas of weakness, especially along cracks or joints in rock layers, and then solidifying. When they are exposed at the surface in an outcrop—whether as igneous, sedimentary, or metamorphic rocks—they are often seen as a ribbon of igneous rock snaking through country rock.

A sill forms in the same way as a dike, but in this case it follows only horizontal weaknesses in the rock. Most often sills are seen as intrusions along different rock layers. It is often difficult to tell the difference between a sill and lava flow deposit. In most cases, a sill can be distinguished by the presence of metamorphosed rock on the top and bottom country rock layers (caused by the heat of the magma intrusion). A buried lava flow is usually associated with metamorphosed rock only on the bottom country rock layer.

What is a **pegmatite**?

When hot magma pushes through the crust from a deep magma chamber, it puts a great deal of pressure on the overlying rock. This pressure creates cracks within the rock, allowing magma to push its way in and intrude as dikes in the rock layers. Some of the magma also partially melts parts of the overlying crust. This partial melting and the escape of volatile gases as the magma cools create a unique rock called a pegmatite. These rocks contain crystals that measure from less than an inch (2 centimeters) to 16 feet (5 meters) across.

Why is **composition** used to identify **igneous rocks**?

Composition is another good way to identify igneous rocks. As the term implies, composition refers to the elements within the magma that eventually cool to form certain minerals. Although there are intermediate compositions, there are two basic types of igneous composition: mafic and felsic. Both are indicative of where the magma formed.

In general, origins of mafic and felsic igneous rocks have a great deal to do with plate tectonics. (For more on plate tectonics, see "The Earth's Layers.") Mafic igneous rocks are those associated with crustal spreading. These sites usually produce basalt if the magma erupts at the surface, and gabbro if the material remains in the magma chamber. (These two rocks have the same composition, but differ in terms of texture.) Mafic rocks also include silicate minerals and rocks high in heavier elements. They are usually dark in color and are relatively heavy. The name mafic is taken from the "ma" in "magnesium" and "fic" from the Latin word for "iron" (derived from *ferrum*). In actuality, however, mafic rocks are rich in calcium and sodium. Some common rock-forming mafic minerals include olivine, biotite mica, and plagioclase feldspar.

Felsic igneous rocks refer to those associated with the crustal collision of plates (compression) and subduction. At these sites on Earth, the rock is dragged into the crust and remelts, resulting in a magma that becomes enriched with the lighter elements. Felsic magmas produce rocks such as granite (if intrusive; this is the most common felsic rock), or rhyolites (if extrusive). They also include silicate minerals and rocks with a lower percentage of heavy elements. Thus, they are high in the lighter elements, such as silica, oxygen, aluminum, and potassium. Felsic is derived from the "fel" in "feldspar" (the orthoclase or potassium feldspars) and "sic" in "silica." They are usually light in color and are lighter in weight than mafic rocks. Common rock-forming felsic minerals include quartz, muscovite mica, and orthoclase feldspar.

There are rocks in between malfic and felsic, too, which are called intermediate magmas and are also associated with crustal plate collisions and subduction. These magmas produce rocks such as diorite (if intrusive) or andesite (if extrusive). They form from the intermediate magmas produced during the conversion from mafic to felsic rocks.

How do the majority of **sedimentary rocks form**?

The majority of sedimentary rocks are formed by two major processes active at the Earth's surface: deposition and precipitation (collectively called sedimentation). To produce sediment, the surface of our planet is worn down by physical means, such as wind, rain, moving ice, chemical weathering, gravity, organic processes, and other less obvious ways, such as weathering from the sun's radiation. This produces rock "waste" ranging from boulder-sized rocks to fine sediment.

Although sedimentary rocks can include larger rock fragments cemented together, such as a conglomerate, the majority are composed of fine sediments. These small particles are transported by water and eventually deposited at river mouths, in lakes, in the oceans, or on land. Such deposits can cover great expanses, and for thousands to millions of years, more and more layers of material build up. Eventually, they solidify, mainly from the pressure of the above sediment layers. The end results are sedimentary rocks consisting of various-sized particles cemented together by certain mineral matter called the matrix.

What are the **types of sedimentary deposits**?

Sedimentary rocks form in three ways and are usually classified as three types of deposits. The following lists these types of rocks. (Note: The origins of some rocks, such as limestone, chert, and dolomite, are sometimes difficult to differentiate between chemical and organic processes):

Clastic—Sedimentary rocks that form by the mechanical accumulation of rock fragments are referred to as clastic deposits. These include sandstones, conglomerates, and shales. The most common constituent of clastic sedimentary rocks is quartz.

Chemical—Sediments deposited through precipitation from solution (or particles that drop out of a liquid solution) are called chemical deposits. For example, evaporites are chemical sedimentary deposits.

Organic—Deposits involving activity by living organisms to create sedimentary rock are called organic. They include coal and oil shale.

115

This beach at St. Joseph Peninsula State Park on the Gulf of Mexico in Florida's panhandle is noted for its strikingly soft, white sand and windswept dunes. *AP/Wide World Photos.*

What is an **unconformity**?

As sediment accumulates on the Earth's surface, it eventually forms layers of sedimentary rock. But there are times when the sediment stops accumulating (such as in a shallow basin when sea level drops) and the existing sediment and underlying rock are stripped off by such erosional agents as ice, wind, and water. Tons of material is removed, along with that part of the rock record. When we see an outcrop with such a break in a sequence of layers, it is called an unconformity.

There are three types of unconformities: In a *disconformity* the layers above and below the break are horizontal; an *angular unconformity* occurs when the overlying layers are not parallel to the base layers; and, finally, a *nonconformity* has igneous or metamorphic rocks as the bottom layer with a sedimentary rock layer on top.

What is the process of **diagenesis**?

Diagenesis is the process that most sediment goes through after deposition, in which it becomes compacted and cemented to form a solid rock mass. Most clastic fragments undergo diagenesis. For example, fine-grained clay and quartz sediments most often form shale; coarse grain sediments consisting mostly of quartz form sandstones (although they may also contain feldspars and other rock fragments); and poorly sorted clastic rocks include conglomerates and breccias. The amount of time it takes for

Are all beach sands made of quartz?

No, not all beach sands are made of quartz. In fact, there are many types of sands, although the ones we're most familiar with are mostly quartz. A "clean" sand consists of about 90 percent quartz; a "dirty" sand has more than 10 percent other material and/or silt mixed in.

Some sands contain relatively no quartz particles at all. For example, you can find light-green olivine sands in Hawaii, which is a definite indication of volcanic activity. There are black sands on the islands, too, which are made of basalt and other volcanic materials. The White Sands National Monument in New Mexico isn't quartz, either, but rather a mix of gypsum grains. And some sands in Florida and the Florida Keys are made of sand-sized fragments of broken shells and skeletons of marine organisms that have been crushed and fragmented by shoreline surf.

the rock to solidify varies depending on what landforms are present, the materials deposited, and the climatic conditions.

What are some **sedimentary rocks** that form from **chemical solution**?

Chemical sedimentary rocks form as they precipitate out of concentrated solutions. For example, carbonate rocks are either directly or indirectly precipitated by carbonate-secreting organisms in shallow oceans.

Evaporites are another type of deposit from chemical solution (usually in shallow marine environments) and include salt (also called halite, table salt, or sodium chloride), anhydrite (calcium sulfate), and gypsum (an anhydrite with water molecules attached). As seawater evaporates, it concentrates high amounts of these various salts, creating evaporite deposits.

You can create a kind of precipitated sedimentary rock in your own backyard. Simply put water, a little dirt, and dissolved table salt in a pan; let the sun evaporate the liquid, and voilà! You have a salt deposit. But you will have to wait a very long time for the actual rock to form. It will take thousands of years for the salt deposit to solidify, assuming conditions are right.

How do scientists **differentiate sedimentary rock particles**?

Scientists measure the grain size of sedimentary particles to better understand sediments and how they accumulate, as well as to differentiate between rocks such as claystone, sandstones, and siltstones. The following lists some sediment grain size classifications:

117

Brilliant waves of gypsum sand form a dune on a 275-square-mile area in the White Sands National Monument in New Mexico, making it the world's largest field of gypsum dunes. *AP/Wide World Photos.*

Particle	Diameter (inches)	Diameter (millimeters)
boulder	> 10	> 254
cobble	2.5 to 10	64 to 254
pebble	0.15 to 2.5	4 to 64
granule	0.07 to 0.15	2 to 4
sand	0.0025 to 0.07	0.06 to 2
silt	0.000015 to 0.0025	0.00038 to 0.06
clay	< 0.000015	< 0.00038

What is the **difference** between a **conglomerate and a breccia**?

In general, the biggest difference between a conglomerate and a breccia is the rounding of the particles within each rock's matrix. In most cases, a conglomerate is composed of rounded particles (usually pebbles and sand cemented together). These particles became rounded by abrasion and collision in water (and sometimes by ice). This makes sense, since most of these types of sedimentary rocks are found along the channels of ancient rivers.

Breccia particles are more angular, indicating that they formed much quicker after the pieces were broken and cemented together. For example, breccias are often found in association with volcanic areas, where angular fragments have been cemented together with volcanic ash or lava. Limestone breccias contain angular pieces of limestone in a matrix of calcium carbonate that most often form in a rapid depositional environment.

What are the various **types of metamorphism**?

There are several types of metamorphism, including some that overlap each other:

Contact or thermal metamorphism—Contact or thermal metamorphism, as the term implies, occurs when the intense heat of an intruding magma changes the surrounding rock. This type of metamorphism is restricted to a zone surrounding the intrusion called a metamorphic or contact aureole. Outside this zone, the rocks are not affected by the intrusive event. For example, shale becomes slate, phyllite, or schist, when the minerals are aligned from pressure. But with contact metamorphism, the shale is baked by an intrusion, becoming hornfels (if fine-grained) or granofels (if medium- or coarse-grained).

Dynamic metamorphism—Dynamic metamorphism occurs when a rock is subjected to high pressure. This type of metamorphism is usually localized in fault zones. The resulting rock is crushed, and there are few resulting mineralogical changes.

Regional metamorphism—This is the most common type of metamorphism and usually occurs over a large area. The rock changes by an increase in pressure (often causing changes in the texture, such as the lining up of minerals) as well as temperature (which usually changes the rock's minerals). For example, the Taconic Mountains of New York and New England were formed by an ancient tectonic plate collision that created metamorphic rocks. Conversely, some rocks affected by regional metamorphism also change due to decreases in temperatures and pressures, which is called retrograde metamorphism.

Are there other **less common types of metamorphism**?

Yes, there are several other types of metamorphism that are less common or that represent crossovers with the more common types. *Hydrothermal metamorphism* involves hot solutions or gases percolating through fractures, causing mineral changes in neighboring country rock. This is common in basaltic rocks; rich ore deposits are often found in association with this type of metamorphism.

A variation of regional metamorphism is called *burial metamorphism,* in which the rock becomes metamorphosed in response to burial under deep layers of sediment. Temperatures at depths of several hundred feet can be greater than 572°F (300°C). New minerals grow within the rock, but in most cases it does not have the appearance of being metamorphosed.

Cataclastic metamorphism is caused by mechanical deformation, such as when two rock bodies slide past one another along a fault zone. The friction from this motion generates heat, deforming and crushing the rock. Although this may seem to be common because of all the fault zones on Earth, it actually isn't; it only occurs under certain special, narrow shearing conditions along a fault.

What causes **impact (or shock) metamorphism**?

Impact (or shock) metamorphism occurs when rocks are altered by great explosions or impacts. Such huge explosions usually occur with very large volcanic eruptions. (For more about volcanoes, see "Volcanic Eruptions.") But that's not the only cause. Impact metamorphic rocks are also produced from massive impacts resulting from meteorites, comets, or asteroids striking the Earth. (For more information about meteorites and Earth impacts, see "The Earth in Space" and "Geology and the Solar System.")

In both cases, ultrahigh pressures are generated in the impacted rock, producing certain minerals that form at very high pressures. For example, the immense pressures associated with impacting space bodies produce two silicate oxides—coesite and stishovite—that are only found in association with craters. They also produce certain features within the "ultra-metamorphosed" rock: essentially, microcracks in mineral grains called *shock lamellae* and larger features called *shatter cones*, which resemble crumbly, pointed ice cream cones within the rock layers.

What is a common way to **categorize metamorphic rocks**?

Metamorphic rocks are often divided into foliated and nonfoliated rocks—criteria that are based on their appearance. *Foliated rocks* have a banded or layered appearance because the minerals within the rock are in parallel alignment. They include schist, gneiss, and slate. *Nonfoliated rocks* include marble, hornfels, and quartzite and do not have banding. They are composed of only one predominant mineral with equally sized crystals.

What are **pelitic rocks**?

When shale is metamorphosed, it produces pelites or pelitic rocks. Shale originates as a rock composed of clay and quartz minerals. If the metamorphism is not intense, recrystallized minerals such as muscovite line up perpendicular to the stress direction and form slate. If the intensity of the metamorphism increases, phyllite forms; at even higher intensity, schist forms; finally, with even higher temperatures and pressures, gneiss is the eventual result.

FAMILIAR ROCKS

What are some **common minerals** found in **igneous rock**?

There are numerous igneous rock-forming minerals, with the vast majority belonging to the silicate group in which silica and oxygen are the basic elemental components. The following lists a few of the more common ones:

Feldspar—Feldspars include the mafic variety (plagioclase), which are mostly striated, and the felsic variety (orthoclase), which usually have no striations. (For more information about mafic and felsic rocks, see above.)

Quartz—Quartz is a hard mineral with no cleavage (ability of a mineral to break along flat planes; for more information about cleavage, see "All about Minerals"), and its crystals come in hexagonal (six-sided) shapes. It is known for its conchoidal fracture (it breaks in a way similar to glass). Quartz is the last mineral to form in a felsic (or granitic) rock, and is commonly found as the filling (matrix) between other minerals.

Mica—Mica is well known for breaking into thin sheets, since it only has one perfect cleavage (for more about cleavage, see "All about Minerals"). It can range from translucent (such as muscovite mica, a felsic mineral) to black (such as biotite mica, a mafic mineral).

Olivine—This mineral is most often seen as small, green to black crystalline-like chunks dotting basalts from places such as Hawaii.

Pyroxene—Pyroxene includes a group of minerals that are green to black in color. Members of this group include enstatite (found in gabbro and mafic diorites). A rare, igneous rock composed exclusively of pyroxene minerals is called pyroxenite.

Amphibole—Amphiboles are another group of minerals that are typically black in color. They include hornblende, a common mineral in diorites.

What are some **common igneous rocks**?

The following lists some common igneous rocks:

Granite—Most unweathered pieces of granite are so coarse that you can see the individual mineral crystals: It consists mostly of feldspar (usually pink

121

orthoclase), quartz (a glassy mineral), mica (sparkling as blackish flakes), and traces of other minerals, such as iron ore. Granites can differ, depending on the chemical nature of the feldspars. For example, a feldspar rich in potassium tends to give the granite more of a white color; a feldspar rich in sodium and calcium would give the rock a pinkish color. (Feldspars, incidentally, represent 60 percent of the Earth's crust.) If the entire continental crust were melted down, mixed, and cooled, the resulting rock would be granite. Two famous landmarks made of granite are El Capitan in Yosemite National Park and the Mt. Rushmore carvings of four presidents in South Dakota.

Basalt—Basalts are fine-grained, extrusive igneous rocks that form from lava flows. They are made up of olivine, pyroxene, and some feldspar, mica, and apatite. Most basalts have a ropy texture on the outside and have a brittle, cinder-like interior. They include aa and pahoehoe lavas, as well as volcanic bombs. (For more information about volcanic materials, see "Volcanic Eruptions"). Many basalts also contain minute bubble holes (called vesicular basalt), in which gases failed to escape before the lava solidified; if the basalts are millions of years old, these holes may be filled with minerals such as calcite. Some of the best places to see basalt include the Hawaiian Islands, Iceland, and the Galapagos Islands. One of the largest ancient basalt flows known (except those on the Moon) were created in west-central India about 65 million years ago: The Deccan Traps covers more than 250,000 square miles (650,000 square kilometers) and is more than 6,500 feet (more than 2,000 meters) thick.

Rhyolite—Rhyolite is an extrusive volcanic igneous rock related to granite. It has the same minerals as granite, but the grains are much smaller. The majority of these rocks are formed in lava flows.

Diorite—Diorite contains equal amounts of dark and light minerals, giving it what is called a salt-and-pepper appearance. The minerals are mostly plagioclase feldspar and hornblende; biotite and quartz may also be present. It is often found in dikes (dykes).

Andesite—Andesite is also related to granite and is named after the Andes Mountains in South America, where it was first studied. It has the same minerals as granite, but contains less silica and no quartz.

Pumice—Pumice forms when lava cools too fast to allow gases to escape, creating a rock filled with holes. If the rock is a light color, it is called pumice; if it is darker, it is called scoria. Pumice is known for its ability to float.

Gabbro—Gabbro is a coarse-grained, dark rock often used as building material. The coarse-grained minerals include mostly plagioclase feldspar and quartz; it is usually found in sills and dikes.

What is a **xenolith**?

A xenolith is a fragment of another type of rock imbedded in an igneous rock. Xenoliths form when the pre-existing country rock is pried apart by the force of a magma intrusion. While some of the rocks are melted down or assimilated into the intrusion, others exists as pieces called xenoliths within the magma. The majority of xenoliths range in size from pebbles to small boulders. Some of the most amazing xenoliths can be gigantic when more resistant country rock is buoyed up within thick magma.

What are the **various types of obsidian**?

As volcanic magma rapidly cools, it often forms obsidian, a rock that resembles glass (it is also called volcanic glass). This quick cooling prevents the magma from forming coarser-crystalline granite. Most obsidian is black, reflecting the elements within the magma, but not all are this color. The various types of obsidian are:

Mahogany obsidian—The rust color in this obsidian is caused by the presence of iron oxide, the substance commonly known as rust.

Sheen obsidian—This obsidian's gold or silver iridescence is caused by tiny gas bubbles or very small crystals of mica, feldspar, or quartz within the rock.

Snowflake obsidian—Markings within this obsidian look like small white snowflakes on a black background. The white inclusions are actually crystobalite, a variety of quartz.

Rainbow obsidian—Rainbow obsidian shows a variety of colorful bands. These are actually formed by small inclusions of minerals such as feldspar, topaz, quartz, and tourmaline.

What are some **common sedimentary rocks**?

There are many sedimentary rocks. Here are some of the most common:

Limestone—Limestone usually forms in the shallow waters of continental shelf areas (most formed between 100 to 500 million years ago) and tends to cover very broad areas. They are made of calcite, and because it is a sedimentary rock it can also be mixed with other minerals, depending on the content of the water when it was formed. As a rule, limestones are any rock with more than 50 percent carbonate minerals. They are formed primarily from accumulated calcareous skeletons of organisms, including shells and coral fragments, or a combination of the two. Limestones can also be formed from the precipitation of calcite from solution. (For more information about limestone, see below.)

Sandstone—The color and texture of sandstone varies greatly, but in general, the majority of these rocks contain medium-grained, well-sorted minerals.

What are some "hot spots" for finding unique rocks?

Certain rocks appear to be unique to a specific location; in other words, they are "hot spots" for those rocks. The following lists only a few found in various regions of the world (of course, since all naturally occurring rocks differ, one can say that they are all unique!):

- Chrysanthemum stone flowers are natural occurrences of celestite crystals in a limestone matrix. They are found in Liuyang County, Hunan Province, China.

- One rock found in the East Kimberley region of western Australia is the zebra stone. This rock actually looks like the hide of a zebra, complete with a whitish background and blackish-red "stripes." In actuality, it is a fine-grained siliceous argillite (a type of siltstone or claystone). The reddish caste is from iron oxide that occurs naturally in the rock.

- Ophiolites are an unique rock type formed only at mid-ocean ridges.

- Brazil is the place to discover amethyst geodes, semi-rounded chunks of rock filled with purple crystals.

- Africa is the spot to find tiger's eye (part of the quartz family).

- Ulexite is a natural mineral from borax mines, such as the one in Boron, California. This calcium-sodium hydrated borax has a unique fiberoptic quality—when polished on one end and put on a book or picture, the ulexite projects the image to the top of the rock.

- Along the edge of certain granite bodies, rocks become altered, forming rich deposits of magnetite, an iron-bearing magnetic mineral. For example, some rocks from mountains such as the Adirondacks in New York contain magnetite. Such rocks were often used in early compasses. (For those of us using compasses near such rocks, it will only deflect the compass needle when you are very close to the rock and not when standing above the ground at a normal distance.)

- Some of the most unique rocks come from space and are called meteorites. The are usually found by chance in random locations around the world. These rocks are thought to originate from the asteroid belt, the Moon, and even Mars. (For more information about meteorites, see "The Earth in Space.")

Most consist of quartz, often accompanied by feldspar, mica, or other minerals. The grains are most often cemented together by silica, calcite, or iron oxides; quartz grains with a matrix of silica are often referred to as "pure sandstones" (and often labeled as quartzite, the same name as the metamorphic rock containing nothing but quartz). They originate from shallow ocean deposits or wind-blown deposits in arid regions.

Conglomerate—Conglomerates consist of rounded pebbles with diameters around an inch (2 millimeters), cobbles, and even boulders, all in a matrix of fine- or medium-grained minerals. The fragments may be made of any sedimentary, igneous, or metamorphic rock, depending on the source rock, but quartz and other harder rocks are usually tough enough to form the conglomerates. These rocks usually originate along ocean, lake, or river shores in shallow water with vigorous water movement, which is necessary not only to round the pebbles but also to move the large rock fragments.

What are some **major types of limestone**?

There are numerous types of limestone (they make up about 15 percent of the Earth's sedimentary crust), but the most common include the fossiliferous (shelly) limestones, each of which is classified by the kinds of fossils they contain. For example, crinoidal limestone contains fossils of crinoid, small plants that lived about 500 million years ago in shallow seas; coral limestones are formed from corals that also lived in shallow oceans.

Another common type of limestone is chalk. These white, yellow, or gray deposits are very fine-grained, porous, and compact. Chalk is actually a very pure limestone consisting of calcite. Any coloring may be from fossils found within the deposit, or small amounts of silt and mud. Chalks are most often formed from the remains of free-swimming microorganisms that fell to the ocean floor, creating a fine-grained calcareous mud. It forms in deeper ocean regions, where there is little deposition of other sediments. Although most chalks accumulated in the Cretaceous period over 65 million years ago, there are still chalk-like deposits forming today.

What are the various **types of sandstone**?

There are various types of sandstone, depending on the composition of the rock. For example, *arkose* is a sandstone that has more than 25 percent feldspar; the original sediment usually comes from weathered granite. *Greensand* is a coarse sandstone, with bits of feldspar and mica in it, as well as large crystals of glauconite, which is a green magnesium-iron mineral. It most often forms on the bottom of the ocean. *Seat earth* is a very pale, granular sandstone containing root fragments. Most are found in former sandbank areas in which vegetation once grew, and are often found with a bed of coal capping the rock. *Graywacke* is a coarse, unsorted sandstone with angular

How is limestone used?

Limestone is used for a variety of purposes, especially in the construction industry. In particular, many limestones contain light fossils in a dark matrix, which can be very beautiful when polished for buildings or monuments. Crushed limestone is used for construction and often roadways and driveways. About 71 percent of all crushed stone produced in the United States comes from either limestone or dolomite. Limestone mixed with clay and water forms cement, and when further mixed with sand it forms mortar. Some limestones have also been mislabeled as a marble. For example, Forsterly marble is actually a limestone imbedded with corals.

As a source of lime, limestone is also used to make windows, paper, plastics, carpets, glass, and mirrors; and it is used in water treatment and purification plants. In the steel-making process, limestone mixed with the impurities in iron creates slag. And, as many gardeners know, pulverized limestone added to soil not only neutralizes the soil's acidity (plants need specific acid or basic soils to grow well) but also helps increase plant nutrient uptake and the presence of beneficial soil organisms.

fragments of quartz, as well as other magnesium-iron minerals and finer clays and mud. It most often forms as continental shelf sediments collapse down the continental slopes and into deeper waters.

How do you tell the difference between **mudstone, clay, siltstone, and shale**?

Each one of these rocks—mudstone, claystone, siltstone, and shale—represents a certain sediment size. So for most field geologists, one of the ways to tell the difference is to chew on the rock to test the grittiness! It is *not recommended* that the reader try this method—just viewing the rock will give you a "taste" of its true identity.

There are other ways to tell these fine-grained sedimentary rocks apart, too. In general, if a rock is easily split into brittle sheets, it is shale; in addition, fossils are more readily seen in shales. In fact, fossils of marine animals are found in deep-sea shales, whereas fossils of freshwater shellfish and plants are found in shallow-water shales.

There are also other distinctions to define clay, mudstone, and siltstone. If the rock has little internal structure, but is slippery when wet, it is clay; if it breaks into lens-like flakes, it is mudstone. Finally, the "larger" grained rocks (those with grains that can be distinguished with the naked eye) are most often siltstone, which is usually formed in oceans, in lakes, or within glacial deposits. Similar to sandstones, siltstones may often show evidence of bedding and ripple marks.

What are some **common metamorphic rocks**?

There are many types of metamorphic rock. The following lists some of the most common:

Slate—As sedimentary rock shale is metamorphosed, it turns into argillite, and then slate. (Similar to its original rock, slate's minerals are mostly invisible to the naked eye.) As argillite is affected by heat and pressure, it causes some clay minerals to chemically change into flakes of muscovite mica and chlorite. The flakes grow, orienting their surfaces mostly parallel to each other and perpendicular to the direction of pressure. This creates the major characteristic of slate: its ability to easily break along cleavage planes (these are often erroneously thought of as bedding planes, but they are not). In turn, it is this characteristic that has long made slate a perfect material for roof shingles.

Phyllite—As slate becomes more metamorphosed, the minerals combine into somewhat larger grains to form phyllite. The clay minerals change to mica, making phyllite similar to a mass of minute mica flakes. It also contains some quartz; and small amounts of orthoclase feldspar are created as the rock experiences more heat and pressure.

Schist—Often the next phase after phyllite is a schist, in which the mica, orthoclase, and quartz molecules combine into larger masses, giving the rock its characteristic layered look (foliation) with flaky or fibrous minerals large enough to be seen with the naked eye. The various types of schists are more easily identified by their texture than their composition, and originate as igneous as well as sedimentary rocks. Examples of schists include biotite schists (black biotite mica flakes with streaks and blobs of quartz and orthoclase); greenschists (the green mineral chlorite dominates); and blueshists (the bluish mineral glaucophane dominates, along with several other minerals). Garnets also occur as red crystals in some schists; they are formed from a recombination of minor elements in the original rock. In fact, garnet crystals are a good indication that a rock is metamorphic.

Quartzite—Quartzite is a layered rock that originates from metamorphosed sandstone. The great heat and pressures literally melt the minerals, which then recombine as the rock cools. But the rock does not change in composition (or, usually, in color) when sandstone turns to quartzite. Instead, the grains just become more tightly packed together.

Gneiss—Gneiss most often originates as a granite (or its close relatives). Like schist, gneiss is identified by its texture, not its mineral composition. Most gneisses are banded, but are more coarse and grainy than schists, and the minerals are not as flaky. Gneiss grains show a streaked pattern, indicating how the minerals flowed and melted as they were metamorphosed.

How does **marble form**?

Quite simply, marble forms from the metamorphism of carbonate rocks, mainly limestone. When limestone and dolomite (the parent rocks of marble) undergo an increase in temperature and pressure, their constituent minerals of calcite and dolomite do not change. This creates marble made up of the original carbonates, but it is a much denser rock.

But the process is not always that simple. If the original carbonate rocks contain impurities, such as silicate minerals, metamorphic reactions take place. For example, *marl* is a mix of carbonate materials and silica that creates minerals called calcsilicates when metamorphosed, including calcite, dolomite, diopside, tremolite, garnet, wollastonite, chlorite, diopside, brucite, and/or serpentine. These mineral impurities create the attractive and decorative lines and patterns within marble that are easily seen when it is polished.

MINING MINES

Where are the **deepest mines**?

The diamond and ore mines of South Africa are probably the deepest in the world. They are so deep that high temperatures and pressures are a problem, often making it dangerous for humans to work in these mines.

What are the **deepest holes** ever drilled?

Since the 1950s, humans have tried to reach into the mantle of the Earth, which, if ever accomplished, would definitely qualify as the deepest holes ever drilled. No ore or mineral mine can reach such depths, however. One of the first attempts to dig to the mantle was conducted by the United States. Called Project Mohole, the plan was to drill through the shallow crust under the Pacific Ocean just off Mexico and into the mantle, but lack of funding caused the project to be abandoned in 1966. Next came

An aerial view of a diamond mine pit dug deep into the Siberian permafrost near the town of Mirny in Russia. Miners have worked in this mine since diamonds were discovered in the Russian Arctic in the 1950s. *AP/Wide World Photos.*

the Deep Crustal Drilling Program in 1966, which also ended quickly. Still another program, called the Moho Project, developed into the Deep-Sea Drilling Project, and the more modern Ocean Drilling Project. The results of these drilling programs were a substantial number of holes—usually they were no deeper than 33,000 feet (10,000 meters)—but no record-making depths were achieved.

On land, the Bertha Rogers well in Oklahoma might be the deepest in the United States. This gas well stopped at 32,000 feet (9,753 meters) when it struck molten sulfur. To date, it apparently still holds the deepest hole status on the North American continent.

But the winner of the deepest hole ever drilled on Earth belongs to Russia, where there is an over-40,000-foot-deep (just over 7 miles [12 kilometers]) hole that was dug on the Kola Peninsula by what was then the Soviet Union government. It took 5 years to drill 4.3 miles (7 kilometers) and another 9 to drill the additional 3.1 miles (5 kilometers). Lack of funding—but not trying—ended the twenty-year project in 1989, at a cost of more than $100 million. Although the Russians were unsuccessful at reaching the upper mantle, they were able to grab the title for deepest hole in the Earth.

What are **native elements** and why are they **important to mining**?

The native elements are the minerals we refer to as metals (native refers to the composition being one element). These elements include gold (Au), silver (Ag), and copper (Cu), all of which are important metals dug around the world.

129

Where does **gold originate**?

Gold deposits originate in various ways, but mostly they exist in fractures and fault zones in association with schists and other rocks. Gold may also be eroded and re-deposited as sedimentary placer deposits. Also called placers, sedimentary placer deposits are accumulations of valuable metal that are concentrated in sediments on top of the ground, in stream sediments, or in beach materials. They form as the result of mechanical and chemical weathering of gold-bearing rock, which is then transported a short distance by water or wind. Gold is also commonly associated with silver and the base metals (copper, lead, and zinc) found in quartz veins deposited from hydrothermal (hot water) fluids in fault zones.

What makes **gold** so **important**?

Gold is a very precious metal. For most cultures, both past and present, it has been of great economic importance. Many nations base the value of their currency on the amount of gold in their reserves (although such gold stashes being "equal" to the amount of paper money is less important now). And gold has historically been the only universally accepted medium of exchange.

What makes gold so valuable? The most important reason is its rarity, although other characteristics make it valuable, too. For example, pure gold is very soft—almost too soft to use in jewelry, unless it is combined with another metal such as silver or copper; it measures about 2.5 to 3 on Mohs' Scale of hardness (for more information about Mohs' Scale, see "All about Minerals"). Gold is an excellent conductor of electricity, too, which is why it is so sought after by the electronics industry.

It is heavy—about 19 times the weight of the same amount of water—and is the most malleable and ductile of the metals. Gold can be easily pounded into thin sheets (one ounce can cover an area of about 160 square feet [15 square meters]), or drawn into a wire more than 37 miles (60 kilometers) long without breaking. This is why gold leaf church domes and other coverings were once so popular; they allow structures to look lavishly rich while only using a thin coating of gold.

Gold is also almost completely insoluble in nature; in other words, very few liquids will affect it in its natural form. Even oxygen and water do not corrode or rust this metal. (Only a mixture of hydrochloric and nitric acids (aqua regia), and some cyanide solutions will dissolve gold.) Therefore, gold is one of the most inert and stable elements known to humanity.

Are there **other types** of **gold**?

Yes, in a way, there are other types of gold, although most have little natural gold in them. For example, white gold, a substitute for platinum, is an alloy of gold with platinum, palladium, nickel, or nickel and zinc. Green gold is usually an alloy of gold with silver. Both are used in the jewelry field. Alloys of gold with copper are a reddish yellow and are used for coinage and jewelry.

How is **gold found**?

Gold is discovered in several ways. It is often found in quartz veins or lodes, but is so finely disseminated in the rock that it is often not visible. One of the most common ways of extracting such natural gold from mineral ores is by mechanical mining. After extraction from the ground, gold-containing ore is processed to remove all the impurities; then it is further separated from other metals using chemical processes, such as the amalgamation process—the process in which valuable metals are extracted from ore. When pure gold has been fully extracted, it is then made into ingots.

Probably the most dramatic type of gold discovery comes when one uncovers a gold nugget, although less than two percent of the world's gold comes in nugget form. Gold is also found as dust, grains, or flakes. A popular way the public collects gold—short of buying it from gold dealers—is to pan for gold dust, grains, or flakes found in alluvial placer deposits (small pieces of gold that are eroded from rocks in higher regions and work their way into rivers). The idea behind panning is that gold is much heavier than most other minerals. Therefore, shaking a pan with water and sediment in it often leads to the discovery of gold flakes at the pan's bottom. This is an enjoyable pastime for many people visiting areas in which gold is found in river sediments, such as in North Carolina or California.

Finally, gold can be found in large amounts in seawater. But because it is so difficult to extract, recovering it in this manner is more expensive than it's worth. Gold also occurs in compounds, notably telluride minerals. But again, the extraction cost is prohibitive.

What is the **largest nugget of gold** found to date?

Some of the richest gold nugget regions in the world are in rural western Australia, with Victoria providing some of the best ever found. But because gold has been mined for centuries—in Australia and elsewhere—it is difficult to know for certain what the largest nugget ever discovered is. The following lists some of the possible (if you can believe them) contenders:

Welcome Nugget—The Welcome Nugget was discovered in June 1858, at Bakery Hill, Ballarat, Victoria, Australia. Twenty-four people claimed to have unearthed the specimen, which weighed 2,217 ounces (138.56 pounds or 63 kilograms), although there are disagreements as to its actual weight. After refining, the nugget yielded about 100 ounces (2.83 kilograms) of fine gold. But as a result of this refining, the original gold nugget no longer exists and is lost to history.

Welcome Stranger Nugget—Another Australian contender, this nugget was found in 1869 in Moliagul, in the Central Victorian Goldfields. It was broken on an anvil in nearby Dunolly in Victoria so it could be weighed at the town's bank. The 156-pound (2,496-ounce, or 70.76-kilogram) gold nugget is claimed by many to be the largest ever discovered. As usual, there are disagreements as to its true overall weight because the discoverers of the nugget may have broken off pieces before it reached the bank.

Hand of Faith—Discovered in Australia, this approximately 61-pound (976-ounce or 27.67-kilogram) gold piece is thought to be the largest nugget still in existence today. It was found at Kingower in 1980 and is the largest gold nugget ever found with a metal detector. It was sold in 1982 for a reputed sum of around one million Australian dollars. It can still be seen on display in the Golden Nugget Casino in Las Vegas.

Where are some of the world's **most productive gold mines**?

Although Australia claims to have some of the largest gold nuggets ever discovered, South Africa holds the record for most gold ever found. In fact, one third of all the world's known gold mined since the Middle Ages comes from just over 2.5 miles (4 kilometers) below the surface in South African mines. Today, South African gold mines supply about 25 percent of the world's yearly production. However, its dominance is diminishing, primarily because of other vast gold discoveries in North America and Australia.

One of the richest gold mines was discovered by Australian George Harrison, who in 1886 crushed and panned some odd looking rock from around Witwatersrand, South Africa. He had uncovered the main gold reef that eventually led to the expansion of Johannesburg; it became the world's richest goldfield, and Johannesburg became the largest mining city.

Is **silver** an important metal?

Yes, silver—a brilliant gray-white metal—is an important material, but not as important as it was decades ago. One reason is that there are so many substitutes now for silver in many applications. For example, digital imaging has made much of the silver used for

How do jewelers define the purity and hardness of a gold piece?

Purity and hardness of a piece of gold—mostly jewelry—is determined by how many carats (or karats) the gold piece contains. The following lists the gold percentages:

- 24 Carat = 100.0 percent pure
- 18 Carat = 75.00 percent pure
- 14 Carat = 58.33 percent pure
- 12 Carat = 50.00 percent pure
- 10 Carat = 41.67 percent pure
- 9 Carat = 37.50 percent pure

film processing and other image development close to obsolete; medical, dental, and industrial X-rays, which also used silver, have also turned to digital images. Mirrors and other reflective surfaces now use much cheaper aluminum or rhodium. And even stainless steel has replaced silver in such traditional items as ornamental tableware.

But silver still has many useful attributes. It is quite soft and malleable, resistant to corrosion, relatively scarce, and it does not oxidize easily (although the surface does tarnish into silver sulfide, which is why silverware needs an occasional polishing). Most importantly, silver is the best conductor of electricity of all the metals. It is often classified along with gold and platinum as a precious metal.

Where is **silver mined**?

The main source of most silver is in lead ore. It is commonly found in association with copper, zinc, and gold and is produced as a by-product of these metal mining activities. Mexico, Peru, and the United States are currently the major producers of silver.

What is the **largest silver mine** in the world?

To date, the Cannington silver/lead/zinc mine is the largest silver mine in the world, processing more than 25 million ounces (708,738 kilograms) annually, with over 700 million ounces (19,844,666 kilograms) of reserves. The mine is located in Queensland, Australia, 124 miles (200 kilometers) southeast of Mount Isa. Although this is the largest silver mine, Mexico and Peru remain the two largest silver producers.

What are some **economically important ores** found in the **oceans**?

There are several areas of the oceans that contain economically important ores. For example, hydrothermal mineral deposits on the ocean floor—marine areas in which rising magma at ocean ridges or subduction zones create fluids enriched with minerals—contain such materials as copper, iron, zinc, manganese, nickel, lead, silver, and gold.

Phosphorite deposits, used primarily in fertilizers, are also an economic commodity found in oceans. Phosphorite is most often deposited where cold, nutrient-rich waters are brought to the surface. As the waters rise and grow warmer, chemical reactions cause the phosphorous to precipitate out onto the ocean floor. Phosphorite deposits off the eastern coast of the United States are said to amount to about several billion tons.

Another well-known type of deposit are manganese nodules, which are rounded spheres made of almost pure manganese (along with traces of iron, nickel, and cobalt) that are found mostly on abyssal plains in all the oceans. These deposits form by the very slow deposition of minerals (about 0.004 inch [1 millimeter] per million years) from seawater.

What is **uranium** and why is it **mined**?

Uranium is the only commercially produced radioactive material. In its natural form, it is found as three isotopes (^{238}U, ^{235}U, and ^{234}U) with ^{235}U being used to generate electricity through nuclear fission. The main uranium-bearing minerals currently mined are uraninite (UO_2), pitchblende (a mixed oxide, usually containing U_3O_8), brannerite (a complex oxide of uranium, rare earths, iron and titanium), and coffinite (uranium silicate). The world's major producers of uranium include Canada, Australia, and Niger.

Uranium is used mostly for electric generation, but other uses include radioisotopes for medical use, preservation of food, and treatment of plants for better yield and disease resistance. In about 26 countries, uranium is used to generate more than 25

The Bonneville Salt Flats southeast of Wendover, Utah, is the home of the Bonneville Speedway, where several land speed records have been set. However, the 28,000 acres that comprise the flats is deteriorating, the salt stripped away by mining and erosion. *AP/Wide World Photos.*

percent of the electricity. It is also used in the manufacture of nuclear weapons, although its generation for such weapons has been curtailed in the past decade.

How is **uranium mined and processed**?

Uranium is mostly mined using one of three methods: open pit (or open cut), underground mining, or *in situ* leaching (which is not as common). In the open pit (literally, an open hole in the ground from which uranium is extracted) and underground mining methods, the ore is gathered, then crushed and milled into a fine powder.

In general, uranium is processed in the following sequence: Chemicals are added to dissolve the uranium, producing slurries of waste and uranium. Sulfuric acid is mixed in, allowing the uranium to separate from the solid part of the slurry. The uranium is then extracted from the other elements and converted to a solid uranium concentrate that is called yellowcake (U_3O_8) because of its distinctive yellow color. The yellowcake is further converted and enriched, dried, and packed for transportation to refineries. The refined uranium concentrate is processed to form hard ceramic pellets that are eventually used in nuclear reactors.

Where is the world's **largest source of uranium** to date?

The Olympic Dam mine in Australia has been in operation since 1988. It is Australia's largest underground mine and is known as the world's largest known uranium resource.

Why is it so important to mine **salt**?

Salt, which is also know as halite, table salt, or sodium chloride, is readily soluble in water. In addition, it is considered to be nearly limitless in supply and inexpensive to produce. This means that it is useful for a variety of applications, including cooking, food preservation, and chemical production. For example, in the United States, table salt for food accounts for around 1 percent of all salt used; 40 percent is used in the chemical industry (mainly for producing chlorine and caustic soda); 40 percent to de-ice winter roads (mainly in the northern states); and just under 20 percent in various manufacturing sectors (rubber and other goods), agriculture, and food processing.

How do **salt deposits** develop?

Sodium chloride is readily dissolved in seawater, and its presence is very profuse. As seawater evaporates in closed places, such as a lagoon, halite (and other minerals) precipitate out, and the crystals sink to the bottom. Over time, these deposits form great beds of rock salt that are often folded and uplifted, exposing the beds to weathering processes. Eventually, they can dissolve, forming brines (water and salt mixtures) that either percolate into the ground or make their way into oceans; they can even collect in salt lakes, such as the Great Salt Lake in Utah.

One other type of salt deposit is associated with oil deposits: the great buried salt domes, some of which are close to a mile (1.6 kilometers) in diameter at the peak. These mushroom or plug-shaped features (called diapirs) usually have an overlying cap rock. They form because of the buoyancy of salt, which flows upward, pushing the layers of sediment into a dome. (They can also form sheets, pillars, or other structures.) Around the domes, usually along the sides where "holes" (or traps) are created, oil and gas can form and get sealed in by the thick salt. This is why many oil companies often look for salt diapirs in their search for petroleum.

How is **salt mined**?

There are several ways in which salt is mined, depending on where the deposit is located. Some salt deposits are mined through the evaporation of sea water, which is why it is called sea salt. The salt is naturally formed from salty water, called brine, which evaporates and creates the halite crystals.

Such salts are also produced by controlled evaporation of seawater or of brines in salt lakes. In this technique, the water is pumped or drained into shallow ponds. Solar evaporation will eventually (especially in an arid climate) concentrate the salt, causing it to crystallize on the pond floor. This process is used around San Francisco Bay, the Great Salt Lake in Utah, and elsewhere.

But occasionally salt is not exposed at the surface, lying underground instead. One method of extracting underground salt is to pump hot water into the salt deposit to dissolve it. The resulting salty water is then pumped to the surface and allowed to evaporate in a process called solution mining. This method is often used at ancient, underground salt domes, such as the ones found in Louisiana and Texas.

Salt mined from solid layers in the ground is called rock salt. It is most often extracted using traditional mining practices of digging with heavy machinery and carrying away the rock for processing. The former salt mines just outside of Salzburg, Austria, and one that is currently near Cayuga Lake in central New York, are examples of this type of salt mine.

What **countries** mine the **most salt**?

Although salt is produced in almost every country on Earth, the United States is the largest producer, providing about one-fifth of the world's supply. It is followed by China, Germany, India, and Canada. In many other countries, coastal collection of salt is used locally as a product of seawater evaporation. In the United States, rock salt accounts for about one third of all the salt produced; solution mining provides about one half of the total. The rest of the salt supply comes from the evaporation of ocean water and lake brines, and a small amount comes from salt crust on dry lakes.

How are **sand and gravel used**?

Sand and gravel are used mostly in construction. This includes road building, concrete production, landscaping, driveways, glass manufacturing, sand casting in foundry operations, snow and ice control, fill, and many other uses.

How are **sand and gravel mined**?

Sand and gravel are mined in various ways—mostly as open pits along ancient rivers. These deposits form as river channels change, with the sand and gravel particles left behind in the old bed. For example, the Susquehanna River in upstate New York has several gravel pits, as does the Platte River Valley in Nebraska; both contain thick gravel deposits left over from rivers that flowed through the areas before the ice ages some 10,000 years ago.

The sand and gravel in these areas are mined in generally the same way. First, the vegetation and top soil are cleared to form a pit. Because the water table is so close to the surface near the river, the pit fills with water, becoming what is called a sandpit lake. The pit is then dredged with machines that extract the sand, gravel, and water. The slurry is then sent to a separator that sorts the sand and gravel according to size, with each different size used for specific applications.

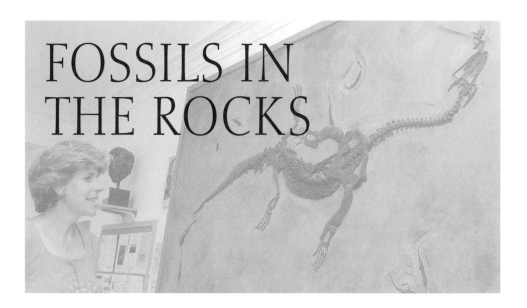

FOSSILS IN THE ROCKS

FOSSIL HISTORY

What are **fossils**?

Fossils—the word comes from the Latin word *fossilis,* or "something dug up"—are any preserved remains or imprints of living organisms (usually ancient animals and plants), such as bones, shells, footprints, or leaf impressions. Most fossils are found in sedimentary rock; others can be found imbedded in igneous rock, or changed by heat and pressure in metamorphic rock. Some fossils resemble the original organism's shape; but those buried deep under sediment layers may be folded, crushed, or twisted from the great pressures. Usually, the hard parts of an animal or plant, such as bones, teeth, shells, seeds, or wood, can all become fossils, but in rare instances soft tissue can be preserved as well.

The word fossil was first used in 1546 by the German scientist Georgius Agricola (Georg Bauer, 1494–1555). During this time, the word fossil was used to describe all minerals and metals dug up from the ground. The word is now used to describe objects obviously formed in the past by living organisms. In general, fossils are usually considered to be objects older than 10,000 years; younger objects are often classified as subfossils.

What are the **three types of fossils**?

The following lists the three major types of fossils:

Body fossils—Body fossils are the preserved skeletal or structural forms of an organism, and in some rare instances, the preserved soft tissues. They are the most common type of ancient preservation we think of when someone mentions the word "fossil."

Trace fossils—Trace fossils are evidence of biological and behavioral activities of organisms while they were alive, including tracks, trails, burrows, leaf imprints, and skin imprints.

Chemofossils—Chemofossils are chemical signatures of life in rock. They may be organic chemical compounds (also called biomarkers) indicative of certain organisms, or trace elements that are the result of biological processes.

How did **trace fossils** form?

Trace fossil are evidence that animals ran, walked, crawled, burrowed, or hopped across land, usually in soft sediment such as sand or mud. For example, dinosaurs walking along a river left their footprints in soft sand; small animals dug branching tunnels in the mud of a lake bed in search of food. These traces then filled with sediment and were buried under layer upon layer of sediment over millions of years, eventually solidifying into rock. When the rock is exposed and collected, the trace evidence of the organisms can be seen. Some of the most famous examples of trace fossils are footprints in hardened sediment, such as the humanlike ones found in East Africa and the numerous dinosaur tracks in the Connecticut River Valley.

What are some of the **oldest fossils** found to date?

The oldest fossils—bacteria found in western Australia—are thought to be single-cell organisms about 3.5 billion years old. In 2002, scientists also discovered one-billion-year-old marks in Australian sandstone that may have been made by wormlike life. They might be the tracks of the oldest multicelled, mobile organisms, but to date, associated fossils have been found. So far, the oldest fossil hard parts (for example, shells) of organisms are from close to 600 million years ago, when organisms first began to develop skeletons, shells, and other hard body structures.

What is **paleontology**?

Paleontology is the study of ancient life (mostly through fossils). The word comes from the Greek word for "ancient life." Paleontologists study all types of fossils—marine and terrestrial, plant and animal. This close look at fossils allows us to understand what ancient life was like over a billion years ago.

There are many divisions of paleontology. A paleobotanist studies fossil flora (plants). A paleozoologist studies ancient fauna (animals), a study that can also be broken down into invertebrate paleontology, or the study of invertebrates (animals with no backbone), and vertebrate paleontology, or the study of vertebrates (animals with a backbone). Paleopalynology is the study of ancient pollen and spores (and in marine environments, single-celled algae); paleoichnology is the study of footprints, tracks, and traces left by ancient organisms (animals and plants).

Fossils also aid other scientific studies. For example, in geology fossils help scientists determine the relative age of rocks. Paleobiologists also use fossils for information about how life has evolved, or what lifeforms have descended from other lifeforms.

What were some **early theories** about the origin of **fossils**?

One of the first reports we have of fossils came from the Greek philosophers Xenophanes (570–475 B.C.E.) and Pythagoras (582–500 B.C.E.), recorded before 500 B.C.E. The two philosophers independently concluded that fossils were the remains of once-living animals and plants, but the organisms were no longer present in the known world, which for them meant mostly ancient Greece.

Herodotus (485–425 B.C.E.), a Greek philosopher and historian who lived about a century after Xenophanes and Pythagoras, studied tiny fossils embedded in the sandstones of the Egyptian pyramids. He concluded that an ocean once covered what is now the Libyan desert. A century after Herodotus, the Greek philosopher Aristotle (384–322 B.C.E.) drew his own conclusion concerning fossils: He proposed that fossils were the failures left behind when life spontaneously generated from mud.

What did the **Chinese** believe **dinosaur fossils** were?

The Chinese have been collecting dinosaur fossils for over 2,000 years, but they mistakenly identified the pieces as the remains of dragons, a prominent cultural symbol for the Chinese. Although modern scientific evidence has dispelled this notion, many still believe that ground up "dragon's teeth" have medicinal healing properties.

Who was the **founder of modern paleontology**?

French naturalist and anatomist Baron Georges Leopold Chretien Frederic Dagobert Cuvier (1769–1832) is considered the founder of modern paleontology. As a scientist at the French National Museum of Natural History, he was the first (in 1812) to propose that certain animal fossils were representative of forms that had completely died

Who discovered the origin of fossils in the Alps?

One of the greatest scientists, inventors, and artists of all time, Leonardo da Vinci (1452–1519) was the first to uncover the true reason behind fossils in the Alps. The Italian Renaissance scientist discovered marine fossils in the Apennines in the late 1400s. He correctly speculated that the fossils were once-living organisms that were buried at a time before the mountains were raised. "It must be presumed," he said, "that in those places there were sea coasts, where all the shells were thrown up, broken, and divided." In other words, he believed that where there was land, there was once ocean; the regions subsequently uplifted, exposing the remains of sea creatures.

Although da Vinci probably could have been called the father of modern paleontology because of his observations, it was not to be. As with many of his ideas, da Vinci hid his notes, keeping them from those who would accuse him of heresy. After all, the main belief at the time—that the fossils were carried to the mountaintops by the Great Flood—was based on the Bible. Still others thought the fossils grew on their own within rock layers.

out or became extinct. Cuvier also tried to reconcile his finding with the Bible by proposing a "catastrophe theory" to account for the extinctions.

What is the connection between **continental drift and fossils**?

There is a definite connection between continental drift—or how the continents move around the planet—and fossils. In particular, the theory of continental movement is based on the idea that identical rock types and associated fossils are found on continents now widely separated by oceans. For example, the Lystrosaurus was a mammal-like reptile that lived during the Triassic Period and behaved much like the modern hippopotamus. Fossils of this reptile are found in modern Africa, Antarctica, and India. This is seen as direct evidence that these landmasses were once linked together. (For more information about plate tectonics and continental drift, see "The Earth's Layers.")

FORMING FOSSILS

What is **taphomony**?

Taphomony is the study of the processes that lead to the formation of a fossil. In other words, it is the study of the conditions under which plants, animals, and other organisms become altered after death to be preserved as fossils.

A visitor looks at a Ceresiosauraus, displayed on a wall of the Fossils Museum in Meride, at the foot of Monte San Giorgio. The Swiss region on the southern shore of Lake Lugano is famous for its abundance of well-preserved reptile and fish fossils dating from the mid-Triassic Period, around 240 million years ago. *AP/Wide World Photos.*

What **conditions** are necessary to **form a fossil**?

In general, the process of fossilization starts when sediment, such as mud or sand, covers a dead organism. Overall, it is believed that fossilization takes well over 10,000 years because most younger bones show little or no mineralization (for more information about mineralization, see below).

It is difficult for something to become a fossil. Most remains are scavenged by other animals before becoming weathered by wind, water, and sunlight. The soft tissues not eaten—such as skin, eyes, muscles, and internal organs—rot away at various rates, depending on the climate, leaving only the bones and teeth to become fossilized, which is why these parts are most often fossilized.

Burial by sediment (sand or mud) is the next and most crucial step in the process of becoming a fossil. It reduces the amount of oxygen that would otherwise decay the bones. If sediment does not bury remains, the pieces are often broken and scattered by such events as flash floods. As more sediment accumulates over millions of years, it buries the remains deep within the ground. Eventually, due to pressure from the overlying layers, much of the water is forced out of the sediment. Minerals in the groundwater (such as carbonate, silica, and iron oxides) help to slowly cement the grains together, turning the sediment into rock in a process called lithification.

143

How do **bones and other hard parts** undergo **mineralization** to become fossils?

Entrapped animal parts—mainly bones and teeth—must undergo some form of mineralization to become fossils. Bones are composed of inorganic minerals and organic molecules (especially proteins and fat). Most of the organic components are eventually consumed by bacteria. What remains are brittle, microscopically porous bones. Water percolating down through the soil above dissolves mineral salts, some of which are precipitated out into the porous areas of the bones. These minerals are usually calcium carbonate (limestone), silica, or iron compounds. Over time, the parts then become a form of rock themselves.

The actual rate and type of bone mineralization depends on the type and chemistry of the sedimentary environment surrounding the buried bones. The following lists several ways in which this occurs:

Recrystallization—In this process, the bone or other hard biomatter converts to a new mineral or to coarser crystals of the original mineral. For example, a mineral found naturally in bones called apatite (calcium phosphate) may recrystallize.

Permineralization—Many bones, shells, and plant stems have porous internal structures that become filled with mineral deposits. In the process of permineralization, the actual chemical composition of the original biomatter might not change. But the bones themselves fossilize, as minerals such as calcite (calcium carbonate) fill the spaces in the bone structures.

Dissolution/replacement—In dissolution and replacement, groundwater (especially acidic water) dissolves the part of the organism that is trapped in sediments; it might simultaneously deposit a mineral such as silica, calcite, or iron in its place.

Carbonization—Carbonization leaves traces in the rock when the temperatures and pressures of burial cause the liquid or gaseous (volatile) components to be squeezed out, leaving a film of carbon.

How does **wood** become **petrified**?

Wood becomes petrified through the process of dissolution and replacement. This occurs when water that contains dissolved minerals such as calcium carbonate ($CaCO_3$) and silicate permeates it. Over thousands of years, the original plant is replaced or enclosed by these minerals—mainly silica—seemingly turning it to stone. Often, the plants' original form is retained, allowing scientists to study the structure of extinct organisms.

What are some examples of **carbonization and carbon films**?

The process of carbonization often produces films that preserve the outlines of animals' bodies, such as those found in the famous Burgess Shale layer of British Columbia. And animals are not the only organisms that can leave carbon films behind: The leaves of some plants have been preserved as carbon layers, revealing an outline and hint of a vein pattern and often allowing the original plant to be identified.

How are **fossils found** at the surface?

The only way to find fossils—short of digging a huge hole—is for the ancient remains to be exposed on the surface. Most fossils found in sedimentary rock have been uplifted by the movement of the Earth's crust (tectonic activity). Another way fossils are exposed is by the action of erosion, especially by wind, water, and ice that carve out the soft sedimentary rocks in which fossils reside. But if the fossils are not discovered in time, the same agents that expose the pieces will eventually destroy them. One prime example of this occurred in the northern regions of the United States, where huge

A section of the rock-lined shore edge in Jamestown, Rhode Island, illustrates how the curvature seen in the rocks shows years of ripple sand and how it was compressed over time. Geologists examine the tissued layers of the rocks and hunt for ancient fossils in its petrified folds. *AP/Wide World Photos.*

glacial ice sheets that receded about 10,000 years ago after the ice ages wore away millions of years worth of rock along with countless numbers of fossils.

Are there **gaps** in the fossil **records**?

Unfortunately, there are many gaps in the fossil records, with whole eras or evolutionary stages missing. The loss of these precious fossils is most often the result of erosion. The action of water, ice, wind, and other erosional agents wears away layers of rock and the embedded fossils. Gaps can also be caused by mountain uplift that physi-

It is hard to estimate how many organisms became fossils over time. In fact, the late United States paleontologist Stephen J. Gould (1941–2002) estimated that 99 percent of all plant and animal species that ever lived are already extinct—and most of these have left no fossil evidence of their existence. The fossils we do find are therefore representative of only a small fraction of the animals and plants that have ever lived on our planet. This seemingly sparse fossil record leaves a considerable gap—and bias—in our knowledge about the history of life on our planet.

There are good reasons for this bias. One is because not all organisms' hard parts survive equally well. Light-weight bones with relatively large surface areas deteriorate more quickly, and thus are less often fossilized. Small, delicate bones, such as those of birds, are more likely to be crushed, eroded, or carried away from the rest of a skeleton by running water, storms, or even winds. Thicker, heavier bones survive much better, giving the fossil record a bias toward organisms that have this type of bone.

Another reason is the rate of fossilization. The best chance for fossilization comes when sediment covers an organism just after it dies. This protects the decaying organism from scavengers and predators, and from possible chemical erosion (such as from acidic water). The soft parts quickly decay, leaving behind the hard bones and teeth, the pieces with the best chance of surviving as fossils.

One other bias in the fossil record results from the paleontologists themselves. All areas of the globe have not been equally searched. Because of the inaccessibility of some regions, such as central Asia and Africa, these fossil records are poorly represented (and understood) compared to those from Europe and North America.

cally destroys the fossils. Hot magma from volcanic activity can also bury and change the rock and associated fossils.

Why is it difficult to **categorize** newly discovered **fossils**?

There are many reasons why it is difficult to categorize newly discovered fossils. In particular, many times there are few fossils—sometimes only one—to determine the species or even family of a new discovery. In other words, there is a poor representation of how the species truly looked. Other fossils might only be partially preserved, thus making it difficult to determine the true identity of the organism.

In addition, there are two different, opposing approaches to identifying the species represented by a fossil (even new species found in modern times). Scientists who use the

What's the largest coprolite found to date?

It is called the pride of Saskatchewan, Canada: "Scotty" the Tyrannosaur, a *Tyrannosaurus rex* fossil excavated near the town of Eastend in 1994. More excitement followed only four years later, when a large, whitish-gray lump about 1.5 feet (0.5 meters) long was discovered a few miles away from Scotty's bones. Although Upper Cretaceous Period crocodiles and small theropod fossils have been discovered in the area, only a Tyrannosaur could have produced a coprolite that large. It is considered to be the largest single feces ever recorded from any carnivore—fossil or living—and the first *T-rex* coprolite ever found. The shattered bone and other material within the coprolite suggest to many scientists that tyrannosaurs pulverized large quantities of bone, along with flesh, skin, and organs, and did not just "gulp and swallow" their prey.

typological approach believe that if two fossils look even slightly different, they must be from two distinct species. This reveals an emphasis on minor differences. In contrast, those who use the populationist approach accept that individuals in all populations of organisms normally have at least minor differences. When they encounter fossils that are similar, but not identical, they tend to lump them into the same species. Currently, the populationist approach to defining species has become the dominant one today.

What are **coprolites**?

Coprolites are the petrified feces of ancient animals. Scientists study the scat (droppings) of ancient animals as it can sometimes give definitive knowledge about the animal's diet and how its digestive systems worked. Coprolite—Greek for "dung rock"—is the scientific name coined by Dr. William Buckland in 1829. A good environment that allowed for the preservation of coprolites was along flat banks (floodplains) of rivers, areas in which the deposited feces could dry slightly before rapidly being buried by sediment from a river flood. Although coprolites are fairly common fossil objects, dinosaur coprolites are rare.

What are **gastroliths**?

Gastroliths, also called stomach or gizzard stones, are smooth, polished stones often found in the abdominal cavities of the skeletons of dinosaurs, such as the Apatosaurus and Seismosaurus (both plant-eaters). Scientists believe the stones tumbled around in the stomachs of the giant animals, helping to grind tough vegetable matter such as twigs, leaves, and pine needles. This is similar to how some animals, especially birds, use small stones today. For example, emus eat tough, fibrous food and need the stones in their digestive tracts because they have no teeth.

Volunteers carefully excavate around the skulls of two sabertooth cats and a dire wolf in the sticky asphalt of Los Angeles's Page Museum at the La Brea tar pits. Oozing in a pit thirteen feet below the heart of the city lies a treasure trove of Ice Age animals, including mammoths, mastodons, and land sloths, which literally sank into history. *AP/Wide World Photos.*

Have any **soft parts** of organisms been found as **fossils**?

Yes, there have been some "soft parts" found as fossils, but very few. The main reason is simple: Soft parts easily decay within a short period of time.

But there have been some interesting discoveries. One recent finding gave scientists a good idea about ancient reptiles' hearts: A plant-eating, 66 million-year-old (late Cretaceous Period) *Thescelosaurus* fossil known as Willo had a fossilized heart. This dinosaur was about 13 feet (4 meters) long and about 665 pounds (300 kg) when alive. It was found in South Dakota and is currently exhibited at the North Carolina Museum of Natural Sciences in Raleigh. Scientists realized after years of painstaking research that this fossil might be the most important dinosaur in the world. Finding the heart attached to the dinosaur's skeleton—and fossilized in iron and sulfur—was an amazing discovery.

The finding also changed scientists' theories about dinosaurs because the heart showed that Willo was warm-blooded like a mammal, rather than a cold-blooded reptile as expected. Of course, not everyone agrees. Some people claim the heart is actually a concretion—a hard, stone-like mass found in sedimentary rock. But the apparent heart in Willo's chest contains complex structures, whereas lumps of mud like concretions do not.

What are the La Brea tar pits?

The Rancho La Brea tar pits are some of the most famous in the world. Located in southern California, the tar pits are excellent sites for fossilization, filled with pools and deposits of asphalt. Tar pits form as crude oil seeps to the surface; the small amount of oil associated with the seep evaporates, leaving behind a heavy tar (asphalt) in pools that easily trap unsuspecting animals.

La Brea in particular has yielded rich collections of fossil vertebrate bones, wood, insects (even insect larvae), plants, mollusks, and sundry other organisms from a time span between 40,000 to 8,000 years ago. Since 1906, more than one million bones have been recovered, representing over 660 species of organisms, including 159 kinds of plants. Of the Pleistocene (Ice Age) vertebrates, 59 species of mammals and 135 species of birds have been identified. Based on the number of animal fossils discovered in the muck, scientists estimate that, on the average, this represents about 10 animals caught every 30 years.

The fossil bones first pulled from the La Brea pits were thought to be from some unlucky cattle that became entrapped in the tar. By 1901, the first scientific excavation of the pits took place, revealing a treasure of ancient animals and plants. The most common large mammal fossils pulled from the pits are the dire wolves. The second most common are fossils of the *Smilodon,* the most famous of the saber-toothed cats. Besides these cats, a number of the larger animal species found at La Brea no longer exist in North America, including native horses, camels, mammoths, mastodons, and longhorned bison. Even today, about 8 to 12 gallons (32 to 48 liters) of tar bubble to the surface, occasionally trapping all types of organisms—from birds and reptiles to small and large mammals—especially on warm days from late spring to early fall, when the asphalt is soft and sticky.

Are remains always **preserved in rock**?

No, not all remains are preserved in layers of rock. While the majority are, there are also other materials that hold fossils, including tar, peat, ice, and the resin of ancient trees (amber). A few fossils have been found pickled in swamps, dried in desert caves, or desiccated and wind-buried in deserts. In any of these conditions, even soft body parts can be well-preserved indefinitely.

Human body remains are also often found preserved as "fossils" in places other than rock. For example, the mummies of ancient Egypt could be considered fossils and contain both soft and hard body parts preserved by extreme, continuous dehydration. Bodies found in cold, stagnant swamps or bogs—and submerged for thousands of years—have also been found in remarkably good condition. The lack of oxygen (anaerobic conditions) preserved their soft tissues, literally halting most decay.

What are **molds and casts**?

Molds and casts are types of fossils. When a dead organism is buried, it often decays completely, leaving behind only an impression in the rock in the form of a hollow mold. The hard parts are most likely to leave an impression, although sometimes so can soft parts. If the mold then fills with sediment, this can also harden, forming a corresponding cast.

What are **pseudofossils**?

Pseudofossils are patterns in rocks that may be mistaken for fossils. For example, some branching or dendritic patterns (called dendrites) found in layers of sandstones and other sedimentary rocks often look like plants. In actuality, they are caused by the percolation of chemicals in rock fissures; the stain from the chemical (often iron or manganese oxide) forms the plant-like pattern. Another example are stromatolites, which are concentric rings formed by ancient blue-green algae.

A large segment of chalk lies at the foot of the famed white cliffs of Dover beneath the South Foreland lighthouse in England. *AP/Wide World Photos.*

Sometimes rocks are not stromatolites but layers of calcite and serpentine minerals that are probably of volcanic origin.

What are the most **common fossils** found on Earth?

The most common fossils on Earth are pollen grains (the male reproductive bodies of advanced plants) and shells. Pollen grains are usually seen under a microscope; most shells are larger and more obvious.

Fossil pollen is found in many rock layers around the world. Produced during the reproduction of plants, these small and mobile cells enable plants to disperse and interact over long distances. Because of this, they have a thick, external wall (of sporopollenin, a durable organic polymer) and resistance to drying out, thus enabling the pollen to be preserved in the fossil record. Scientists have used the pollen record to determine what the ancient climate and environment were like. For example, scientists know that Mongolia had plains 36 million years ago, thanks to studying fossil plant pollen within the region's rocks. Plants have changed little over the past 40 million years, and plant families found today in arid regions are similar to those found in 36-million-year-old Mongolian rocks. This suggests that it was dry in Asia at that time, too.

Shells can be as large as a giant clam shell or as small as plankton. In fact, the chalk and beaches of the White Cliffs of Dover were created from the lime shells of billions of Cretaceous microscopic, single-celled marine animal and plant plankton (coccoliths). The chalk cliffs began to form as the plankton died, the layer becoming ever thicker as their shells fell to the ocean floor between 80 and 65 million years ago.

How did **ancient insects** become trapped in **amber**?

Ancient insects have been found in amber, a resin produced from certain cone-bearing trees, such as pine, fir, and spruce. When a tree was damaged, it produced the resin, a sticky, viscous fluid. If an insect was traveling up the tree bark and encountered the resin, it could become entangled in the blob. (Fossilized feathers, pieces of bark, and at least one small frog have also been found.) If the resin was buried (before or after the tree fell), it eventually produced fossil amber.

Insects preserved in amber have been discovered in certain Baltic Sea coast regions and various islands in the West Indies. But in reality, fossilized insects, in amber or rock, are rare. The chance that amber would form in conjunction with the meeting of the damaged tree and insect was not high. In addition, although more than half of the millions of species alive today are insects, of the 250,000 species known from fossils only 8,000 are insects. Fewer insects meant there was less of a chance of catching one in the ancient resin.

DATING FOSSILS

How is a **fossil's relative age** determined?

The relative age of a fossil is based on its position within the surrounding rock layer. The most common way of looking at relative age is called superposition, which is based on the fossil's position in a stacked sequence of sedimentary rock layers. The ages are relative: Fossils in the lowest layers are older than the fossils in the upper layers. This method has been used for centuries to determine the approximate age of rocks and fossils. (For more information about ages of rocks, see "Measuring the Earth.")

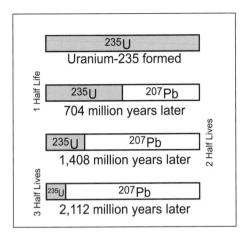

1 Half Life

235U
Uranium-235 formed

235U | 207Pb
704 million years later

2 Half Lives

235U | 207Pb
1,408 million years later

3 Half Lives

235U | 207Pb
2,112 million years later

The radioactive decay of Uranium-235 into lead (Pb-207). The half-life of Uranium-235 is 704 million years.

What are **index fossils**?

Index fossils are those that form a "pattern" throughout geologic history. These are organisms that existed during limited periods of geologic time, making them perfect to use as guides to date the rocks in which they are found. For example, ammonites were common during the Mesozoic Era (245 to 65 million years ago), but became extinct at the end of the Cretaceous Period (during the Cretaceous-Tertiary [K-T] extinction event that killed off the dinosaurs). Other index fossils include brachiopods (during the Cambrian Period, 540 to 500 million years ago), with some still surviving; graptolites from the Cambrian Period to the early to mid-Carboniferous Period (360 to 320 million years ago); conodonts from the Paleozoic (540 to 245 million years ago) and the Mesozoic (245 to 65 million years ago) eras; and trilobites from the beginning of the Paleozoic (about half the Paleozoic fossils are trilobites) to the late Permian Period (248 million years ago).

How do you find a **fossil's absolute age**?

The absolute age (or direct dating) of a fossil is determined by knowing the age of the surrounding rock. The most common way is by radiometric dating, a method that relies on the natural radioactivity of certain elemental isotopes. Different isotopic elements—such as uranium, rubidium, argon, and carbon—have varying (but constant) rates of radioactive decay. This constancy provides a "radiometric clock," allowing scientists to analyze rock samples and measure the relative ratios between the "parent" and "daughter" isotopes. The ratios reveal the age of the rock layer, and thus, the age of the associated fossil. As with many scientific measurements, although the method is called "absolute," the dates are not exact and there is room for potential errors.

What are some **absolute dating methods**?

The following lists a few of the radiometric and other dating methods used to determine the absolute age of rocks and fossils:

Thermoluminescence and electron spin resonance—The atoms in minerals are continuously being bombarded by radioactivity present within the ground. This excites electrons that subsequently become trapped in the minerals' crystal structures. Geologists use thermoluminescence and electron spin resonance to determine the number of excited electrons present in the

Professor Jacques Millot, left, and assistants, are shown with a preserved coelacanth in Paris, France, in 1956. The lobe-finned fish, previously thought to be extinct, was caught off the coast of East Africa at a depth of more than 1200 feet. *AP/Wide World Photos.*

minerals of a rock. Thermoluminescence uses heat to free the trapped electrons, while electron spin resonance measures the amount of energy trapped in a crystal. This data is compared to the actual rate of increase of similar excited electrons. Geologists can then calculate how long it took for the excited electrons to accumulate within the minerals, determining the age of the rock and, thus, the age of the fossils.

Uranium-series dating—Another radiometric method used to determine the age of a fossil is called uranium-series dating. This measures the amount of thorium-230 present in limestone deposits. Uranium is present when these deposits form, but almost no thorium. The rate of decay of uranium into thorium-230 is known, so once the amount of thorium-230 in a limestone rock has been determined, the age can be calculated.

Tree rings and varves—Tree rings (the growth rings found in fossilized and living trees that indicate years) and varves (seasonal sedimentary deposits) are consistent in their growth. Thus, some scientists also use these patterns to determine the absolute age of rocks and fossils.

What are **carbon dating techniques**?

Carbon dating techniques measure the age of organisms that are relatively young in terms of geologic time. This method is based on the fact that there are two stable isotopes of carbon, ^{12}C and ^{13}C, with the former being slightly lighter than the latter. Living organisms will preferentially absorb the ^{12}C, as it takes less energy. A larger than normal ratio of ^{12}C to ^{13}C in a sedimentary rock indicates that these amounts have been altered by living organisms, and therefore life was once present.

Since the decay of carbon isotopes does not occur over a very long period, carbon dating can be used to determine the age of more recent fossils, especially those younger than 45,000 years. When nitrogen-14 (^{14}N) in the atmosphere is bombarded by cosmic rays, carbon-14 (^{14}C) is produced, drifting down towards the surface, where plants absorb it from the air. Animals take the ^{14}C into their bodies by either eating the plants directly or eating other plant-eating animals. The death of an organism stops the intake of ^{14}C, which begins to decay to ^{14}N. The half life of ^{14}C is 5,730 years, meaning that in this time period one half of the ^{14}C will change into ^{14}N. In another 5,730 years, half of the remaining amount of ^{14}C will change into ^{14}N, and so on. By measuring how much C-14 is left in the fossil—and how much has decayed—geologists can determine the approximate age of the fossil.

What are **"living fossils"**?

Living fossils are modern animal and plant species that are almost identical to their ancestors that lived millions of years ago. In fact, many of these species were discovered and identified as actual fossils long before they were found as modern living organisms. For example, one species of modern ginko tree survives intact from the Triassic Period (approximately 220 million years ago). One of the earliest true flowering plants, the magnolia, existed some 125 million years ago during the Cretaceous Period; the horsetail was present during the Devonian Period almost 380 million years ago.

And it is not just plants that are found today as living fossils, but animals as well. The modern brachiopod *Lingula* is almost identical to its ancestor that lived during the Devonian Period. Other animals include the tuatara, a reptile species from the Triassic, and the didelphids, marsupials, including the modern opossum, that lived at the end of the Cretaceous Period. Insects such as cockroaches and dragonflies are also living fossils that evolved during the early Carboniferous Period (approximately 350 million years ago).

DINOSAUR, HUMAN, AND OTHER FAMOUS FOSSILS

What did some **ancient cultures believe** about **dinosaur fossils**?

People have been finding and digging for fossils for hundreds—perhaps thousands—of years. The "dragon" bones found in Wucheng, Sichuan, China, over 2,000 years ago were probably dinosaur fossils. And the Greek and Roman ogre and griffin legends might have originated with dinosaur fossil discoveries.

When was the **first dinosaur fossil** collected and described?

In 1676, a huge thigh bone was found in Oxfordshire, England, by Reverend Robert Plot. Although Plot scientifically recorded the bone, he also claimed it was from a humanlike "giant," not anything like a dinosaur. By 1763, British naturalist R. Brookes studied the fossil, also concluding it was from a giant and giving it the name *Scrotum humanum.*

The first dinosaur to be truly scientifically described (and the first theropod dinosaur discovered) was a Megalosaurus, which was also found in Oxfordshire, England. (It is interesting to note that the first dinosaur found was an Iguanodon, but it was named and scientifically described after the Megalodon.) The fossil's genus was named in 1824 by British fossil hunter and clergyman William Buckland (1784–1856), who studied the creature's lower jaw and teeth; Gideon Mantell (1790–1852) gave the scientific type species name, *Megalosaurus bucklandii.* No one knew it was a "dinosaur" yet, the word having not yet been invented.

What are some **famous dinosaur fossil discoveries**?

There are numerous famous dinosaur fossil discoveries. The following lists some of the more interesting ones:

- The first dinosaur fossil discovered in the United States was a thigh bone found by Dr. Caspar Wistar in Gloucester County, New Jersey, in 1787. The bone has since been lost, but more fossils were later found in the area.
- In 1877, the first dinosaur bones were pulled out of a rock formation that would prove to be one of the most prolific in the United States. Named after a nearby Colorado town where the first dinosaur bones were discovered, the world-famous Morrison Formation is where most of the fossils were found in the middle green siltstone beds and in the lower sandstones.

155

- Perhaps the most famous fossils are of dinosaurs preserved in the act of nesting on their eggs. Located in the Gobi Desert, Ukhaa Tolgod proved to be an incredibly rich area for dinosaur, mammals, and other animal skeletons. The fossils discovered here are mostly uncrushed, often complete, and preserved in exquisite detail.
- In an extremely rare discovery in 2001, a 77-million-year-old, nearly complete duck-billed dinosaur fossil was found in Montana with much of its skin and muscle tissues preserved by the mineralization process.

A Neanderthal skeleton reconstructed from casts of more than 200 Neanderthal fossil bones stands in the foreground with a modern version of a *Homo sapiens* skeleton behind. *AP/Wide World Photos.*

Is there a **connection** between **dinosaurs and birds**?

Most likely, yes. Many scientists now believe there is a connection between birds and dinosaurs. One of the first birdlike fossils to be found was the *Archaeopteryx lithographica,* which literally means "ancient wing from lithographic limestone."

Thought by many paleonologists to represent the oldest bird yet discovered, the first fossil was found in 1855 in the Solnhofen quarry in southern Germany. This fossil laid in sedimentary rock from the upper Jurassic Period, but it was not recognized as a bird until 1970. Over the years, six more fossil skeletons have been uncovered and are dated between 125 to 147 million years old. The fossils of these birdlike creatures—about the size of a present-day crow—were used by some early paleontologists to help substantiate the theory of evolution.

What happened to the Homo neandertalensis?

More fossil bones of the *Homo neandertalensis* have been found in relatively good condition compared to earlier humanlike species—mainly because they are younger and the Neandertals (also seen as Neanderthals—both spellings are correct and in common usage, even among paleoanthropologists) buried their dead. On the average, they were 5.5 feet (just under 2 meters) in height, had short limbs, and were well-adapted to living in a cold climate. Attached to their thick and heavy bones were powerful muscles. The Neandertal's brain was larger than that of living humans, with its shape longer from front to back and not as rounded in the front.

The anatomy of Neandertals and modern humans (*Homo sapiens*) seem so similar that in 1964 it was proposed that Neandertals were not a separate species from modern humans, but the two were actually subspecies: *Homo sapiens neandertalensis* and *Homo sapiens sapiens*. Popular throughout the 1970s and 1980s, this naming convention has since been dropped in favor of the two-species classification. Either way, Neandertals represent an extremely close evolutionary relative of modern humans.

What happened to the *Homo neanderthalensis*? No one has yet figured out this mystery. It is interesting to note that the Cro-Magnons arrived about 40,000 years ago in Europe, a region already populated with Neandertals. The two populations coexisted for as much as 10,000 years, then the Neandertals were either wiped out or assimilated into the Cro-Magnon groups.

In modern times, many scientists see the *Archaeopteryx* as marking the transition between dinosaurs and birds, providing proof that birds descended from dinosaurs. But other scientists are not ready to jump on the bird-as-dinosaur bandwagon, especially because there have been so few such fossils found around the world.

What were some of the first **hominid fossils discovered**?

One of the first hominid (humanlike) fossils was discovered in 1856 and belonged to the *Homo neandertalensis*. The discovery of a skullcap and partial skeleton in a cave in the Neander valley near Dusseldorf, Germany, signaled the first recognized fossil human form. *Homo neandertalensis* lived about 200,000 to 30,000 years ago. In 1893, the first fossil of *Homo erectus* (known as "Java Man") was uncovered in Indonesia. Similar fossil remains were subsequently found throughout Africa and Asia, making it the first wide-ranging hominid with a skeleton very similar to that of modern humans, although it was thicker and heavier. *Homo erectus* was probably the first hominid to use fire and lived from about 1.6 million to 200,000 years ago.

What are thought to be some of the **evolutionary steps** leading to **Homo sapiens**?

Although the road to discovering how *Homo sapiens* evolved is a long and fiercely debated one, there are still some fossil hominid bone discoveries that do give some evidence. Based on these fossil finds, scientists have determined some of the previous human-like species that existed for the past approximately 4 million years. The following lists some of the names given to these fossils and their approximate ages (note: this listing of names and dates continues to be highly debated):

Species	Years Ago
Ardipithecus ramidus	4.4 million
Australopithecus anamensis	4.2–3.9 million
Australopithecus afarensis	3.9–3.0 million
Australopithecus africanus	2.8–2.4 million
Australopithecus garhi	2.5 million
Paranthropus aethiopicus	2.7–1.9 million
Paranthropus boisei	2.3–1.4 million
Paranthropus robustus	1.9–1 million
Homo rudolfensis	2.4–1.9 million
Homo habilis	1.9–1.6 million
Homo ergaster	1.8–1.5 million
Homo erectus	1.6 million to 200,000
Homo heidelbergensis	600,000 to 200,000
Homo neandertalensis	200,000 to 30,000
Homo sapiens (and *Homo sapiens sapiens*)	100,000 to present

What are some of the more **famous human fossils**?

There are numerous human fossils that have captured the public's attention. For example, "Lucy"—from the species *Australopithecus afarensis*—is one of the most famous and dates to about 3.2 million years ago. Her remains were found in Hadar, Ethiopia—and she is considered one of the oldest human fossils ever found. Another famous human fossil is the "Taung child," an *Australopithecus africanus* that lived at least 2.4 million years ago. These fossil remains were found in Taung, South Africa. Still another human fossil is the "Peking Man," a member of the *Homo erectus* that lived about 500,000 years ago. His remains were found in Zhoukoudian, China.

What are some **other famous fossils**?

There have been many famous fossils around the world. Here are just a few of the old and new discoveries:

Have there been any fossils from Mars?

A 4.5 billion-year-old rock, labeled meteorite ALH84001, is believed to have once been a part of Mars and to contain fossil evidence that primitive life might have existed on the red planet more than 3.6 billion years ago. The rock is a portion of a meteorite dislodged from Mars by a huge impact about 16 million years ago. It subsequently fell to Earth in Antarctica 13,000 years ago. The meteorite was found in 1984 in Allan Hills ice field, Antarctica, by an annual expedition of the National Science Foundation's Antarctic Meteorite Program. It is currently preserved for study at the Johnson Space Center's Meteorite Processing Laboratory in Houston. (For more information about this special meteorite, see "The Earth in Space.")

Trilobites—One of the most famous fossils, trilobites date back 550 million years ago and were the first animals to develop eyes. The largest fossils of this "armored" animal can measure over 2 feet (almost a meter) in length.

Burgess Shale—Discovered in 1909, the Burgess Shale animal fossils comprise more than 140 species in 119 genera, with the majority of species being bottom-dwelling (benthic) organisms. In addition to specimens with the usual hard skeletal material fossilized, the excellent preservation of the rock layer has resulted in a large number of soft-bodied organisms being discovered.

Chengjiang mud beds—Many Burgess Shale-type fossils have also been found in lower Cambrian Period deposits near the town of Chengjiang in the Yunnan Province of China; they are about 15 million years older than the Burgess Shale deposits. To date, more than 10,000 specimens have been discovered.

Wormlike fossils—In 1998, researchers discovered what appeared to be worm-like animals in rock over 1 billion years old—almost twice as old as any other multicellular life yet discovered. This finding has lead many scientists to look once again at the origins of multicellular organisms, which are typically thought to have originated around the Cambrian Period around 600 million years ago.

Are there any **fossil parks in the United States**?

Yes, there are many National Parks known for their fossils. The following lists the major ones administered by the National Parks Service:

- Agate Fossil Beds National Monument, Nebraska
- Badlands National Park, South Dakota
- Channel Islands National Park, California

- Death Valley National Park, California and Nevada
- Delaware Water Gap National Recreation Area, Pennsylvania and New Jersey
- Dinosaur National Monument, Colorado and Utah
- Florissant Fossil Beds National Monument, Colorado
- Fossil Butte National Monument, Wyoming
- Grand Canyon National Recreation Area, Arizona and Utah
- Grand Canyon National Park, Arizona
- Guadeloupe Mountains National Park, New Mexico and Texas
- Hagerman Fossil Beds National Monument, Idaho
- John Day Fossil Beds National Monument, Oregon
- Petrified Forest National Park, Arizona
- Theodore Roosevelt National Park, North Dakota
- Yellowstone National Park, Montana, Idaho and Wyoming

FOSSIL FUELS

FOSSIL FUELS BASICS

What are the **major fossil fuels**?

Oil, gas, and coal are the major fossil fuels. They were formed over millions of years as a thick blanket of sediment covered organic matter. As more layers of sediment fell on top, increasing the pressures below, energy-rich compounds of hydrogen and carbon (also called hydrocarbons) formed, including oil (also called petroleum), gas, and coal.

What are some of the **first records** mentioning what we now call **fossil fuels**?

The earliest records mentioning fossil fuels go back thousands of years. The following are only some of the earliest references to these materials:

Oil (petroleum)—Most ancient cultures "used" oil in one way or another. Egyptians, for example, preserved mummies with oil (but only for the poor, not the more affluent). In fact, mummy comes from the Persian word *mummeia,* meaning "pitch" or "asphalt," a naturally occurring crude oil that oozed to the surface). Mummeia was also a prized medicine in Europe, but it was limited in supply. The early civilization of Mesopotamia used asphalt for construction purposes, especially to waterproof houses and boats. The ancient Greeks and Romans used petroleum as a lubricant and for sealing. There is strong evidence that Native Americans, at least as far back as 1410, used oil for medicinal purposes. They obtained oil by hand digging small pits around active seeps (areas in which oil is exposed at the surface), then lining the hollows with wood. Early European settlers in North America also skimmed oil from seeps, using the petroleum for lamp fuel and to lubricate machinery.

What's the connection between Marco Polo and fossil fuels?

Along the Caspian Sea in modern Azerbaijan, in the Persian city of Baku, the famous Italian explorer Marco Polo (1254–1324) witnessed oil being collected from seeps. He wrote, "On the confines toward Georgine there is a fountain from which oil springs in great abundance, inasmuch as a hundred shiploads might be taken from it at one time. This oil is not good to use with food, but 'tis good to burn, and is also used to anoint camels that have the mange. People come from vast distances to fetch it, for in all countries round there is no other oil."

He also mentioned other materials we now call fossil fuels, including mud volcanoes associated with natural gas leaking into ponds. He actually witnessed the "Eternal Fires of the Apsheron Peninsula," in which natural gas seeps through cracks in broken shale—eternal fires that were (and have been) worshipped for centuries. He also brought back knowledge of coal to Italy from China, referring to it as the "stone that could burn." It was an item never used before in Europe. However, the origin for his reference to coal is questioned, as some historians doubt Polo actually made it to China.

Natural gas—Records of natural gas appear as early as 3000 B.C.E. (some historians say 6000 B.C.E.), as ancient people worshipped "eternal fires" that most likely were produced by seeping natural gas. By 900 B.C.E., the Chinese had discovered natural gas in wells originally drilled for salt brine pools. This gas finally made its first big impact on the public in 1807 when gaslights lit up the streets of London.

Coal—Because coal lies mostly below the Earth's surface, its use through the centuries has been more limited than seeping oil and natural gas. The Chinese dug coal more than 3,000 years ago. They used bitumen (a black, heavy, viscous oil) when building the Great Wall of China. In 1679, French explorers on the Illinois River first mentioned coal in the United States. Mining coal commercially occurred in 1750 in Richmond, Virginia. From 1850 to 1950, it became the most important fuel in the United States.

How much of our **energy in the United States** comes from **fossil fuels**?

More than 85 percent of the energy used by the United States comes from fossil fuels. Oil supplies account for approximately 40 percent of our energy, natural gas about 25 percent, and coal about 20 percent.

Do **fossil fuels develop differently**?

Yes, various fossil fuels developed differently. Scientists believe coal formed from plant debris, while natural gas and oil formed from tiny organisms (plants and animals) that settled to the bottom of ancient seas and rivers. Oil and natural gas formations depended on the amount of time the organisms decayed underground, along with temperature and pressure differences. In general, where the underground temperature was hotter or the pressures were the greatest, natural gas formed.

Where do you **find oil and natural gas**?

Oil and natural gas often appear as seeps at the surface. For example, the "eternal fires" in Baba Gurgur, Kirkuk, in northwest Iraq are petroleum and natural gas seepages that probably were originally ignited by lightning.

But the majority of oil and natural gas reserves are drilled from deep underground. Oil and natural gas are most often found in reservoirs or pools, trapped in rocks containing holes much like a sponge. (Natural gas is either dissolved in the oil or rests on top of the

This oil drilling operation in Colorado features advanced drilling technology, resulting in thousands of gallons of water being saved and a reduction in truck traffic and waste. *AP/Wide World Photos.*

oil as a separate layer.) These sites are most often found under dome-shaped structures called anticlines. In most cases, at the crest of the anticline is a thick rock called a caprock. This dense, nonporous rock traps the oil and gas, keeping both from reaching the surface.

How is an **oil or natural gas well dug**?

A well is a hole drilled or bored into the Earth in order to extract gas or oil (the other fossil fuel, coal, is most often mined). In the United States, the average oil well reaches about 1,650 feet (500 meters) deep. In general, a rotary tool with a tough diamond bit   163 bores deep through the rock layers, typically drilling anywhere from 10 to 20 feet (30 to 60 meters) per hour. A fluid composed of clay, water, and chemicals—called "drilling mud"—is pumped down with the bit to keep it cool. When oil is found, it is extracted with a pump. In order to get out as much oil as possible, drillers pump in other materials—such as water, gases, or chemicals—into the well to keep the pressure high. At the top of the well, a collection of pipes and valves called a "Christmas tree" controls the flow of the oil (or natural gas).

What is a **well blowout**?

A blowout occurs when a well releases an uncontrolled flow of gas or oil (or other fluids) into the air. If the pressure in the gas or oil reservoir underground exceeds the weight of the drilling fluid inside the well hole, a blowout can occur. In the early days of oil drilling (and a typical scene in old movies), it was common to have gushers in which a drill bit hit a pocket of pressurized natural gas, forcing petroleum to the surface. Modern well drillers now have special instruments that continually measure well pressure. They also use a blowout preventer—a special collection of heavy-duty valves on top of the wells—to prevent the blowout from occurring during drilling operations.

OIL (PETROLEUM) PRODUCTS

What is **petroleum**?

The word petroleum comes from the Greek *petros* ("stone" or "rock") and the Latin word *oleum* ("oil"). Petroleum applies to crude oil and oil products in all their different forms.

What is **crude oil**?

Crude oil—a smelly, yellow-to-black liquid—is petroleum in its natural state as it emerges from a well or after passing through a gas-oil separator. To put it another way, it is a naturally occurring mixture of mostly hydrocarbons that exists in a liquid state in underground reservoirs before refining or distillation.

What are **bitumen and extra-heavy oil**?

Bitumen is best described as a thick, sticky form of crude oil—a tar-like mixture of petroleum hydrocarbons. It is so much more viscous than crude oil that it will not flow unless heated; at room temperature, it is similar in consistency to cold molasses. This heavy, black oil must be diluted with lighter hydrocarbons to convert it to an upgraded crude oil, and to make it transportable via pipelines. From there, it can be used by refineries to produce diesel fuels and gasoline.

Extra-heavy oil is another form of crude oil, but it's not as degraded as natural bitumen. (Approximately 90 percent of all extra-heavy oil deposits are found in the

The Trans-Alaska Oil Pipeline snakes across the Alaska tundra under the Brooks Range about 150 miles from Prodhoe Bay, Alaska, carrying crude oil about 800 mile to Valdez, Alaska. *AP/Wide World Photos.*

eastern Venezuela basin.) Drillers often use methods such as steam injection to recover this oil.

How does **crude oil (petroleum) form**?

Although still highly debated, most scientists believe in the organic theory of oil formation: Petroleum is formed from the remains of marine organisms—probably phytoplankton—that died millions of years ago. (Others believe terrestrial plants also formed oil.) This is because certain carbon-containing substances are found in oil that could only have come from such organisms.

After the organisms died, their remains settled to the bottom of the ocean. The organic matter was subsequently buried in mud, and over hundreds of thousands of years, layer upon layer of sediment increased heat and pressure on the deposit. Millions of years later, microscopic globules of oil (and natural gas) were created, with the oil occupying from 5 to 25 percent of the rock volume. These crude oil-saturated rocks were most often a sedimentary rock called shale, which is why rocks containing large amounts of oil are often referred to as oil shales.

A larger petroleum reservoir (or trap) forms when the oil and gas from the shale moves into a porous and permeable rock, such as sandstone or limestone. It is thought that oil can migrate to the trap in several ways: Seawater (more dense than oil) in the rock pushes the oil upward through nearby rock layers, eventually becoming trapped by non-porous rock; or the weight of overlying rock layers squeezes oil into cracks and holes, eventually collecting in nearby rock.

Either way, over time the oil becomes trapped in underground reservoirs. Oil traps include anticlines or dome-shaped rock formations; along faults that shift an impermeable (impenetrable) layer of rock next to a permeable (porous) one that contains oil; in salt domes; and in nonporous rock layers called stratigraphic traps. Petroleum also often seeps to the surface in many oil and natural gas-rich areas around the world.

What is **oil shale**?

Oil shale is shale—a sedimentary rock—that contains oil recoverable by distillation. The shale itself is an inorganic, nonporous rock that contains a certain amount of organic material in the form of kerogen. Many scientists consider oil shales to be a hybrid of oil and coal. Some have more kerogen than an oil-containing rock but less than coal. Two thirds of the world's oil shale reserves are located in the United States. The largest known reserves of hydrocarbons of any kind are the Green River shale deposits in Wyoming, Colorado, and Utah.

The biggest difference between oil shale rock and a rock that contains oil is that oil shale has greater amounts of kerogen (as much as 40 percent) than oil source rock (usually about 1 percent). In addition, oil shale has never been exposed to the high temperatures necessary to convert the kerogen to oil. Oil shale is not used as a solid fuel; it is desirable because of its possible conversion to liquid fuels (usually by heating). If the shale is lean, with about 4 percent kerogen, it can yield about 6 gallons of oil per ton of shale; rich shales with up to 40 percent kerogen can yield about 50 gallons of oil per ton of shale.

What are **oil or tar sands**?

Oil (or tar) sands are deposits of bitumen. While crude oil flows naturally at the surface or is pumped from the ground, oil sands must be mined or recovered in place. Hot water is used to extract the oil sands from the rock; from there, various processes are used to separate the bitumen from the sand and water.

How much liquid does a **barrel of oil** hold?

A barrel is the standard unit for measuring liquids in the oil industry; it contains 42 U.S. standard gallons (159 liters). As for "how much," in terms of money, it is dependent on many factors, especially supply and demand. But because oil means power in this modern world, the predominant factor driving the cost is political, as countries position themselves to dominate and control the flow of oil.

How much of the **United States' energy** come from **petroleum products**?

Together, oil and natural gas supply 65 percent of the energy used in the United States.

According to the United States Energy Information Administration, oil furnishes 40

> ## Why are petroleum products in great demand?
>
> Petroleum products are in great demand because of their versatility: They can be a fuel for heating or power production, or the byproducts can be used to produce everyday chemicals. In many ways, humans have produced so many items that are dependent on petroleum that it has been difficult to find suitable replacements.
>
> There are good reasons for this dependency, especially in the production of petroleum as fuel. Petroleum fuels ignite and burn steadily; they also produce a great amount of heat and power in relation to their weight; and they are easy to handle, store, and transport. In fact, 53 percent of the world's energy requirements are met from oil.

percent of our energy, natural gas 25 percent, coal 22 percent, nuclear power 8 percent, and renewables (hydroelectric, geothermal, wind, and others) about 5 percent.

How are **petroleum products used**?

The list of petroleum products seems endless. They fall into three major categories: fuels such as motor gasoline and distillate fuel oil (diesel fuel), where demand is greatest; finished nonfuel products, such as solvents and lubricating oils; and materials (called feedstocks) for the petrochemical industry, such as various refinery gases.

Almost all our methods of transportation—cars, trucks, trains, motorcycles, and airplanes—depend on fuels made from oil. We use them as gasoline to fuel cars and in the heating oil to warm our homes. Many industries also rely on petroleum to run their plants and factories, and they use lubricants made from oil to keep machinery running in their factories. Other uses for oil are numerous, from producing fertilizer to grow plants to plastics for everything from toothbrushes to hairbrushes.

Over 90 percent of the source material needed to produce thousands of chemicals comes from petroleum. For example, ethanol, styrene, ethyl chloride, butadiene, and methanol are the intermediate chemicals needed in the production of many substances, including plastics (needed for new technologies, such as compact discs and cellphones), synthetic rubbers, refrigerants, fertilizers, resins, solvents, detergents, nylon, explosives, insecticides, dyes, and some medicines and paints.

What are some **refined petroleum fuel products**?

Fuel products—or energy products produced after processing of crude oil and natural gas—account for nearly 9 out of every 10 barrels of petroleum used in the United States. The leading fuel, motor gasoline, consistently accounts for the largest share of total demand for petroleum products: more than 40 percent.

An oil worker climbs a set of stairs amidst a sea of pipelines, a telltale feature of an oil refineries around the world. *AP/Wide World Photos*.

Other petroleum fuels include distillate fuel oil (diesel fuel and industrial and heating oils), liquefied petroleum gases (LPG's, including propane, ethane, and butane), jet fuels (most are a kerosene-based fuel used in commercial airlines and military aircraft), residual fuel oil (used by electric utilities to generate electricity), kerosene (residential and commercial space heating, in water heaters, as a cooking fuel, and in lamps), aviation gasoline, and petroleum coke (often used as a solid fuel for power plants and industrial use).

Who first **distilled kerosene**?

Kerosene was first distilled in 1849 by Canadian geologist Abraham Gesner (1797–1864). This would eventually be a boon for oil drillers (and whales) and a bust for whalers. After all, kerosene eventually replaced whale oil as the illuminant of choice for many streets around the world: In 1857, Michael Dietz invented a kerosene lamp that forced whale oil lamps out of the market.

What are some **non-fuel uses** of petroleum?

Non-fuel use of petroleum is much less when compared with fuel use, but it is still widely used to produce specialized products for the textile, electrical, and metallurgical industries. For example, nonfuel products from petroleum include solvents for paints, lacquers, and printing inks; asphalt for roofing materials, roads, airfields, and other surfaces; petroleum wax (as paraffin) for candy making, candles, matches, and furniture polishes; and even petroleum jelly (petrolatum) used in medical products.

What is the story behind leaded and unleaded gasoline?

In the early 1920s, major car companies wanted to make car engines work more efficiently. To solve this problem, scientists at one company added a lead compound (tetraethyl lead) to gasoline. This additive eliminated the rackety sound that indicates a poorly running engine, raised the octane levels, and provided lubrication to vital engine parts. The resulting fuel was sold under the name Ethyl—what we now call leaded gasoline.

For about 50 years, leaded gasoline was still used in cars, even though for almost the same amount of time people had warned of the possible health side effects. It was reported that several scientists initially working on formulating the leaded gasoline had died from its effects. In the 1960s, evidence mounted, showing that airborne lead was a serious health hazard, causing a wide range of illnesses in adults and especially affecting younger children's neurological development, growth, and intelligence. By the 1970s, federal restrictions governing the lead content of motor fuels came into effect. Lead was also removed from gasoline because it damaged emission-reducing catalytic converters that began appearing in cars in the early 1970s. On December 31, 1995, tetraethyl lead was banned from all gasoline in accordance with the Clean Air Act Amendments of 1990.

More than 90 percent of cars on the road today were designed to use unleaded fuel. For cars that still need leaded gas, such as vintage vehicles, the "replacement" includes fuel additives equivalent to low-lead gasoline—or the owners need to make mechanical modifications to keep the car running with unleaded gas.

What does gasoline **octane measure**?

Octane is a measurement of a fuel's resistance to "engine knock," or the metallic sound that emanates from an engine running on too-low octane. The most common octane levels are 87 (regular), 89 (mid-grade), and 93 (premium). Octane enhancers are added to increase the octane readings. For example, ethanol—a high octane, water-free alcohol from the fermentation of sugar—is often used as a blending ingredient in gasoline to produce higher octanes.

If the octane in your car's tank is too low over a long period of time, the engine knocking can eventually damage pistons and other vital engine parts. The correct octane level depends on many factors, including vehicle age and mileage, driving habits, climate and geography, and individual vehicle requirements. The listing of your car's minimum octane needs is usually found in the owner's manual.

Why do **gasoline prices fluctuate**?

Currently, gasoline prices fluctuate because refiners pay more for the crude oil extracted from the ground. The reasons are many—from political and social strife to natural catastrophes. For example, in 2002, a barrel of crude oil cost about $12 United States dollars; a year later, the cost had risen to about $36 dollars per barrel. In this case, prices rose for three main reasons: Venezuela, the United States' fourth largest supplier of crude oil, was experiencing social and political unrest, including recovering from a strike by oil industry workers. Weather also played a part, as the winter of 2002–2003 was much colder than normal (both in the United States and Europe), so the demand for crude oil for heating fuel increased. And finally, there was uncertainty in the world, as one of the major oil producing regions became involved in war when the United States entered into a conflict with Iraq. This action affected the continued supply of crude oil from the Middle East, and prices again rose.

When was the **first oil drilled**?

The first record of oil being drilled is from China: In 347 C.E., oil wells were dug using bits attached to bamboo poles. These holes were about 800 feet (244 meters) deep.

How has **oil been used historically**?

One of the first records of oil use dates back to the 1500s. Oil seeping to the surface in the Carpathian Mountains was used to light street lamps in Poland. Some of the first oil sands were mined and the oil extracted at Pechelbronn field in Alsace, France, in 1735. In the United States, one of the first records of oil production comes from Pennsylvania, but this was an unwanted byproduct in which brine wells seeped oil during the search for salt. Probably the first modern oil well was drilled in Asia in 1848, on the Aspheron Peninsula northeast of Baku, by Russian engineer F. N. Semyenov. The first oil well drilled in North America was in 1858 in Ontario, Canada; the next year, the first commercial oil well was dug in the United States.

Who drilled the **first oil well in the United States**?

In August 1859, Edwin Drake drilled the first oil well, a hole 69-feet (21 meter) deep, in Titusville, Pennsylvania. Almost overnight, the small town turned into an oil boom town, marking the beginning of Pennsylvania as an oil state. Drake initially chose the site because of the many active oil seeps in the area. Numerous wells had already been dug in the area in the search for salt (often found in association with oil seeps) and drinking water. After striking oil, the drillers abandoned the wells, because at that time the black liquid was considered useless.

What was the **initial reason** for the **commercial oil drilling**?

The first commercially drilled oil wells were not dug for industry fuel, transportation, or plastics. Instead, during the 1850s, drillers hoped that the "rock oil" would be used

What was the deepest oil well ever drilled?

The answer to this question is difficult, since there are several claims to "world class" fame! It is known, however, that the deepest oil wells ever drilled to date are over 20,000 feet (6,096 meters) deep. One in Pecos County, Texas, reached 25,340 feet (7,724 meters) in 1958 before being abandoned. Others who claim the "world's deepest oil well" status don't even come close: Some people say Wasco, California, has the world's deepest oil well at 16,004 feet (4,878 meters). In reality, this is close to the average oil well depth of about 10,000 to 16,000 feet (about 3,000 to 5,000 meters).

commercially as a fuel for lamps. Oil at that time had also been refined and sold commercially for one of its byproducts: kerosene.

How much oil comes from **offshore drilling**?

About one third of the world's oil comes from offshore wells. The main drilling areas include the North Sea, Arabian Gulf, and Gulf of Mexico. These wells are usually drilled from fixed platforms—huge frameworks that are often as tall as skyscrapers. Wells are drilled from semi-submersible structures or tankers; the beginning of the well on the ocean floor is often over a half mile (1 kilometer) or more below the water's surface.

Where does the **United States' oil originate**?

About half of the oil Americans consume is produced in the United States; the rest is imported. Of the oil we import, 51 percent comes from other nations in the Western Hemisphere: 21 percent from the Middle East (where 75 percent of the world's oil reserves are found around the Persian Gulf), 18 percent from Africa, and 10 percent from other countries.

What are the current predictions for **future oil consumption**?

The United States Energy Information Administration predicts an average rise in world oil consumption of about 3 percent each year for the next 20 years. With more demand for petroleum products—not to mention the growth in our global population—this is only an estimate. Many people hope this growth rate can be slowed somewhat by developing alternatives to petroleum products and using alternative energy sources, such as wind or solar power.

EXPLORING NATURAL GAS

What is **natural gas**?

Colorless, and in its pure form odorless, natural gas is a naturally occurring mix of hydrocarbons in a gaseous state often found among petroleum deposits. It is mainly composed of methane, a simple compound made od one carbon atom surrounded by four hydrogen atoms. Once considered to be worthless—and thus discarded—it is now considered by some to be the "perfect" fuel, as it is convenient, efficient, and relatively clean, generating no ash and little air pollution. (Unlike coal and oil, natural gas emits less carbon dioxide for a given quantity of energy consumed.) Some countries (not the United States) still doubt its worth, eliminating the gas by burning it in fires large enough to be seen in space.

How much **natural gas** is **used by ordinary Americans?**

Natural gas currently provides one-fourth of all the energy used in the United States. Nearly half of all the energy used for cooking, heating, and for fueling other types of home appliances is supplied by natural gas. Because the gas is odorless, gas companies add a smell similar to rotten eggs that allows the homeowner to detect a gas leak in or around the home.

What are **natural gas liquids (NGLs)?**

Natural gas liquids (NGLs) are a collection of hydrocarbons; they are constituents of natural gas recovered as liquids in separators, field facilities, or processing plants. They mostly include ethane, propane, and butane, along with small quantities of non-hydrocarbons. (Propane and butane are sometimes referred to as liquefied petroleum gases.) NGLs are used in oil refining and petrochemical manufacturing. For example, ethane is used to make polyethylene; butane has a number of industrial and petrochemical uses; and propane is widely used as a fuel, as is butane.

How does **natural gas form?**

Natural gas is most often found in association with crude oil (as a "layer" on top) or dissolved in the oil. It is thought to have formed the same way as petroleum: from the decay of marine organisms that are then subjected to pressure.

Who dug the **first natural gas well?**

In 1821, William A. Hart drilled a 27 foot (9 meter) well to increase the flow of a natural gas surface seep in Fredonia, New York. This was the first intentionally dug well of its kind (smaller collections of natural gas were already being used to fuel lamps).

How was **natural gas** used in **the 1800s?**

Natural gas was used almost exclusively as fuel for lamps in the 1800s, with most pipelines bringing gas to light city streets (by the 1890s, electricity replaced gas street lamps). In 1885, Robert Bunsen invented the "Bunsen burner," a burner that mixed air with natural gas, thus paving the way for gas to be used to heat buildings and for cooking.

A natural gas plant in Lake Maracaibo in western Venezuela. The severe poverty that characterizes this South American nation has long been an anomaly, as Venezuela sits atop the biggest proven oil reserves outside of the Middle East. *AP/Wide World Photos.*

How does **natural gas** get from underground **to your home**?

In general, natural gas is first extracted from a reservoir (either with or without the associated crude oil). Because natural gases often contain other types of hydrocarbons, such as butane, ethane, and propane, the gases need to be processes to create a more "clean" methane. The extracted gases travel through pipelines to a refining plant, where non-methane hydrocarbons, water, and other impurities are removed. Some of the unwanted, extracted hydrocarbons are recovered to eventually be used as fuel, such as propane for camping stoves and barbecues.

At this point, a chemical called mercaptan is added to give the odorless methane the distinctive smell associated with natural gas. From there, natural gas flows through the long, continuous network of pipes—mostly made of steel and measuring from 20 to 42 inches (51 to 107 centimeters) in diameter—that eventually lead to the smaller pipes to your home.

What was one of the **first natural gas pipelines** built?

One of the first extensive pipelines—a 120-mile (193-kilometer) long line from gas fields in central Indiana to Chicago, Illinois—was built in 1891. The major push for natural gas pipelines came after World War II, especially in the 1950s and 1960s, when thousands of miles of pipeline were built throughout the country. In the United

What will eventually be the longest undersea gas pipeline?

The longest undersea gas pipeline might be completed by October 2007. When done, it will be 746 miles (1,200 kilometers) long and run from the Ormen Lange gas field to Great Britain. The Ormen Lange field is the largest undeveloped gas field on the Norwegian Continental Shelf, located some 62 miles (100 kilometers) northwest of the Møre and Romsdal coast at a water depth of some 3,300 feet (1,000 meters).

But there are other contenders, including China, who want to run a 1,988 mile (3,200 kilometer) pipeline from Kazakhstan's mammoth Kashagan oil field (one of the world's most prolific oil and gas fields, set to start producing in 2006) to China. With the need to satisfy the world's seemingly insatiable appetite for gas (and oil), the competition to build such pipelines will only continue.

States, there are currently more than a million miles of pressurized, underground natural gas pipelines. If laid end-to-end, this length would stretch to the Moon and back again two times.

What are the **deepest gas wells** drilled to date in **North America**?

The deepest gas well ever drilled in North America measures 31,441 feet (9,583 meters) deep. Located in the Anadarko Basin of Oklahoma, drilling of the so-called Lone Star 1 Bertha Rogers well (which once held the "deepest hole on Earth" status) was stopped when it struck molten sulfur. To date, the Bertha Rogers and Long Star Baden wells—which are also in Oklahoma and reach 30,050 feet (9,159 meters) in depth—are reportedly the only wells in North America that reach below 30,000 feet (9,144 meters).

What is **sour gas**?

Sour gas is a naturally occurring gas containing more than 1 percent hydrogen sulfide (H_2S) and found in deep, hot, high-pressure natural gas deposits. It is typically identified by a strong "rotten eggs" smell. In most cases, sour gas is recovered and converted to elemental sulfur, which is used to manufacture fertilizers, paper, steel, and other products.

DIGGING FOR COAL

What is **coal**?

Coal is a brownish or black solid formed by the partial decomposition of vegetable matter (with no access to air). A combustible, organic rock, it contains more than 50 percent carbonaceous material by weight. It is composed largely of carbon, hydrogen, oxygen, and nitrogen, varying amounts of sulfur, and traces of other materials, ranging from aluminum to zirconium.

Coal's complex range of materials, qualities, and quantities varies greatly from deposit to deposit and depends on many factors: the coal's original types of vegetation; the depth at which the deposit formed; the temperatures and pressures exerted at that depth; and the length of time the coal has been forming.

Coal also varies depending on how much the original plant material changed into carbon. The amount of vegetative material needed to create a coal bed is huge: On the average, it takes a layer of dead plant material 10 feet (3 meters) thick to create a 1-foot-thick (0.3 meters) layer of coal.

What are the various **coal types**?

Coal is classified into four different types (ranks), based on the amount of heat produced. As coal increases in rank it becomes harder: Bituminous coal is harder than lignite, and anthracite is the hardest. The higher the carbon content, the higher the rank given to the coal and the cleaner the energy generated. The following list provides details of the major coal types:

Anthracite—Anthracite is the hardest, purest type of coal; it is a brittle, lustrous, jet-black substance with a high luster. Almost pure carbon (90 percent) it is the highest rank of coal and creates the most heat and very little smoke. Most anthracites are more than 200 million years old. In the United States, it is primarily mined in northeast Pennsylvania and used mostly for residential and commercial heating.

Bituminous— Bituminous coal is the most common type of solid fossil fuel. It is a soft, dense, black (or dark-brown) material with well-defined bands of bright and dull material throughout. Most bituminous coals are more than 60 million years old. It is mined chiefly east of the Mississippi River, and it is mainly used for steam and power in manufacturing, and to make coke.

Subbituminous—Subbituminous coal is dull and black (or dark brown), and is often referred to as black lignite. Ranking between a bituminous and lignite, it is used for generating electricity and space heating. It can be soft and crumbly or jet black and hard. Subbituminous coal formed several million years after lignite. It is mined in the western United States and used primarily as fuel for steam-electric power generation.

Lignite—Lignite ("brown coal") is the lowest rank of coal. It is brownish-black, only about 65 percent carbon, and has a high moisture content. It also provides the least amount of heat of any coal—and the most smoke when burned. Lignite forms from peat after a few million years of burial. It is mined

Is coal being formed today?

Yes, in many ways, the early stages of coal formation still continues today as trees and plants in swampy areas die and decay to form peat. If these bogs are allowed to continue without human intervention (such as development to change the environment of the bog), and sediment continues to bury the peatlands, coal could, theoretically, eventually form. Of course, don't wait around—it will take millions of years for this to occur.

in Montana, North Dakota, and Texas and used mainly as fuel for steam-electric power generation.

What is **peat**?

Peat is a soft, dark-brown or black deposit created from the partial decomposition of vegetative matter (a variety of plants and occasionally trees) in waterlogged marshes and swamps. Peat is formed very early in the decomposition process, usually after a few hundred or thousand years. It is often considered a type of soil, and in many countries, such as Ireland, it is dried and used as a fuel.

How did most modern **coal deposits** form from **peat**?

Peatlands (or boglands) have been around since before the dinosaurs. The Carboniferous Period occurred from about 360 to 286 million years ago, when the land was covered with swamps filled with huge trees, ferns, and other large, leafy plants. When the plants died, they sank to the bottom of the swamp, forming a dense layer of vegetation. In these environments, there was minimal oxidation, so plant material did not decay completely. Instead, bacteria partly removed the oxygen and hydrogen in the organic material during decomposition, concentrating the carbon. (Before the bacteria completely destroyed the plant material, they were killed off by acids produced in the decomposition process.)

This process slowly created peat, the precursor to coal. Over thousands of years, these beds of peat became buried by sand, clay, and other mineral deposits—all of which eventually formed sedimentary rock. Water was squeezed out of the rock as the weight from sediment above increased. Millions of years later, heat and pressure increased the peat compaction even more to form coal.

How is **coal mined**?

Coal is mined using a variety of methods. One of the most well-known (and notoriously dangerous) is by tunneling out deep mine shafts, with coal miners entering the depths in elevators. Another method is strip mining, in which huge steam shovels strip away the top layers of rock above the coal seam. The overburden, or the rock and soil layers covering a coal seam, is removed using large equipment; it is either used to backfill areas previously mined or hauled to dumping sites. The coal is then shipped by train or ship. Some coal is even transported through pipelines. To do this the coal is ground up and mixed with water to make a slurry, and is then pumped many miles through the conduits to be used as fuel for power plants and other factories.

What is **coke**?

Coke is a hard, dry carbon substance produced by heating coal to a very high temperature in the absence of air. It is used in the manufacture of iron and steel. Metallurgical coal—or coking coal—is the type of coal converted to coke for use in manufacturing steel.

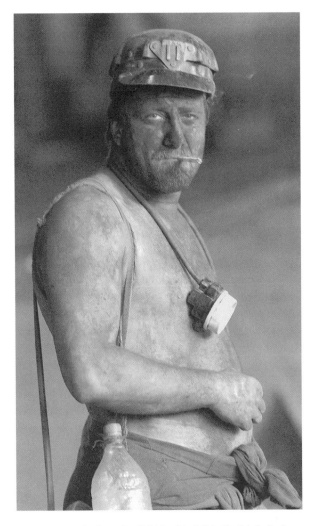

A Romanian coal miner after finishing his shift in the Petrila Coal Mine, west of Bucharest. This miner is one of about 2,000 coal miners who work more than 3,000 feet bellow ground level for a monthly average salary equivalent to about $200. *AP/Wide World Photos.*

What are **gasification and liquefaction**?

Gasification is a group of processes that turns coal into a combustible gas by breaking apart the coal using heat and pressure and, in many cases, with hot steam. Liquefaction is a process that converts coal into a liquid fuel similar to crude oil.

How is **coal used**?

Coal has been burned for many centuries, mainly as a source of heat and energy. The earliest known use of coal was in China. Coal from the Fu-shun mine in northeastern

China may have been used to smelt copper as early as 3,000 years ago. According to Marco Polo, the Chinese called coal the "stone that could burn."

Modern uses for coal include electrical generation (about 38 percent of the total electricity produced around the world comes from burning coal); steel production; and heating for industrial processes and home/commercial use. Its gases, oil, and tars can be extracted and used in the manufacture of numerous products, ranging from gasoline and perfumes to mothballs and baking power.

How common is **coal use in the United States**?

The United States has more coal than any other fossil fuel—and more than any other single country in the world. About one quarter of all the known coal deposits are in the United States, with large deposits found in 38 out of 50 states. About 10 percent of United States coal is exported to other countries.

In the United States, coal is primarily used to generate electricity. It is burned in power plants to produce more than half of the electricity used in the entire country. The amount of coal used to generate electricity for an average household is surprising: To generate the electricity for an electric stove for a year, for example, about half a ton of coal is used; for an electric water heater, about two tons of coal a year is used; and for an electric refrigerator, about a half ton a year is used.

What **other countries** around the world have substantial **coal deposits**?

It is interesting to note that the countries with the most coal deposits are not the countries with the most oil or natural gas reserves. The main coal countries besides the United States are China, Russia, Australia, and India.

PRESENT PROBLEMS, FINITE FUTURE

What is the **cleanest burning fossil fuel**?

Most people agree that natural gas is the cleanest burning fossil fuel, although it is not perfect. The greenhouse gas amounts released from burning natural gas are significantly lower than emissions from wood, coal, and oil. When natural gas is burned, it produces mainly water vapor and carbon dioxide. Although methane in natural gas is a more potent greenhouse gas than carbon dioxide, the actual amount released during processing is small—less than 1 percent of all the methane released by other industries. About 80 percent of greenhouse gas emissions from the use of natural gas occur at the final destination: your pilot light.

Natural gas processing plants are another story. Although most plants try to recover the majority of the sulfur compounds in raw gas, some is released into the

> ## Will oil, gas, and coal last forever?
>
> The Earth has a great deal of oil, gas, and coal reserves—or so it seems—but these are finite resources. It is estimated that at today's usage rates, all the world's coal and oil will be burned in 60 years and its natural gas in 220 years.

atmosphere as sulfur dioxide. In addition, processing plants use energy and often release volatile organic compounds (VOCs) into the atmosphere. (VOCs are substances containing carbon and different proportions of other elements, such as hydrogen, oxygen, fluorine, chlorine, bromine, sulfur, or nitrogen; these substances easily become vapors or gases.) Other problems include odors from the plant, groundwater contamination, waste product disposal from processing, and land-use questions, such as building or drilling on unique or endangered wildlife habitats.

What is the **Strategic Petroleum Reserve?**

The Strategic Petroleum Reserve is a fuel reserve that is held underground in Texas and Louisiana. Equal to about 300 days of crude oil imports from the Middle East, the purpose of the reserve is mainly to provide fuel in case of a national emergency. By law, the SPR can be used only if there is a serious loss of supplies that might result in a major adverse impact to the national economy or safety.

How does burning **fossil fuels** affect the **environment**?

Fossil fuels consist primarily of hydrocarbons. When burned, the carbon combines with oxygen to yield the greenhouse gas carbon dioxide. (The amount depends on the carbon content of the fuel.) Over 75 percent of carbon dioxide emissions come from the use of coal and petroleum fuels. For each unit of energy produced, natural gas produces half as much carbon dioxide as coal, and petroleum fuels about 75 percent as much as coal. And although the industrial sector is the largest energy user, the transportation sector—cars, trucks, train, planes, etc.—emits nearly as much carbon dioxide because of its almost complete dependence on petroleum fuels.

In the United States alone, nearly 85 percent of greenhouse gas emissions come from the burning of fossil fuels. Methane from landfills, coal mines, oil and gas operations, and agriculture represents an additional 10 percent of annual emissions (on a carbon equivalent basis, meaning that the methane gas has been converted to the equivalent of either carbon or carbon dioxide based on the gas's global warming potential). Other significant sources include manmade gases, such as hydrofluorocarbons (HFCs), perfluorocarbons (PFCs), sulfur hexafluoride (SF_6), and nitrous oxide (N_2O), the last being a compound emitted by combustion of fossil fuels as well as released by nitrogenous fertilizers and by certain industrial processes.

White plumes of water vapor rise from the two stacks of the emission scrubber system at the Tennessee Valley Authority's Cumberland Fossil Plant near Clarksville, Tennessee. The scrubbers reduce sulfur dioxide emissions that cause haze and acid rain. *AP/Wide World Photos.*

Greenhouse gases are known to affect the environment and are thought to be a contributor to climate change around the world. It is thought that the increase in greenhouse gases cause the sun's radiation to become trapped, creating a greenhouse-like environment. Such conditions can raise the average global temperatures, changing weather patterns, climate, and sundry other global phenomena. (For more information about global climate change, see "Ice Environments.")

What causes **acid rain**?

The use of petrochemicals does have many drawbacks. One in particular is acid rain. This is rainwater with an abnormally acidic pH balance of between 2 and 5, which is 100,000 to 100 times more acidic than distilled water. The acidity is caused by the burning of oil, gas and coal, all of which release large quantities of sulfur- and nitrogen-bearing gases that create the acidity in the rain. The resulting acid rain is harmful to plants and soil, changing their pH. Acid rain falling into bodies of water can also change the pH, harming aquatic flora and fauna. Furthermore, acid rain can corrode outdoor structures, such as sandstone buildings and stonework.

Where does **acid rain fall**?

Acid rain probably began around the turn of the 20th century, when the industrial revolution took hold on the world. Today, many sections of our planet have been affected.

The major areas affected include the following: The northeastern United States (caused by high numbers of factories and power plants in the Midwest); southeastern Canada (from factories in the Toronto-Hamilton region); central Europe and Scandanavia (from British and other European factories); and parts of Asia, specifically India and China.

How does burning **fossil fuels** affect the **ozone layer**?

The ozone layer (a region in the stratosphere that protects organisms from the sun's harmful ultraviolet radiation) is affected by the burning of fossil fuels. Scientists know that there are holes in this layer. One is about 1.5 times the size of the United States (although it does vary in size) and is located above the Antarctic continent. It usually lasts for about three months, starting in the beginning of the Southern Hemisphere's summer. Smaller holes occur over the Northern Hemisphere's populated mid-latitudes and above the Arctic. Chemical reactions that naturally create and destroy the ozone are out of balance in these holes.

Although the main culprit is thought to be chlorine-containing chemical compounds, the burning of fossil fuels also creates problems. In particular, such burning pumps huge amounts of the greenhouse gas carbon dioxide into the atmosphere. The carbon dioxide, along with other greenhouse gases such as methane and nitrous oxide, traps heat and warms the lower atmosphere. This causes the stratosphere to cool, speeding up the chemical reactions that cause the ozone holes.

A resident stands in front of her oceanfront house in Bodega Bay, California, where El Niño-driven storms eroded the cliffs below, threatening to send her house and several others plunging into the Pacific Ocean. *AP/Wide World Photos.*

Physical weathering—The agents responsible for physical weathering include gravity, wind, tumbling rocks, and moving water, all of which mechanically affect rock. They are responsible for the development of fractures in rocks, generating fragments of rocks, and creating sediment. The development of rock fragments and sediment is most often caused by *abrasion,* which is the grinding of rock by friction and impact as fragments are transported in rivers and streams, under and around glaciers, and the movement of sand by winds.

Chemical weathering—Chemical weathering takes place when rocks react with chemicals in solution, essentially decomposing the rocks and soil by a chemical reaction. This usually occurs in water that is rich in carbon dioxide, which is, in turn, produced mainly by the decomposition of plants. For example, limestone caves are weathered in this way. (For more information about cave formation, see "Exploring Caves").

Biological weathering—Biological weathering is caused by organisms and other vegetation that break down rock either physically or chemically. It includes a large range of organisms—from bacteria to plants to animals. For example, lichens play an important part in weathering because they are rich in chelating agents, which trap the metallic elements of the decomposing rock. Some lichens live on rock surfaces (epilithic), some actively bore into the rock's surface (endolithic), and still others live in the hollows and cracks in the rock (chasmolithic).

Why do people confuse erosion and weathering?

Although they seem to be the same processes, a common misconception is that erosion and weathering are synonymous. But there is a definite difference: Erosion happens when a rock particle (most likely loosened by a weathering process) moves away from its original rock. This may be caused by gravity, air, water, or ice. Weathering is a process that originally caused the particle to loosen. One easy way to remember the difference is that if physical or chemical forces loosen a particle and it stays where it falls, it is weathering; but if the particle starts moving, the process (once it moves) is termed erosion.

Which is more important: **chemical or physical weathering**?

Physical weathering is more obvious than chemical weathering, and both operate together, even to the point of assisting one another. But there is no general agreement about whether chemical weathering is more important than physical weathering. Many say it's chemical because, for example, if a rock is chemically broken down, rainwater, rivers, and streams can carry the elements to the oceans, thus adding to the plethora of material in seawater. Others say it's physical weathering because, after all, the creation of landforms through physical weathering defines an environment.

What is **sediment transport**?

After weathering loosens rock, the resulting sediment has to go somewhere. As the phrase indicates, sediment transport is the term used to describe the natural agents—ice, wind, moving water, or gravity—that cause material (sediment or any loose material) to move from one place to another. This can happen either at the surface, such as with the movement of mud by a flood, or it can occur near the Earth's surface: for example, the transport of silt high into the air during a dust storm.

What are **frost action** and **frost heave**?

Frost action is a type of physical weathering. Water commonly collects in cracks and wedges everywhere on Earth. In a cold region—especially during the winter and early spring—water seeping within cracks can pry apart rocks as it freezes (water expands by 9 percent as it turns to ice) and thaws. This can take anywhere from a few years to hundreds of years, depending on the rock type and the amount of precipitation in a region.

Frost heave is caused by a similar process. Water in the soil creates ice lenses that expand, causing the ground to rise. The amount of heaving depends on how much water is available to freeze. If more water is added, the lens can continue to grow.

Such heaving affects roads, sidewalks, and foundations. It can also affect a garden, heaving up rocks buried in the soil.

How are **clastic and non-clastic rocks** connected to weathering?

Clastic rocks are those made of preexisting rock fragments formed from physical weathering. Non-clastic rocks are those formed as the product of chemical precipitation or organic activity. Most non-clastic rocks eventually break down into rock fragments, forming clastic rocks.

What is the **rate of chemical weathering**?

The rate of chemical weathering is most often determined by various factors, especially climate. In general, it is accelerated by the presence of warm temperatures and moisture. Minerals—singly and in rocks—react with chemically active reagents such as water (the most important), carbon dioxide, oxygen, and organic acids. These reactions form new minerals and/or dissolved elements that originate from the minerals. In most cases, the minerals become new chemical substances, many of which are much softer and more susceptible to agents of physical weathering than the original material.

Climate is not the only factor; rock type is also important. Some rocks, such as limestone or feldspar, may be more easily affected by chemical erosion than a basalt or quartz, for example.

What role does **oxygen** play in **chemical weathering**?

Oxygen plays an important role in chemical weathering because it readily combines with other elements to form new chemical elements. This process is called *oxidation,* a form of chemical weathering in which the oxygen anions combine with a mineral's cations to break down or soften the original material. (Ions are atoms, molecules, or compounds that carry either a positive [cation] or negative [anion] electric charge.) For example, iron and water form iron oxide—or rust—because of the oxidation process.

Why is **hydration** important in **chemical weathering**?

Hydration is another form of chemical weathering, in which hydroxide (OH) anions combine with the rock's mineral cations to break down and soften the original mineral. For example, hydration causes feldspar to turn to clay.

What role do **natural acids** play in **chemical weathering**?

There are many natural acids in our environment; most of them not strong enough to affect us. But they can, over time, change rocks and minerals. Called *dissolution,* this form of chemical weathering occurs when minerals within a rock are dissolved by the natural acidic reaction in the environment. For example, natural rain water is slightly acidic because carbon dioxide combines with water in the atmosphere to form carbonic acid.

In addition, as rain water percolates through soils, plants contribute additional acids, such as humic acid. These natural acids can disrupt the atomic structure of many minerals. The two most important plant influences are the root tips, which exude acidic solutions (a pH as low as 2, where neutral pH is 7); and dead vegetation, which decomposes, forming more organic acid and carbon dioxide that goes into solution.

Are there **other types of acids** in our environment?

Yes. There are also human-made acids present in our environment. One group in particular is the sulfuric acids caused by industrial emissions, which is also known as "acid rain." In certain sections of the world, industry and power plants add sulfur dioxide and nitrogen (nitrous) oxides to the atmosphere, creating compounds that are many times more acidic than natural rain.

Perhaps a more precise term than "acid rain" is acid deposition, which has two parts: wet and dry. Wet acid deposition includes acidic rain, fog, mist, and snow; dry includes acidic gases and particles. In the latter case, the dry particles can be washed from surfaces and vegetation by rainstorms. This, in turn, makes the combination more acidic than the rainfall alone, causing more chemical weathering in the region.

And, of course, it is not only the rocks that are affected. According to the Environmental Protection Agency, acid rain and the dry deposition of acidic particles contribute to the corrosion of metals (such as bronze) and the deterioration of paint and stone (such as marble and limestone). These effects seriously reduce the value of buildings, bridges, cultural objects (such as statues, monuments, and tombstones), and cars. One prime example is the Jefferson Memorial in Washington, D.C. The monument's marble surfaces have developed a rough "sugary" texture because calcite grains are being loosened, continually dissolving as they are exposed to acid in the rain.

What is **carbonation**?

Carbonation is another chemical weathering process. It is especially active when the environment has abundant carbon dioxide. This leads to the formation of carbonic acid, a result of carbon dioxide gas reacting with water in the atmosphere.

What is **solution weathering**?

Solution weathering is yet another form of chemical weathering. It occurs when minerals dissolve in water (go into solution). This happens because some types of rock are easily dissolved in rainwater. Weathering by solution typically produces rather smooth, scalloped surfaces. For example, soft calcite and gypsum often show evidence of solution weathering.

A view of Half Dome from the valley floor of Yosemite National Park in California. The hiking trail to the summit of Half Dome is one of the most famous and scenic day hikes in the United States. Serious rock climbers climb straight up the face, sometimes faster than hikers coming up the trail. *AP/Wide World Photos.*

What types of **rock are prone** to **chemical weathering**?

Certain rocks are easily affected by chemical weathering. For example, minerals such as feldspar, olivine, and hornblende react with chemicals (called reagents) to produce various other types of minerals. The silica, aluminum, and iron ions from these minerals often help generate secondary rocks such as quartz, muscovite mica, and hematite. Potassium, sodium, calcite, and magnesium ions from these minerals usually dissolve and are carried away in solution.

Not all types of rock are susceptible to chemical change. For example, quartz is resistant to chemical decay and remains unaltered. But quartz can still be weathered out of its parent rock by physical weathering; erosion then takes over, battering, rounding, and reducing the quartz in size as it is transported by currents and/or waves in rivers, lakes, and oceans.

What is **exfoliation** and how is it caused?

Exfoliation is a form of physical (mechanical) weathering that occurs when curved plates of rock are stripped from the parent rock, resulting in exfoliation domes, dome-like hills, and rounded boulders. The reason for huge exfoliation domes is still under investigation. One theory is that differences in pressure at the Earth's surface caused rock to expand and easily break into the exfoliation domes. Some rocks experience exfoliation domes and domelike hills more readily than others. For example, granites

Does weathering take place on other planets?

Yes, weathering takes place on all the other planets, as well as on their satellites and on asteroids. In other words, no body in the solar system seems to be exempt from the effects of weathering, except possibly the gaseous giants, Jupiter and Saturn, both of which have no actual "solid" surface.

But not all planetary weathering is similar to Earth's. For example, on Mercury and the Moon, physical weathering takes place not from running water or wind, but from the bombardment of the sun's radiation and from small micrometeorites that continually strike their surfaces. Venus also has a different type of weathering: sulfuric clouds "rain" acid on the planet's surface, which causes a great deal of chemical weathering. Add to that surface temperatures that hover around 900°F (482.2°C), and there is no doubt that physical weathering occurs on Venus, too.

Mars is probably the planet on which physical and chemical weathering phenomena most closely resemble those on Earth. The "red planet" is home to huge dust storms and frosts that no doubt cause physical weathering. It is thought that this planet contains underground water or frozen ice that could have caused not only physical weathering at one time (Mars may have been warmer at one time, making free-flowing rivers possible), but also chemical weathering.

are known to weather in this way. Some good examples of large exfoliation sites include Half Dome in Yosemite National Park, California, and the granitic Idaho Batholith.

Exfoliation of boulders looks similar to that of domes and hills, but it is on a much smaller scale. These boulders become rounded by concentric shells of rock peeling off, similar to the way layers peel from an onion. Mechanical and chemical weathering are at fault: the rock breaks down because of constant temperature changes from both daily and seasonal heating and cooling. Along with the physical weathering, chemical weathering can cause certain minerals at the rock's surface to change. For example, the feldspar in granite boulders is often chemically converted to clay, which occupies a much larger volume and helps to split the rock apart.

Do **humans cause any weathering** on the Earth's surface?

Most definitely, humans do cause chemical and physical weathering to occur on our planet. As mentioned above, human activities have produced a type of chemical weathering called "acid rain." Also, uncontrolled destruction of forests can add to chemical weathering, as more acids from decomposing vegetation affect soils and groundwater, and the lack of groundcover accelerates erosion.

Humans have also contributed greatly to physical weathering on the Earth's surface. For example, road building moves an estimated 300 trillion tons of rock and soil throughout the world each year, resulting in a form of physical weathering as the rock and soil are dug out of the ground. Coal, metals, and minerals are mined in open pits, creating physical weathering also. (See also humans and mass wasting below.)

What is **differential weathering**?

You have probably seen this in many roadcut/outcrops in your travels: layer upon layer of rock that are all weathering at different rates, making the outcrop look like an uneven stack of flat rocks. Called differential weathering, this occurs when the layers in an outcrop contain more than one type of rock. For example, certain ancient marine environments may deposit separate layers of sand and silt, creating an outcrop of sandstone and shale. As these two types of rock weather, the result is often differential weathering in which the sandstones are more resistant to weathering than the shales. Examples of such outcrops can be seen in upstate New York and western Newfoundland.

MASS WASTING

What is **mass wasting**?

Mass wasting—also called mass movement—is a collective term and covers many different phenomena. Overall, it refers to the usually slow (but sometimes rapid) downslope movement of rock, soil, regolith, and sundry debris under the influence of gravity. In most cases, this mass movement occurs when a certain threshold (also called a geomorphic threshold) is reached. Gravity then pulls the material downhill. For example, the excess weight of rainwater-saturated soils can eventually cause them to slide down a steep mountain slope. Other times, mass wasting can be caused by isolated geological events, such as earthquakes.

Definitions of the various types of mass wasting—from debris flows to landslides—are highly debated; many overlap, as the following questions illustrate. But in general, the divisions have a few characteristics that are dissimilar.

What is **regolith**?

Regolith is literally the very loose top portion of the Earth's surface. It consists of all materials above bedrock, such as soils, sediments, and weathered rock.

The term regolith is more often used in reference to other members of our solar system, including the Moon, asteroids, and certain planets and satellites. On our Moon, the regolith is the overall soil horizon, a collection of particles fragmented by the impacts of large and micrometeorites over time. In the case of the Moon (although the sun's rays do affect the surface), there is no true weathering of the lunar regolith like there is on Earth.

Why is **mass wasting important** for geologists to understand?

Mass wasting is very important for geologists to understand because slides and other ground failures included in mass wasting processes cost more lives, money, and property damage each year than all other natural disasters combined. And the incidence of such movements appears to be increasing.

Disaster occurs when mass wasting directly affects humans. For example, a slope in nature evolves to a certain balanced angle, and material moves downhill at a controlled rate. If the conditions change, then this balance may be upset, and a sudden mass movement can take place.

Because few studies have been conducted on mass movements, many areas continue to be affected by sudden slides and flows. Of even more concern are the people who continue to build and live in places that are likely to be affected one day by avalanches of rock, mud, and debris.

What is the **angle of repose**?

The angle of repose refers to the angle at which loose, unconsolidated materials—such as gravel, soil, and sand—can sits at rest; this is usually about 25 to 40 degrees. A slide or slump often occurs if the materials go past the angle of repose.

What are the main **types** of **mass wasting**?

Although many scientists believe there are two types of mass wasting, slow and fast, these movements can be subdivided into four types based on how fast the material moves down a slope:

High velocity events—High-velocity events include *falls and avalanches,* in which the material is dry and loose, and *flows,* in which material (soil and/or rock) is loose but highly saturated with water and/or air.

Low velocity events—Low-velocity events include *landslides* (although some can occur rapidly, such as those caused by an earthquake), which are displacements of somewhat consolidated, wet (but usually not saturated) rock, and soil along a slope. *Creep* is the gradual, more steady movement of surface soil, in which individual particles move by small-scale processes such as freeze-thaw or wet-dry events. Creep is the low-velocity event that causes trees and telephone poles to lean downslope at an angle over time.

What is a **talus slope**?

Talus slopes are the collection of materials—usually in the shape of an upside-down cone—that result from avalanches that have caused dry material to fall down a slope.

What are the various **types of landslides**?

When most people mention mass wasting, they think of landslides—an often rapid displacement of rock and soil. They occur all over the world, and in the United States cause almost $2 billion in damages each year. Slides are categorized into two types: *translational* (movement along a plane) and *rotational* (movement along a concave surface, such as at the base of a hill).

What are the **major factors** that cause **mass wasting**?

There are several major factors that contribute to mass wasting. These include the following:

Gravity—Gravity is, of course, the force that pulls on everything, including soil, dust, and debris, especially downslope.

Shape of the slope—The shape of a slope has a great deal to do with mass wasting, especially when the slope has an extreme angle.

An aerial view of the landslide and remains of flattened ski lodges at the Australian resort of Thredbo taken July 31, 1997. Landslides occur all over the world, and in the United States, cause almost $2 billion in damages yearly. *AP/Wide World Photos.*

Water—Oftentimes, depending on the material found on slopes, water can affect its strength. If saturated enough—and if the right rock, minerals, and fragments are present—water can create a mass movement of material.

Sensitive soils—Some soils are more sensitive than others when it comes to mass movements. For example, certain clays easily slide down slopes.

Triggering events—Some mass movements occur over time after a certain threshold is reached, but others are triggered by sudden events. For example, earthquakes are often responsible for upsetting the balance of soils along a slope, with the sudden movement literally shaking soils off the slopes.

What are the "falling rock" zones seen along highways and country roads?

We've all see them: a sign that says "falling rock" along a highway or a country road. In most cases, the sign means just what it says: Rock will possibly fall or has fallen from high above the road, usually on a road cut into a cliff or mountain.

Falls—sudden, nearly vertical movements of large or small rock chunks—occur in many places around the world. The rock can be loosened as a result of road construction or other human-made excavations, from the freeze-thaw action in cold climates, or even as a result of plant growth or animal burrowing. When the fall hits the ground below, material will either bounce, roll, or fall flat, and this is clearly a major hazard to anyone walking, driving, biking, or otherwise traveling in a fall zone. Because it is difficult to judge when or where a rock will fall, it is always smart to pay attention to such signs along highways.

How is **water** important in certain **mass wasting events**?

Water can be important in some mass wasting events by contributing to the soil and rock instability. In particular, water adds weight; more weight leads to instability of soils, rock, and debris along a slope. Water also tends to lubricate loose material.

More or less water can even change the characteristics of a material. For example, clays like kaolinite expand and become slicker when mixed with a great deal of water; when dry, the clay shrinks. Both conditions can contribute to mass movement of the material downhill.

What is the **difference** between **falls and slides**?

There is a definite difference between falls and slides, although, like all mass movements, the differences are not always agreed upon. The following lists some general distinctive characteristics:

Rock fall—A rock fall occurs when a piece of rock on a steep slope becomes dislodged and falls down the slope.

Debris fall—A debris fall occurs in the same way, but includes a mix of soil, rocks, and regolith. Both types of falls form a talus slope, or an angled collection of debris at the bottom of the fall area.

Rock slide and *debris slide*—Both slides (often termed landslides, though this term is not specific enough) occur when blocks of rock or massive amounts of loose material, respectively, slide down a preexisting slope. Both types of

193

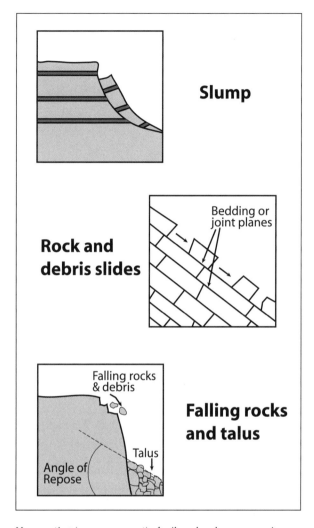

Mass wasting (mass movement) of soils and rocks can occur in several ways, such as through slumps, rock slides, or falling rocks.

slides create a pile of talus at the bottom of the slide. These are among the most destructive of mass movements and are most often triggered by excessive rain, melting snow, or earthquakes.

What are **slumps**?

Slumps are masses of soil, rock, or rock fragments that most often move slowly along a curved, rotational surface. At the upslope end of the slump, one or many *scarps* (small, crescent-shaped cliffs) form. A mass of material forms at the bottom (toe) of the slump. These mass movements are sometimes seen along highways in which graded soil along the road sides is a little too steep; they are often triggered by excessive rains.

What are **sediment flows**?

Sediment flow is the general term applied to the movement of mostly saturated rocks, soil, and debris that flows down a slope. It can be broken down into two types, depending on the amount of water present: *slurry flows,* which contain between 20 and 40 percent water; and *granular flows,* which contain up to 20 percent water.

What are the various **types of slurry flows**?

The following lists the most common slurry flows:

> *Mudflows*—Mudflows contain a highly fluid mix of loose sediment and water and are the consistency of wet concrete. These types of flows are very wet, usually forming after heavy rains in loose sediment. As the flow travels

downslope, it gathers more loose sediment, which often speeds up the movement and contributes to its destructiveness. Mudflows are commonly found in volcanic areas, where they are called *lahars;* they often form when volcanic activity melts mountaintop snows, with the resulting meltwater mixing with volcanic deposits on the mountainside. These flows are very quick, destructive, can travel long distances, and usually follow established drainage valleys or even gently sloping stream beds.

Solifluction—Solifluction occurs in areas where the soil remains mostly saturated with water or frozen throughout the years. This movement of the top soil (regolith) is measured in fractions of an inch per year.

Debris flows—Debris flows occur at high velocities, often as the result of heavy rains saturating the soil and regolith. They sometimes start with slumps, then gather more speed to flow downhill, often forming large lobes at the foot of the flow.

What are the various **types of granular flows**?

The following lists the most common granular flows (*creep,* which was defined earlier in this chapter, is also a type of granular flow):

Earthflows—Earthflows form in humid areas along hillsides, most commonly after a heavy rain or melting snow. Fine-grained materials, such as clay and silt, mix with the water. They can move anywhere from less than a fraction of an inch (1 millimeter) to over 10 feet (3 meters) per day, but are usually long-lived, lasting days or even years.

Debris avalanche—Debris avalanches occur after the complete collapse of a mountain slope. These very high velocity flows are huge, carrying tons of rock, soil, and debris downslope and traveling for considerable distances, even along a relatively gentle slope. They are most often triggered by larger events, such as volcanic eruptions or an earthquake.

Can **mass wasting** occur in **frozen regions or oceans**?

Yes, mass wasting can occur in the coldest regions of the world, including in oceans. In particular, areas covered by permafrost (semi-permanently frozen ground) can experience mass wasting. When the "active" surface layer thaws in the summer, the material can flow in a process called solifluction (see above).

In the oceans, mass wasting most often occurs on steep slopes, and includes slumps, debris flows, and landslides. Material may move down a slope if there is an overaccumulation of sediment, especially in submarine canyons (for more information about submarine canyons, see "Geology and the Oceans"), or even from pressure

Do humans contribute to mass wasting?

Yes, humans do contribute to mass wasting. If a slope is artificially over-steepened by humans—such as by roadcuts or excavation of minerals—material will become unstable. This instability often leads to movement of materials downslope.

Humans are not only direct agents of mass wasting, they can also accelerate the process inadvertently. For example, scarification is the movement of soil and rock that results from digging in such places as a pit mine, trench, or road cut. Humans can also overload a slope by adding more material, piling soil, rock, and debris until it reaches a threshold and begins to move downslope. Again, mining and debris from roadcuts can add to such an overload of material. More importantly, humans can remove material that supports a slope, or even remove vegetation that prevents a slope from experiencing too much erosion.

It is estimated that humans move about 40 to 45 gigatons of soil and rock per year. In comparison, mountain building moves about 34 gigatons per year; rivers transfer about 14 gigatons of sediment per year; sedimentation in the deep oceans moves about 7 gigatons; glaciers about 4.3 gigatons; wave action and erosion about 1.25 gigatons; and wind about 1 gigaton per year. This illustrates that humans clearly do have an impact on mass wasting and weathering on our planet.

due to the generation of methane gas (caused by the decay of organic material). Further mass movements occur if there is a triggering event, such as an underwater volcanic eruption and/or earthquake.

Can **mass movements** occur in **ancient sediments**?

Yes, mass movement—especially slides—can occur in ancient sediments. For example, on April 27, 1993, a landslide severely damaged three homes near the town of LaFayette in the Tully Valley, 15 miles (24 kilometers) south of Syracuse, New York. This slide occurred not only because of melting winter snowpack, but also because more than 7 inches [190 millimeters] of rain fell that April. The Tully Valley slide was the largest in New York in the past 75 years, creating a 1,500-foot- (450-meter-) wide scarp and a 1,800-foot-long (600-meter) landslide.

This earthflow consisted of mainly red lake clay deposits of glacial origin that were covered by various thicknesses of glacial till. The area was ripe for such a slide. After all, it was located around former glacial lakes, and such deposits, particularly lake clays, are generally highly susceptible to landslides.

WIND–BLOWN SANDS AND DESERTS

How does **wind** contribute to **rock erosion**?

Winds slowly erode rock in three ways: saltation, in which the particles of sediment (such as sand) move by essentially hopping along the ground; suspension, in which the particles of rock (such as silt) are so light they become suspended in the air and move along with the wind; and rolling, in which the winds roll the sediments along the ground.

What are the **major deserts** on Earth?

With the exception of Europe, every continent on Earth has a desert. The following lists the major deserts around the world:

Continent	Deserts
Africa	El Djouf, Libyan, Namib, Nubian, Sahara, Tenere
Asia	Dashi-i-Lut, Gobi, Kara Kum, Kyzyl Kum, Rub al Khali, Syrian, Takla, Thar
Australia	Great Sandy, Great Victoria
North America	Great Salt Lake, Mohave, Sonoran
South America	Atacama, Patagonia
Antarctica	Antarctica (almost the entire continent can be considered a desert)

A red willow stands in the desert in Ejina Qi of north China's Inner Mongolia Autonomous Region. The effects of wind erosion are clearly visible in the sand as China's vast western area faces the danger of continuous expansion of the desert due to driving winds and drought. *AP/Wide World Photos.*

197

Are there different **types of sand dunes**?

Yes, there are different types of sand dunes, which are collections of sand and smaller material that form arcs as the winds blow. The following lists those most commonly seen sand dunes that are found anywhere from your local beach to the great Sahara Desert:

Transverse—broken-crescent shaped dunes that contain a great deal of sand

Barchan—crescent moon-shaped dunes that contain little sand

Longitudinal—long, tear-dropped shaped dunes that are formed by steady winds

Star dune—multi-faceted dunes fromed when winds from many directions push up many sides of the dune

Parabolic—large arcs in the shape of a parabola that mark a transition between transverse and longitudinal dunes

How much of the **Earth's land** surface is covered by **desert**?

Deserts occupy between 15 and 20 percent of all the land surface on the Earth. Here, the sun bakes the desert sands so that daytime temperatures can range from 70°F (21.1°C) to upwards of 100 to 120°F (37.8 to 48.9°C) during the summer months. During winter, however, temperatures can dip below freezing, so a desert is not defined by how hot a location is; rather, it depends on the amount of rain a region receives—less than 19.7 inches (50 centimeters) per year. This is why some high-altitude or cold regions, such as Antarctica, can be called deserts, even though they are not in warm climates.

What is **desert pavement**?

Many deserts are covered with stony material called desert pavement. This is created when the wind blows fine material away, leaving the coarse material behind.

What are the almost-trianglar-shaped rocks often found in the desert?

These wind-scoured rocks are called *dreikanters,* after the South African word meaning "three corners." Sometimes called *ventifacts,* they are common in the United States' southwest deserts and are formed by winds that erode the fragments into angular pieces. Ventifacts found in the northern United States formed during the Pleistocene tens of thousands of years ago, when vegetation was sparse and wind-blown sand and silt was more abundant.

Where are the **highest sand dunes** in the United States located?

The Great Sand Dunes in Colorado are the highest sand dunes in the United States. The winds blow sand toward the mountains here, but they can't carry the particles over the highlands. As the air funnels through a low pass in the mountains, it leaves behind tall sand dunes.

Where are the **largest sand dune fields** in the United States located?

As strange as it may seem, Nebraska has the largest sand dune fields in the United States. The Sand Hills exhibit almost every type of dune and cover much of the western part of this state. They were formed during the Pleistocene and are made up of glacial debris that eroded out of the Rocky Mountains.

What is **cross-bedding**?

Cross-bedding can be seen on many windblown or river sedimentary deposits. These beds show a criss-crossing of patterns that were formed by river channel currents that changed course or when winds blew in various directions.

SOILS

What is the **definition of soil**?

Soil is a mixture of minerals and rock fragments, water, air, and sediment derived from the weathering of bedrock; it is also defined to be material modified to the point that it will support life (in most cases). Also included are pieces of degraded organic matter (mostly from overlying vegetation). This definition excludes materials such as volcanic ash, alluvial sand and gravel deposits, beach sand, and glacial deposits that do not contain organic matter or that display the weathering features described below. Layers of soil are divided into *soil horizons,* with each distinguishable by characteristic physical or chemical properties (also see below).

What are the major **factors** that influence **soil formation**?

There are several factors that influence the formation of most soils. They include the following:

Time—Time, as with everything in nature, is important in the formation of soils. In general, the more time, the greater the buildup of soil.

Climate—The climate of an area determines the type of weathering taking place to form the soil, as well as the thickness of the soils.

Source material—The source material (or parent rock) determines the soil chemistry, and thus the soil's fertility. Soils can either come from the rock located in the region (from erosion of the parent material), or from material transported to the region (for example, sediment that travels from mountain slopes).

Plant and animal life—Plant—and, to a certain extent, animal—life contributes to soil composition. Organisms furnish materials such as organics (which also leads to organic acids) and fertilizer to the soils; they also contribute to the breakup of the soils (for example, from the burrowing of animals such as earthworms and moles), allowing air and water to infiltrate.

Slope—The slope of soils also contributes to its formation by helping to dictate the amount of erosion that takes place and how much water is retained in the soil.

How are **soil horizons** divided in a temperate climate?

In a temperate climate, the following table lists the major *soil horizons* (also called zones) present in a cross-section of soil:

Soil Horizon	Comments
O	organic debris (detritus) from dead vegetation
A*	decomposed organic matter, mineral and rock fragments; water moves downward from here
E	fine particles are washed to deeper depths (eluviated) here
B	zone of accumulation of material leached downward from the A horizon; precipitation of hydroxide and carbonate materials causes acids to be neutralized; eluviated clay minerals accumulate here, making the layer quite clayey; it is often stained red or brown by hematite (iron oxide) and limonite
C	partially weathered and fragmented bedrock
R	relatively unweathered, but still fragmented bedrock

* Soil horizon A is often referred to as the "zone of leaching"; this is because the acids produced in the O horizon by the decay of organic matter percolate downward, chemically weathering and removing (or leaching) the layer.

What **types of soils** can form in **different regions**?

Depending on the region, various types of soils can form. For example, *pedalfer* soils contain a high content of aluminum and iron. These rich, clay soils are found in the B horizon. They usually form in middle latitudes in which moderate rainfall carries away soluble material, leaving behind the iron oxides and clays. These soils are often red in color and are the most common type of soil in such areas.

Pedocol soils tend to be thin, with little leaching and humus; these calcite-rich soils occur in arid and semi-arid (low rainfall) areas. Because local rainfall does not penetrate deep enough into these soils, the soluble minerals (especially calcium carbonate) are dissolved from the uppermost layer and redeposited in the lower levels.

There are also *laterites,* or soil that is formed by intense chemical weathering in tropical areas. In the upper zones, the silica and soluble materials in the soil is depleted, which helps to concentrate iron and aluminum.

Finally, *caliche* is a hardpan soil that forms when calcium carbonate and other salts precipitate in the soil after water evaporates; it usually forms in arid environments. In these regions, there is not enough water to break up the caliche, which can result in a very thick layer over time.

What are **paleosols**?

Soils that are buried rapidly, such as from a volcanic eruption, and are preserved in ancient rock layers are called paleosols. Geologists study paleosols in order to understand not only past weathering processes, but also the biogeology of that time period (how organisms affected the soils).

BUILDING MOUNTAINS

HOW MOUNTAINS FORM

What is a **mountain**?

The definition of a mountain is not easy. In general, a mountain is an area in which the ground has been naturally uplifted. It typically has steep sides, a narrow summit, and often reaches great heights. In reality, "mountain" is a relative term and can apply to any land mass that conspicuously jets up above its surroundings. What one person considers to be a mountain may be a hill to another person. For example, the term mountain in Florida often means a rise of about 120 feet (37 meters) in height. On the other hand, in the Colorado Rocky Mountains the tallest mountain, Mount Elbert, is about 14,433 feet (4,399 meters) high, and the tallest mountain in the world, Mt. Everest, is 29,035 feet (8,850 meters) high. In addition, all mountains differ by geologic history, age, origin, rock variety, layers, and structure. Therefore, no two mountains—or mountain ranges—are alike.

How does **tectonic activity** create **mountains**?

The theory of plate tectonics states that the Earth's crust is broken up into a series of large plates that "float" on the planet's moving mantle. It also often explains the building of mountains and mountain ranges: As these plates roam around the Earth, they collide, slide, or pull apart from each other, creating mountains in various ways. (For more information about plate tectonics, see "Through the Earth's Layers.")

In particular, during collisions one plate may slide under another plate, creating volcanic activity and mountain ranges, such as the Andes of South America. Other times, the boundaries force material upward to form mountains—a process called orogenesis—such as the Himalayas and Tibetan Plateau. Where plates separate, 203

Why is Mt. Everest so famous?

Mt. Everest is one of the most famous mountains in the world, probably because it is the highest mountain on Earth, and one of more than 30 peaks in the Himalayas over 24,000 feet (7,315meters) high (Himalaya is a Sanskrit word meaning "abode of snow"). The mountain became famous when Sir Edmund Hillary and Tenzig Norgay climbed to its summit in the early 1950s, becoming the first to officially make it to the peak. Hillary surveyed Mt. Everest at that time and determined the mountain was 29,000 feet (8,839 meters) high, a figure amazingly close to the current reading of 29,035 feet (8,850 meters) confirmed using Global Positioning Satellite technology.

magma (hot, molten rock) adds to the mass of material, often forming mountains such as the underwater Mid-Atlantic Ridge, the largest mountain range on Earth. The plates here have been separating for more than 200 million years, and the mountains are still growing.

What are the **hinterland and foreland**?

When describing the process of mountain building (orogenesis), a geologist needs to know certain characteristics of the mountain groups. Two terms used to describe these characteristics are hinterland and foreland. The hinterland is most often found close to the continental edge, an area in which deformation is usually more severe. The foreland is next to the interior of a continent, in which a continent-to-continent collision has occurred.

What are the **world's highest mountains**?

The Earth has some very high mountains. The following lists the five highest in the world:

Name	Location	Height (meters/feet)
Everest	Nepal/Tibet	8,850/29,035
K2	Kashmir/China	8,611/28,250
Kanchenjunga	Nepal/Sikkim	8,598/28,208
Lhotse	Nepal/Tibet	8,501/27,890
Makalu I	Nepal/Tibet	8,470/27,790

What are **mountain ranges and mountain systems**?

Mountain ranges are groups of aligned mountains—usually in rows—related with respect to composition and origins. A mountain system is a grouping of several ranges

The southern face of Mt. Everest, the world's highest peak at 29,035 feet, is seen above the clouds. Likely the most famous mountain in the world, Everest has attracted a steady stream of climbers since Sir Edmund Hillary and the Nepalese Sherpa Tenzing Norgay became the first to reach its summit on May 29, 1953. *AP/Wide World Photos.*

that seem to be similar in form, alignment, and structure. For example, in North America the American Cordillera system includes the Cascade Range, Sierra Nevada, Rocky Mountains, and Canadian Rockies.

What are the **types of mountain ranges**?

Geologist categorize mountain ranges based on several characteristics. The following lists several basic types:

Valley and ridge range—Valley and ridge ranges contain a series of alternating folds (troughs and crests, respectively called synclines and anticlines) of the rocks covering the surface. The valleys are usually composed of easily eroded rock, such as limestones or shales; the ridges are composed of less resistant rock, such as sandstone. For example, central Pennsylvania's Valley and Ridge represents a good sample of this type of mountain range.

Crystalline upthrust range—When continental crust is highly compressed, it often reaches the surface in the form of brittle block uplifts. When this occurs, the sedimentary rock overlying the crystalline rock becomes folded. For example, the mountain ranges in Wyoming exhibit this type of formation.

Crystalline core range—Along the hinterland of mountain ranges, the crystalline basement breaks into sheets that are pushed toward the foreland by

Air balloons fly over the village of Zermatt, Switzerland, in the Swiss Alps with the Matterhorn's peak in the background. *AP/Wide World Photos*.

tectonic forces. These are mostly metamorphic rocks, crystalline sheets changed by orogenic forces. For example, the Blue Ridge Mountains (southern Appalachians) comprise a crystalline core range.

Plateau uplift—Crustal rocks that lift high above sea level are called plateaus; they do not appear to be mountain ranges until one views the plateaus from the edges. The Colorado Plateau is the largest in the United States; the Tibet Plateau is the largest in the world.

Fault block mountain range—A fault block mountain range forms when the mountains separate from the valley floors by faulting. These faults create a huge displacement in the land, raising or lowering the land in between the mountains. The Basin and Range of Nevada is a fault block mountain range.

Volcanic mountain range—A volcanic mountain range usually forms around subduction zones at plate boundaries, creating a chain of volcanoes as magma erupts over time. The Cascade Range of California and Oregon is part of a volcanic mountain range.

What **mountain range** is still **growing in the Pacific Ocean**?

On the interior of a plate at a hot spot, magma reaching the surface can create a mountain range. One of the largest such mountain ranges is the Hawaiian Ridge. Its sheer size and length go unrecognized because the majority of the volcanoes are underwater. The entire Hawaiian Ridge is only 43 million years old and stretches hundreds of miles across the Pacific Ocean floor, and it is still growing.

What are the names of some **important mountain ranges**?

There are many mountain ranges around the world—too many to mention here. The following lists some of the more well-known and larger mountain ranges:

Name	Location
Brooks Range	Northern Alaska, Northwestern North America
Alaska Range	Northwestern North America
Coastal Mountains	Northwestern North America
Rocky Mountains	Western North America
Appalachian Mountains	Eastern North America
Andes Mountains	Western South America
Drakensberg Mountains	South Africa
Ethiopian Highlands	Eastern Africa
Atlas Mountains	Northwestern Africa
Pyrenees	Between Spain and France
Alps	Southern Europe
Carpathian Mountains	Eastern Europe
Caucasus Mountains	Southern Russia
Ural Mountains	Central Russia
Greater Khingan Range	Northern China
Altai Mountains	Western China
Himalayas	Southern Asia
Great Dividing Range	Eastern Australia
Southern Alps	New Zealand, southern island

MOUNTAIN HISTORY

What did some **early cultures** believe about **mountains**?

Humans have always been fascinated by mountains, viewing them alternatively with awe and fear, depending on the prevailing culture. The ancient Greeks believed mountains were the homes of gods and goddesses; Zeus, Apollo, Hera, and others resided on Olympia, looking down on mortal humans and influencing or meddling in their lives. The Buddhists of Japan and China believed mountains were associated with divinity and built temples and shrines on the peaks. Most Europeans regarded mountains with superstition and fear, an attitude that lasted even until the early 19th century. They believed these high places were the tumors of the Earth, populated by gnomes, demons, and goblins. For them, mountains were places to be avoided.

What were some **early theories** about **mountain development**?

There were many early theories about how mountains originated. In 1545, German geologist Georgius Agricola (1494–1555) proposed that mountains originated by means of earthquakes, water erosion, and sundry other natural phenomena. In 1669, Nicolaus Steno (Niels Stensen, 1638–1686), Danish geologist and anatomist, conducted field studies in Tuscany. He concluded that mountains formed from the shifting of the Earth's strata, either through violent upthrusting, downfall, or slippage. In 1740, Italian scientist (Antonio) Lazzaro Moro (1687–1764) described the history of early Earth, noting that mountains developed because of a central fire within the planet.

By 1830, French scientist Elie de Beaumont (1798–1874) came close to accurately describing mountain formation. He believed that each mountain range formed due to a catastrophe and that mountains formed because of the cooling (and thus contracting) of the Earth's crust. But it was Swiss scientist Alphonse Favre (1815–1890) who showed, in 1880, that mountains can form under lateral compression. His model was simple: By pushing a layer of soft clay and a thick band of rubber, the model's "mountains" formed by folding. Favre concluded that the Appalachians, Alps, and Jura mountains formed in this way.

What is an **orogeny**?

An orogeny is a period of mountain building that occurs as the Earth's lithospheric plates move and collide (the actual formation of the mountains themselves is called orogenesis). During an orogeny, rock can be deformed, thrusted, and folded. Sometimes, an orogeny is accompanied by the intrusion (invasion) of igneous rocks or volcanics; and metamorphic rocks can be created at the site by the intense heat and pressure of the orogeny. (For more about plate tectonics, see "The Earth's Layers.").

The first mention of orogenies was in 1887, when French geologist Marcel Alexandre Bertrand (1847–1907) proposed that Europe had gone through three major orogenies. The results were the Caledonian, Hercynian, and Alpine mountain chains.

When did the **first orogeny** occur?

The first known orogeny occurred approximately 3.96 billion years ago in what is now North America. Evidence for this comes from a gneissic rock layer. Since that time, there have been numerous other orogenies. Geologists separate them into the time periods in which they occurred, including the Precambrian, Paleozoic, Paleozoic to Mesozoic Era Transition, Mesozoic Era, and Cenozoic Era.

What **orogenies** occurred during the **Precambrian**?

The following table lists the known orogenies that occurred during the Precambrian:

Time Period (millions of years ago)	Name	Location (referenced to present location)
3960 (oldest rocks)	Acasta gneiss	North America
3300–2700 (volcanic)	Kalahari craton	Africa
3200–2700 (volcanic)	Baltic Shield	Baltica
3050–2700 (volcanic)	Pilbara block	Australia
2700–2500 (volcanic)	Canadian Shield	North America
2600–2400	Algoman	North America
1820–1640	Hudson	North America
1750–1650	Penokean	North America
1200-900 Rift (volcanic)	Mid-Continent	North America
1000-880	Grenville	North America
600-550	Baikalian	Siberia

What **orogenies** occurred during the **Paleozoic Era**?

During the Paleozoic Era, widely separated landmasses moved together to form one large continent in the north, called Laurasia, and one in the south, called Gondwanaland. Between the Paleozoic and Mesozoic Eras, the two continents came together to form the supercontinent of Pangea (or Pangeae). The following orogenies occurred during this time period (and the transition time between the Paleozoic and Mesozoic Eras):

Time Period (millions of years ago)	Name	Location (referenced to present location)
540–530	Adelaide	Australia
480–460 (early)	Appalachian	North & South America, Antarctica
460–440	Taconic	North America, Baltica
450–430	Caledonian	North America, Baltica
410–380	Acadian	North America
380–350	Antler	North America
380–300	Uralian	Baltica, Siberia Kazakhstania
380–250	Tasman	Australia
350–245	Hercynian	Baltica
325–310	Ouchita	North America
320–220	Allegheny	North America, Africa
250	Cape Folding	Africa
250 (volcanic)	Siberian Traps	Siberia

What **orogenies** occurred during the **Mesozoic Era**?

During the Mesozoic Era, Pangea broke up to form Laurasia and Gondwanaland again. In turn, these two landmasses began to break up into smaller pieces that vaguely resembled our modern continents. The following orogenies occurred during this time period:

Time Period (millions of years ago)	Name	Location (referenced to present location)
200–190 (volcanic)	Eastern N. America	North America
170–160	Rifting (volcanic)	Africa, Antarctica
190–140	Nevada	North America
140–80	Sevier	North America
135–130 (volcanic)	S. Atlantic	South America, Africa
122	Ontong (volcanic)	Java
112 (volcanic)	Kerguelen	South Indian Ocean
110–100	Rajmahal	India

What **orogenies** occurred during the **Cenozoic Era**?

The continents continued their movement throughout the Cenozoic Era, arriving at their current positions. The following orogenies occurred during this time period:

Time Period (millions of years ago)	Name	Location (referenced to present location)
84–50	Laramide	North America
80–60	Andean	South America
65–63 (volcanic)	Deccan Traps	India
62–55	Brito-Arctic	North America, Baltica
57 (volcanic)	North Atlantic	North Atlantic Ocean
55–45	Pyreneen	Baltica
40–5	Alpine	Baltica, Africa
25–18 (volcanic)	Ethiopa	Africa
24–0	Himalayan	India, China
18–15	Columbia River	North America

Have any **ancient mountain ranges disappeared**?

Yes, scientists believe that many ancient mountain chains have disappeared. Many are seen as the eroded remnants of huge batholiths (of volcanic origin) or even belts of highly deformed metamorphic rocks.

What major orogeny is currently taking place?

One of the fastest growing mountain chains—and thus orogenies—is the Himalayas. These mountains are growing by more than a half inch (1 centimeter) a year, a growth rate of 6 miles (10 kilometers) every million years.

MOUNTAIN EVOLUTION

What was the **first model of mountain evolution**?

Published in 1899, the "geographic cycle" was the first comprehensive model of mountain evolution. It expounded a lifecycle for mountains that included a relatively brief birth caused by a violent, sudden uplift, and an old age due to slow erosional processes. This geographic cycle was accepted for almost 100 years, and only recently has it been modified as a result of new research.

What is the **modern model of mountain-building**?

Formulated in 1998, the most recent model of mountain building links the forces of plate tectonics, erosion, and climate, all of which influence each other in a continuous feedback cycle. The complex interplay of these forces results in the mountains (and surrounding regions) continually changing their height and topography.

Plate tectonics starts the process off as the collision of continental plates leads to uplifting. Then, erosion and climate begin to work on the rock, forming the mountainous features we see. Erosional agents include gravity, wind, water, and glacial ice, all of which weather the rock, turning it into increasingly smaller sediment chunks. These pieces move down to the mountain base and are eventually transported away by streams and rivers. The rate at which erosion of mountains occurs is directly related to the presence and extent of these agents, the steepness of topography, the type of rocks present, and the climate.

In turn, the climate becomes a key component of erosion, affecting how fast it occurs and the amount of material removed. But this relationship is complex. For example, a cold climate can lead to the development of glaciers, a very aggressive erosional agent. The Sierra Nevada of the western United States and the Alps of Europe have been extensively carved by glaciers. In Antarctica and Greenland, the ice sheets are frozen to the underlying rock, resulting in little erosion. Another example of this complexity is the influence of a wet climate on erosion. Mountains in this type of climate will normally erode faster. But a wet climate may also enable the growth of plants, which anchor the soil and slow down erosion rates.

What is **isostasy**?

Isostasy refers to blocks of the Earth's crust that are in equilibrium. First proposed in 1889 by the American geologist Clarence Edward Dutton (1841–1912), this model views the blocks of crust as if they were floating stationary in a liquid, held up by the force of buoyancy. To be in equilibrium (isostasy), mountains must have deep, less dense roots than the underlying mantle; this is similar to icebergs in the oceans. But erosion upsets this equilibrium by removing more of the rock on top, diminishing the weight above. To restore balance, the mountains rise in a process called isostatic uplift.

For example, as the miles-thick glacial ice sheets of the recent ice ages moved over the land, their weight caused the continents to subside (sink). After the ice sheets retreated, the continents rebounded. Many are continuing to rebound even today, bringing them back into equilibrium. Small earthquakes occur in the formerly ice-covered Adirondack Mountains of New York—not because of active faults, but due to the continuing rebounding of the mountains.

What are the **three stages** of **mountain development**?

Currently, geologists theorize there are three stages of mountain development: the formative, steady-state, and erosional stages. Each of these stages can be measured in thousands, if not millions, of years. Tectonic activity is dominant in the formative stage, a time in which the rate of uplift is greater than erosion. As the mountains continue to grow, they eventually become high enough to change the surrounding climate, leading to an increase in erosion.

The steady-state stage occurs when the rate of uplift—either due to isostasy or continuing tectonic activity—is the same as the rate of erosion. At some point, the tectonic activity lessens and the rate of uplift slows, allowing erosion to become dominant. The erosional stage occurs as the rock is slowly eaten away. The once tall, pointed peaks diminish and become rounder, or they might even disappear. If the climate changes drastically, or if uplift begins again due to renewed tectonic activity, the mountains may exist for many millions of years, or they might rise up again to their previous heights.

WELL-KNOWN MOUNTAINS AND RANGES

What are the **Appalachians**?

The Appalachians are a wide belt of mountains stretching 1,600 miles (2,600 kilometers) from Newfoundland to Alabama in eastern North America. In the northern part of this range are the White Mountains of New Hampshire, the Green Mountains of

Haze hangs in the valleys in this view from Clingman's Dome on the Tennessee–North Carolina line in Great Smoky Mountains National Park. Scientists believe the Smoky Mountains were so named because of the haze caused by isoprene, a visible hydrocarbon emitted from trees. *AP/Wide World Photos*.

Vermont, the Catskill Mountains of New York and Pennsylvania, and the ranges of the Alleghenies. To the south are the Blue Ridge Mountains of Virginia and North Carolina, and the Great Smoky Mountains of North Carolina. All together, they are essentially a barrier between the center of the continent and the eastern coastal plains.

Why are the **Black Hills and Black Mountains** famous?

The Black Hills were formed by an uplifting of an area that spreads across northeastern Wyoming and southwest South Dakota. The most famous place in the Black Hills is found on the granite side of Mt. Rushmore, where the faces of four American presidents were carved by artist Gutzon Borglum (1871–1941). As for the Black Mountains themselves, they are famous because they are the highest point east of the Rocky Mountains. As part of the Appalachian Mountains in North Carolina, they rise to 6,684 feet (2,037 meters) at Mt. Mitchell.

What are the **Coast Mountains**?

The Coast Mountains form a range that runs parallel to the Pacific coast, from Alaska to British Columbia, for about 1,000 miles (1,600 kilometers). It is mostly composed of ancient metamorphic rock that has been highly eroded by glaciers and rivers. Mount Waddington is part of the range; measuring 13,262 feet (4,042 meters), it is the highest mountain in Canada.

What mountain is the home of the world's worst weather?

The mountain that claims to have the world's worst weather is not in Antarctica but in the White Mountains of New Hampshire: Mt. Washington. This 6,288-foot (1,917-meter) mountain has the most severe combinations of wind, cold, icing, and storminess anywhere in the world. This has been determined by scientists who are always there to take measurements (only once since 1932 has the summit not been manned 24 hours a day, 7 days a week).

The summit lies in the path of the principal storm tracks and air mass routes affecting the northeastern United States; the mountain is so high that it essentially squeezes the air over the top of the summit. This creates extremely high winds (and high windchill temperatures). In fact, the highest wind gust ever recorded on Earth—it was measured at a whopping 231 miles (372 kilometers) per hour—took place on Mt. Washington on April 12, 1934. The average wind speed for the year is 35.3 miles (57 kilometers) per hour; the number of days per year that exceed hurricane force winds (75 miles [121 kilometers] per hour) is 104 days. In addition, the average yearly temperature is about 26.5°F (–3.06°C); and average snowfall is about 256 inches (650 centimeters). Finally, because of its elevation, biologically and ecologically the mountain is similar to a subarctic zone.

What are the **Coastal Ranges**?

The Coastal Ranges are a series of very young, granite mountains running along North America's Pacific coast, from Alaska to southern California. The mountains formed from folding at the North American and Pacific plate boundaries, and include the Olympic Mountains of Washington, the Klamath Mountains in California, and the Diablo and Santa Lucia ranges.

What is the **"hottest spot"** in the **Sierra Madre**?

The Sierra Madre is a mountain system that occupies most of Mexico. The western range runs parallel to the Pacific coast for about 680 miles (1,100 kilometers), while the eastern range converges on the Pacific coast. One of the most famous volcanic mountains exists along this range: the second highest volcano in Mexico called Popocatepetl. This giant volcano stands 17,925 feet (5,465 meters) tall. Only about 45 miles (70 kilometers) southeast of downtown Mexico City, it is about 30 miles (45 kilometers) southwest of the city of Puebla. The name Popocatepetl, meaning "smoking mountain," was given to the volcano by the Aztecs, suggesting that it has been active for a long time. Popo, as it is often called, is built on an older volcano that adds 12,464 feet (3,800 meters) to Popocatepetl's elevation.

Are there any **mountain ranges in Antarctica**?

Yes. The Transantarctic Mountains essentially cut the huge, icy continent in half, and extend from the tip of the Antarctic Peninsula to Cape Adare, a distance of 3,000 miles (4,800 kilometers). Other ranges include the Prince Charles Mountains and the Ellsworth Mountains, which are Antarctica's tallest mountains.

Toward Ross Island in Eastern Antarctica, there is also an active volcanic mountain: Mt. Erebus. This mountain is about 900 miles (1,450 kilometers) from the South Pole and measures about 12,280 feet (3,743 meters) high. Thanks to the clean air, atmospheric conditions, and flatness of the area—not to mention the fact that it is actively smoldering—Mt. Erebus is often visible from 100 miles (160 kilometers) away.

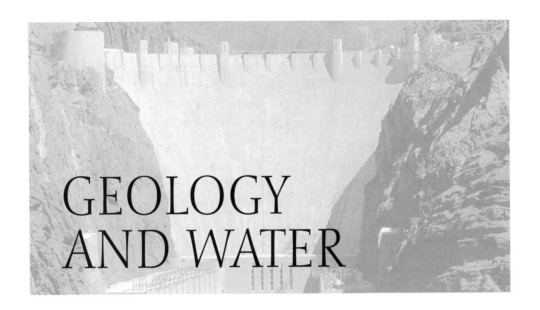

GEOLOGY AND WATER

WATER STATISTICS

Why is **water so important** to the Earth?

Water is one of the most important—if not the most important—features on our planet, especially when it comes to life on Earth. It plays a crucial role in our atmosphere, soil, and underground regime (as groundwater), allowing plant and animal life to flourish. Most organisms are mostly made of water. For example, the adult human body is between 50 and 70 percent water (a child's body is approximately 75 percent water), and our brains are about 75 percent water.

How is **water distributed** on Earth?

Water is also one of the most common molecules circulating in our atmosphere; the volume of the global oceans equals about 330 million cubic miles (1,400 million cubic kilometers) of water. It is estimated that if all the water on Earth were spread evenly over the planet, the resulting layer of water would be almost 2 miles (3 kilometers) thick.

In reality, the distribution of water is not even. It is stored mainly in the oceans, which hold about 97.2 percent of all the water in the world. Of the remaining 2.8 percent, 2.15 percent is found in ice sheets and ice shelves (or polar ice), and in glaciers; 0.62 percent is located in groundwater; 0.011 percent is in saltwater lakes and inland seas; rivers, freshwater lakes, and wetlands hold about 0.013 percent; and soils hold about 0.005 percent of the water. The final 0.001 percent is found in the atmosphere. Humans use less than 1 percent of all the water on the planet—mostly from groundwater, lakes, and rivers—some of which returns to the atmosphere and falls as rainwater, but much of which is not renewed.

Why is water important to geology?

Beside water being important to life on Earth, the movement of water is one of the most important geologic processes affecting our planet. It not only plays a part in shaping the Earth's surface, but also in shaping human lives on the planet. After all, humans have long settled where there is plenty of water, especially along rivers and ocean shores.

Water has carved a "feature-full" world. Lakes form high in the mountains or in flatlands. Running water from mountain glaciers carves deep valleys and carries sediment into rivers that drain into the oceans. Water will always be with us—at least until the Earth is consumed by the sun—which means we can count on about 5 billion years of water carving rock and depositing material on our planet's surface.

What is **runoff**?

Runoff is the water that does not evaporate or sink into the ground after a precipitation event (snow, ice, or rain). Runoff flows downhill in response to gravity, following the contours of the land. Most runoff ends up in rivers, streams, and creeks; some soaks into the soil, eventually contributing to the groundwater.

If the runoff is great enough—especially from a heavy storm event—the water does not infiltrate the soil, but can literally flow in sheets along streets, sidewalks, and gentle slopes. This is called *Hortonian overland flow,* a term first used by hydrologist Robert Horton, who noted the effect. This flow has a dragging effect on the soil and materials underneath, with enough force to move fine clays and coarse gravels, a process often called *sheet erosion.* In arid regions and those with little vegetation to hold soils, sheet erosion can cause considerable wearing away of sediments.

How does **water cycle around the Earth**?

Many people separate the water and hydrologic cycles, but they are intertwined and inseparable. These cycles are driven by the "engine" of the sun, mainly through evaporation and precipitation; the water eventually flows to the oceans or into other water reservoirs.

Simply put, the cycles combine to distribute water as follows: Water is evaporated from the Earth's surface by the energy from the sun. From there, it enters the atmosphere in the form of water vapor clouds (water molecules stay an average of 10 days in the atmosphere). Depending on the temperatures and weather conditions, water vapor then condenses, falling to the surface in the form of precipitation—from rain to sleet and snow. Much of the water reaching the surface evaporates again into the atmosphere, either by sublimation (water turning into a gas) or biogenic processes, such as transpiration in plants. Other precipitation will run from high to low areas on the surface, such as mountains to valleys. This surface runoff will either find its way into rivers, streams, lakes, seas, or oceans (surface water's usual destination), or will percolate (infiltrate) through the soil, eventually reaching the groundwater.

Runoff water continually erodes and deposits sediment on the Earth's surface. Thus, these cycles grind down the higher elevations, drive surface rocks downhill, and gradually deposit the resulting sediment into the oceans. From there, after millions of years, they form rocks that are eventually affected by tectonic forces.

RIVERS AND STREAMS

What is a **channel**?

A channel is simply the bed where a natural stream or water runs—or a clear, defined pathway the water follows. It is also considered to be the deeper part of a river, harbor, or strait. Such areas are often dredged by humans to allow the water to flow more freely and/or to allow ships to better navigate the channel.

What is the **source** of some major **rivers**?

The source of a river is the location where the river begins. Most sources originate in the mountains; others are in lakes or seas; still others come from a combination of places. For example, the source of the Nile River in Egypt is actually two rivers: the White Nile and the Blue Nile, which join together at Khartoum to form the Nile. The White Nile begins at Lake No in south-central Sudan, while the Blue Nile starts in Ethiopia, at Lake Tana.

Most other river sources are more straightforward. The source of the Yangtze River, China's longest river, is in the Himalayan Mountains at Mt. Geladaintong; the source of

How does water get into a river or stream?

There are four major ways water gets into a river or stream. Contrary to popular belief, precipitation (rain, ice, or snow) contributes only a small amount to the overall amount of water in a river. But runoff from precipitation (rainfall, melting ice, and snow) does represent a very large part of river flow. The next major source is interflow—the water that infiltrates the soil and gradually moves toward the stream channel—which is dependent on the soil's structure and the depth to the groundwater table. Finally, groundwater also contributes to the river flow, as underground water naturally migrates toward a river or stream.

The space shuttle *Endeavor* passes over the Nile River in North Africa, the world's longest river at 4,160 miles (6,695 kilometers). *AP/Wide World Photos.*

the Yellow River (Huang He or Hwang Ho), China's second longest, is in Qinghai province. In the United States, the source of the Susquehanna River in New York and Pennsylvania is Otsego Lake in Cooperstown, New York. The source of New York's Hudson River is at Lake Tear of the Clouds on the flank of Mt. Marcy in the Adirondack Mountains. The Colorado River has its source high in the Rocky Mountains, at Grand Lake, Colorado. And the Rio Grande's river source is in the Rio Grande National Forest, San Juan Mountains, Colorado (at the Continental Divide), at a clear, spring- and snow-fed mountain stream that is 12,000 feet (3,658 meters) above sea level.

What are the **five largest rivers**, based on discharge?

Discharge is one way to categorize rivers; it is the volume of water flowing through a cross section of the channel over a certain period of time. The following lists the Earth's five rivers with the most discharge, as measured in cubit feet (and meters) per second:

River	Discharge (feet^3/second)	Discharge (meters^3/second)
Amazon	4,000,000 to 5,000,000	113,267 to 141,584
Congo	1,400,000	39,644
Yangtze	770,000	21,804
Mississippi-Missouri	620,000	17,556
Yenisei	615,000	17,415

What are the **five longest rivers** (and largest, in terms of area) in the world?

The five longest rivers in the world are *not* the same as the five with the most discharge. They are as follows:

River	Length (in miles)	Length (in kilometers)
Nile	4,160	6,695 kilometers
Amazon	3,900	6,276 kilometers
Mississippi-Missouri	3,890 miles	6,260 kilometers
Yangtze	3,600 miles	5,794 kilometers
Ob	3,200 miles	5,150 kilometers

The rivers with the most area (or largest basin) also differ. They are as follows:

River	Area (miles^2)	Area (kilometers^2)
Amazon	2,368,000	6,133,091
Congo	1,550,000	4,014,481
Mississippi-Missouri	1,244,000	3,221,945
Nile	1,150,000	2,978,486
Yenisei	1,000,000	2,589,988

What are the **major parts** of a **river system**?

In general, there are three major parts of a river system. The first is the collecting system, otherwise known as a basin. This is where erosion is the dominant process: The rivers throughout the basin carry materials (called *load*), which vary in amount and type, depending on the surrounding rock, and speed and volume of the stream. In turn, this is based on the slope (or *gradient*) of the stream—the steeper the slope, the faster the flow.

The next major parts of a river system are the tributaries, areas in which deposition and erosion occur and transport of materials (especially sediment) is the most important process. Finally, the last part of a river system is called the dispersing system. This is the region in which deposition of materials is the most important process, such as a delta area.

What are **deltas and alluvial fans**?

Deltas are alluvial deposits—an accumulation of silt, sand, gravel, and other materials transported and deposited by running water—that most often occur at the mouths of some rivers. Deltas should not be confused with alluvial fans, which are deposits that form at the base of mountains where a stream meets a valley floor. Deltas are similar

Brazilian Amazon Basin inhabitants row a boat along the Amazon River. Due to the lack of land routes between villages, rivers become the main communication stream and transportation system in the area. *AP/Wide World Photos.*

phenomenon, but they occur where a river meets the ocean and loses its velocity to deposit its sediment load.

What are the major **types of river valleys**?

There are two major types of river valleys that form as waters carve their way through the Earth's surface. *V-shaped valleys* form when a river is fast or it is actively cutting its way downward through hard bedrock. As the water moves, it carries away most of the sediment associated with erosion. The other type of valley is often called a *U-shaped valley,* which is a much broader, shallower valley that surrounds the main river channel. Because they are less steep, the rivers in U-shaped valleys are much slower, which means they also tend to deposit more sediment in these valleys.

What is a **floodplain**?

Broader, less-steep valleys around rivers usually have expanses of flat areas. These sections, called floodplains, are usually at or near the level of the main river. Floodplains are notorious for flooding during flash floods or incessant rains and runoff. As the water eventually recedes after a flood, it drops a great deal of rich sediment, which is one reason why farmlands on floodplains are such good places to grow crops.

Will rivers ever stop eroding the Earth's surface?

Not likely, unless all water is boiled away from the Earth's atmosphere and oceans! The Earth is dynamic, and part of this dynamism involves rivers continually carving away at the land. For example, the uplifting of land from lithospheric plate movement renews the energy of a river, causing it to erode more vigorously. The movement of an earthquake can change the land, including the channels of rivers. And even a volcanic eruption can create fresh, pristine surfaces on which a new drainage system develops.

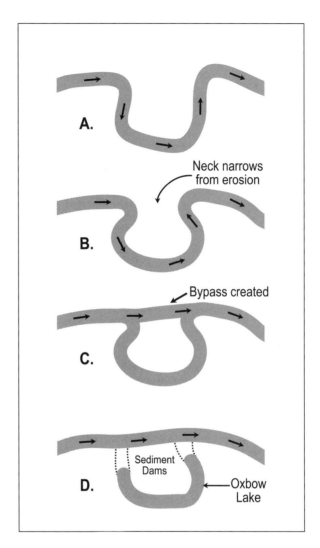

What are **meandering rivers** and **braided channels**?

The main river channel within a floodplain often follows a curvy path, also known as a meandering river. These curves and bends in the river slowly migrate over the floodplain from side to side, depending on the erosional and depositional features of the channels, further leveling the floodplain. If any of these meanders are cut off by sediment deposits or occasional flooding, the resulting curved water reservoir is called an *oxbow lake*.

Where there is a very high sediment load and frequent changes in the water's flow levels, the floodplain often divides into numerous overlapping channels called *braided streams*. These occur when sediment starts to drop out, and the river divides into multiple channels.

The formation of an oxbow lake. Beginning with a meandering river (A), the base of the meander begins to narrow (B) until a bypass is created (C) Eventually, sediments build up between the bypass and the original course of the river until an oxbow lake forms (D).

How do **rivers erode soil and bedrock**?

Rivers erode soil and bedrock in three general ways. First, the river picks up loose particles, carrying them away. This is why soil is so easily lost to streams and rivers. (For more about weathering and eroding streams, see "Wearing Away the Earth.") Second, the particles of sand, soil, and debris act as scouring agents, continually carving the channel and surrounding rock in a process that is most pronounced at waterfalls and rapids. Third, chemical weathering from the water breaks down bedrock by eating away at minerals, especially along joint planes of the rock.

223

What are the various **types of streams**?

There are several types of streams that are categorized based on the flow of water. If a stream is a dry channel throughout most of the year, only bearing water during and immediately after a rainstorm, it is called an *ephemeral stream*. If a stream carries water part of the year and is dry part of the year, but receives groundwater when it is high enough, it is called an *intermittent stream*. And finally, a *perennial stream* is one that carries water the entire year and is fed by groundwater flow.

What are the types of **river or stream drainage patterns**?

There are many different drainage patterns that rivers and/or streams follow. The following lists some of the most common:

The giant Hoover Dam, the largest of the 50 dams controlling the 1,450-mile Colorado River, is shown near Boulder City, Nevada. *AP/Wide World Photos.*

Dendritic—This type of river or stream forms a spreading, treelike pattern, usually in horizontal sediments or in crystalline rocks. Some examples include rivers and streams in the Great Plains and along the Appalachian Plateau.

Rectangular—These rivers and streams form a compact, perpendicular network of channels, usually with the channels predominantly lying in two directions. Examples include rivers that go through joints or faults, such as those in the Adirondack Mountains and Zion National Park.

Pinnate—These rivers and streams are featherlike, closely grouped, and with short tributaries. They usually occur in areas containing easily eroded material, such as those found in parts of the Colorado Plateau.

How do dams affect rivers around the world?

River systems are directly and absolutely changed when dams are built. The main reason is obvious: Dams block the channels, altering the water's direction by decreasing or increasing the amount of water that flows through the channel. In turn, this modifies (or completely changes) the river's erosional and depositional characteristics, thus changing the channel's landscape and affecting the local environment.

Although there are good reasons for dams (mainly to stop flooding in populated areas), it's known that there are often just as many potential problems. One in particular is the erosion that occurs just below the main structure holding back the water. Because sediment is no longer transported within the water (the load is dropped in the reservoir), the water from the spillway often erodes the channel immediately below.

Another problem can also arise: Because there is less sediment load, there is also less of a delta being formed at the mouth of a river. For example, the Aswan High Dam along the Nile River in Egypt was finished in 1966, primarily to provide electricity and irrigation. But the water is dammed up in a lake about 175 miles (280 kilometers) long, and this is starving the Nile delta of sediments. Because of this, the currents in the Mediterranean Sea are carrying away more sediment than the river can replenish and causing the delta to slowly erode away.

One of the biggest problems of happens with the placement of a dam, or many dams, along a larger river that causes low (or no) flow in the lower reaches of that river. For example, the fifth longest river in North America, the Rio Grande, runs through Colorado, New Mexico, Texas, and Mexico. The dams along the river—mostly for irrigation—have contributed to the detriment of the entire river system. With no tributaries to replace water withdrawn for irrigation, the river frequently dries up by the time it reaches Fort Quitman, Texas.

Trellis—Trellis drainage patterns have one dominant direction, with secondary streams perpendicular to the main river. Trellises closely resemble rectangular drainage patterns, but are more elongated (along the main river) and less compact. Examples of this type of pattern occur in tilted or faulted sediments, such as those in the European Alps, and along elongated landforms, such as hill-sized, teardrop-shaped deposits called drumlins, which are formed by retreating glaciers.

Radial—Radial drainage patterns are just what their name implies: rivers or streams radiating from a central point. Examples include rivers and streams on volcanic cones or domes, including such places as Koko Crater, Hawaii, and the Paricutin volcano in Mexico.

225

Parallel—Parallel drainage has many channels that are regularly spaced and parallel to each other. Examples are found in steeply sloping surfaces, such as those found in Mesa Verde National Park.

LAKES AND GROUNDWATER

What is a **lake**?

In general, a lake is a hollow filled with water; most lakes are filled by rivers, streams, and/or springs. A body of water has to be of substantial size to be called a lake, rather than a pond. The size of a lake depends on the size and shape of the hollow, which, in turn, depends on how it was formed, the local bedrock, and even the climate. It is also dependent on the inflow and outflow of water, which may vary with the seasons.

What is a **pond**?

In general, ponds differ from lakes in that ponds are usually smaller. Ponds are also shallow enough for plants to grow from shore to shore—and even sometimes in the middle of the pond; lakes are usually too deep to support rooted plants except along the shore. Another way to define a pond and lake is by sunlight: Since the sunlight plants need generally does not penetrate deeper than 6 feet (2 meters), a pond is often defined as being less than 6 feet (2 meters) in depth and a lake is deeper than 6 feet (2 meters).

There are other differences: In a pond, water temperature is fairly even from top to bottom, changing with the air temperature; and ponds usually freeze solid in the winter in cold regions. In the summer months, water temperatures in lakes are not uniform from top to bottom, with most developing three distinct layers: The top layer stays around 65 to 75°F (18.8 to 24.5°C); the middle layer is usually to 45 to 65°F (7.4 to 18.8°C); and the final, bottom layer stays around 39 to 45°F (4.0 to 7.4°C). Almost all lake-dwelling creatures spend the summer months in the upper layer; in the spring and fall, when the lake temperature is more uniform, they are found throughout the layers of the lake. In the winters, the top layer of a lake usually freezes, with the lower layer remaining unfrozen.

What is the **difference** between **freshwater and saline lakes**?

A lake can be freshwater or saline (salty). A freshwater lake usually forms where there is a free flow of fresh water through the lake to an outlet that eventually leads to either an ocean or sea. Freshwater lakes are important to life on Earth because they contain a great deal of the unfrozen fresh water. For example, Lake Superior in North America contains the most fresh water of any lake in the world; and Lake Baikal in Siberia is the largest freshwater lake by volume, as well as being the world's deepest.

A salt lake usually forms where evaporation is high and the incoming water can't flow out to an ocean or sea. Saline lakes are often used as reservoirs for salt deposits, but they are not as important to life on Earth (whereas oceans are huge saline reservoirs that are very important to life on our planet). For example, the Dead Sea is the world's saltiest lake, and is about 9 times more salty than the oceans.

What are some of the **world's largest lakes**?

Some of the largest lakes are in North America, Asia, and Africa, and they were all formed in different ways. Currently, the top ten largest lakes and their average areas are as follows (note: seas are also considered lakes in this list):

Lake	Location	Area (miles^2)	Area (kilometers^2)	Comments
Caspian Sea	Asia	143,630	372,002	It is a saline lake
Lake Superior	North America	31,700	82,103	The world's largest freshwater lake
Lake Victoria	Africa	26,828	69,485	The world's second largest freshwater lake
Lake Huron	North America	23,000	59,570	
Lake Michigan	North America	22,300	57,757	
Aral Sea	Asia (Russia)	15,500	40,145*	
Tanganyika	Africa	12,700	32,893	It is also the world's longest lake, measuring 420 miles (676 kilometers) long and between 10 and 45 miles (16 and 72 kilometers) wide
Baikal (Baykal)	Asia (Russia)	12,162	31,500	
Great Bear	North America	12,096	31,329	
Malawi (Nyasa)	Africa	11,150	28,879	

* Because of overuse of the Aral Sea for irrigation decades ago, this sea has gone from the fourth largest lake to the eighth largest lake. Its area also changes each year; therefore, this is merely an estimate.

What are some of the world's **deepest lakes**?

The world's deepest lakes are not always the largest. The following lists the deepest five lakes known in the world and their average depths:

Why is Lake Baikal famous?

Lake Baikal, a giant lake measuring almost 400 miles (636 kilometers) long and about 50 miles (80 kilometers) wide, is located in the taiga between Siberian Russia and Mongolia, along a natural divide between the mountains to the north and rolling grasslands to the south. Lake Baikal is the oldest (25 million years) and deepest (5,315 feet/1,620 meters) lake in the world. It is also the world's largest freshwater lake by volume, with some scientists estimating that it contains about 20 percent of the world's unfrozen freshwater reserves. It is so large and deep that all of the combined rivers on Earth would take an entire year to fill it.

Lake	Location	Depth (feet)	Depth (meters)
Baikal (Baykal)	Asia (Russia)	5,315	1,620
Tanganyika	Africa	4,708	1,435
Caspian Sea	Asia	3,264	995
Malawi (Nyasa)	Africa	2,280	695
Great Slave Lake	North America	2,015	614 meters

What are some common **types of lakes**?

There are plenty of lakes all over the world. The following lists some of the more common types:

Tarn—Tarn lakes (also often classified as alpine lakes) are usually associated with glacial regions—areas that have been glaciated in the past. These hollows form as the ice scours out the side of a hill, creating what is called a cirque. (For more information about glaciers, see "Ice Environments"). Tarn lakes from the last ice ages are evident along some slopes of the Adirondack Mountains and from more recent glaciation in the European Alps.

Rift valley lake—Rift valley lakes occur in areas where rock is pulling apart. As a narrow wedge of land drops from the movement, a long, narrow lake often forms. The Rift Valley, which cuts through Africa and Asia, contains the largest group of such lakes, including Lake Nyasa.

Crater lake—Crater lakes form at a volcanic peak. When an extinct volcano collapses, it often leaves a somewhat circular depression; water from rain and runoff eventually fills the lake. Crater Lake in Oregon is a good example of such a lake.

Deflation lake—Deflation lakes usually occur in deserts as the wind blows out depressions in the sand. If the hole reaches the local groundwater table, it can create an oasis. Such features dot the Sahara Desert in Africa.

Oxbow lake—Oxbow lakes are created by a curved river meander that was cut off by sediment or other means, creating a bowed-shaped lake. The Mississippi River shows evidence of past oxbow lakes.

Artificial lake—Artificial lakes are those usually formed by the damming of a river. These lakes are usually created to regulate the flow of a seasonal flooding river, and many times offer a constant supply of potable (drinkable) water or generated electricity.

What is **groundwater**?

Groundwater is just what the word implies: water that flows underground. From the time that water formed on the Earth, it has been endlessly cycling, and part of that cycle is the presence of groundwater. It is also an important part of the hydrological cycle.

Water reaches underground mostly as rain or snow. As it falls or melts, the water percolates down through a connection of pore spaces within the soil and rock. Eventually, the water reaches a level at which all pore spaces and cracks in the soil, sand, and rocks are filled with water. This area is called the saturated zone; the top of this zone is called the *groundwater table* (or water table), which can either lie right at the surface or up to hundreds of feet underground. Precipitation percolating through the soil to the water table helps to replenish—or *recharge*—the groundwater system.

Groundwater is found almost everywhere, but some places, such as under the loose soil and rocks of glacial deposits, have more underground water than others. Most groundwater is naturally stored in huge reservoirs called aquifers that are often used by humans for drinking (potable water supplies) and irrigation. (For more information about aquifers, see below). Groundwater also often feeds into lakes, springs, and other surface waters.

Does the **groundwater table fluctuate**?

Yes, the groundwater table fluctuates up and down, depending on many factors. It can rise because of runoff from heavy rains or melting snow, or an extended period of dry weather can lower the water table.

Groundwater levels are also greatly affected by humans. For example, it is estimated that over 50 percent of the United States population depends on groundwater for drinking each day. In addition, groundwater is one of our most important sources of water for irrigation. In some areas in the United States—and around the world—groundwater is used much faster than it is replenished, causing groundwater levels to fall. The development of areas in which structures such as houses, buildings, parking lots, and driveways

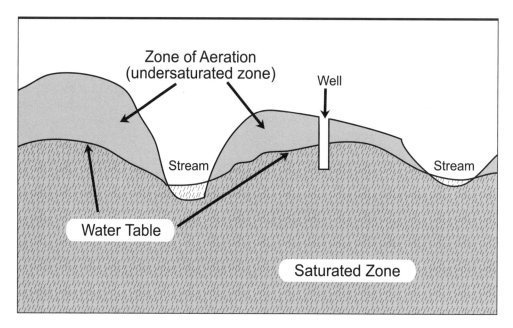

Zone of Aeration
(undersaturated zone)

Well

Stream

Stream

Water Table

Saturated Zone

A simplified depiction of a water table illustrates how water levels can fluctuate underground.

are built can cause the local groundwater level to change. This is because such structures cover the soil, preventing water from refreshing groundwater supplies.

Is **groundwater** important to the formation of underground **caves and caverns**?

Yes, groundwater is very important to the formation of underground caverns and caves. Like rivers, it carries sediment and chemical elements, carving out or dissolving rocks, as well as depositing sediment or minerals. For example, karst topography is greatly affected by groundwater activity, forming some very famous caves and caverns, such as Luray Caverns in Virginia and Carlsbad Caverns in New Mexico. (For more information about karst topography and caves, see "Exploring Caves").

How much **water** is **underground**?

There is a great deal of underground water around the world. It is estimated that less than 3 percent of the world's water is fresh, and around 75 percent of that is frozen in polar ice sheets. Of the balance, about 95 percent is stored as groundwater. That works out to more than two million cubic miles (8,336,364 cubic kilometers) of fresh water stored in the Earth, with 50 percent of that located within a half mile (0.8 kilometer) of the surface.

How are **groundwater supplies used**?

The following list, from the United States Geological Survey (USGS), estimates how groundwater is used in the United States:

What is the dramatic evidence for groundwater under deserts?

Some of the most dramatic evidence for groundwater is found in the world's deserts. Called an *oasis* (plural is oases), these often-vegetated spots in the desert represent areas where the water table is near the surface. Groundwater can easily be extracted from these sites, with the larger oases supporting humans, plants, and local wildlife.

For example, Al Ain (meaning "the spring" in Arabic) in the United Arab Emirates is a fertile oasis city. The city was once a well-known stop on a major south-west route used for centuries by traders. The district known as Buraimi—once part of now-divided Al Ain that belonged to neighboring Oman—is often found on ancient maps as a place in which food and water could reliably be found in an otherwise barren region.

- Groundwater supplies account for 22 percent of all freshwater withdrawals
- Groundwater supplies account for 37 percent of agricultural use (mostly for irrigation)
- 37 percent of the public water supply withdrawals are from groundwater supplies
- 51 percent of all drinking water for the total population comes from underground water
- 99 percent of all drinking water for rural populations comes from groundwater supplies

What is an **aquifer**?

Most groundwater is stored in, and moves slowly through, layers of soil, sand, and rock called an aquifer. This feature is essentially a pool, with the water held underground by an impenetrable layer of hard rock or clay. Aquifers are usually made of sand, gravel, sandstone, or easily fractured rocks like limestone. These materials are called *permeable* because they allow the water to flow through large connected spaces.

Most aquifers have accumulated water over thousands of years or longer, with some water in the desert regions dating back over 40,000 years. In the majority of cases, groundwater either naturally comes to the surface through springs or flows into lakes and streams, following the contours of the land (most groundwater flows toward rivers and streams). Aquifers are also the source of wells for potable water supplies, irrigation, and sundry other water uses. Although the majority of water is brought up with the use of pumps, some wells do not need a pump. These *artesian wells* have natural pressures within that force the water up and out the well, often like a geyser.

What is a **well**?

Related to groundwater, wells are simply holes drilled into an aquifer. A pipe is put into the well, with a pump used at the top to pull the water out of the ground. A screen is usually used to filter out unwanted particles that can plug up the pipe; in addition, most commercial and residential owners of wells use additional filters inside the home to further purify the well water. Water wells come in all sizes, depending on the soil and rock and how much water is being pumped.

Is **pollution** a problem with **groundwater**?

Yes, groundwater pollution is a big global problem. In particular, areas polluted by human activities often lead to polluted groundwater supplies. Underground water can be polluted in many ways, such as leaching of pollutants from landfills, septic tanks, junkyards, chemical spills, mining sites, and leaking underground gas or storage tanks (called *source pollution*). It can also be polluted by less obvious means (called *non-point source pollution*), such as runoff from agricultural fields (that carry fertilizers and pesticides) and parking lots and roads (that carry oil, gas, and sundry other pollutants, including salt from winterized roads).

Why is this such a problem? The main reason is that so many people rely on groundwater for their drinking water. In the United States, over 50 percent of the population, including almost everyone living in rural areas, relies on groundwater supplies. Source pollution is one of the major culprits. For example, it is estimated that over 10 million storage tanks—tanks that can corrode, crack, and leak over time—are buried in the United States alone, and many of these are already leaking. Also, in the United States, there are more than 20,000 hazardous waste sites, with more being "discovered" every year. These sites are particularly insidious, causing groundwater contamination from barrels or other containers that crack and leak. Tragically, sometimes the actual contents of the containers are not known because of bad record keeping, companies moving away, or people abandoning barrels on empty lots.

It is not easy to clean up this type of pollution, either. The pollutants in underground water are difficult to discover and track, because the speed at which groundwater flows depends on the size of the soil, rock pores, and cracks, as well as how the spaces are connected. These conditions often cause the water to change direction, something that can't be easily observed from the surface. The polluted water has to be pumped out, but with little knowledge of where and how deep the pollutants are located in the underground labyrinth of rock, soil, cracks, and crevices, it is nearly impossible to extract all the contaminated water.

Are there any **other problems** associated with **groundwater**?

Yes, there are additional problems with groundwater not necessarily associated with pollution. In particular, when wells pump out too much groundwater, the ground itself can sink. Groundwater depletion has caused the land to subside as much as 4 inches (10.16 centimeters) in parts of the Mojave Desert in southern California, according to the United States Geological Survey (USGS). The subsidence, or sinking, of the land occurred between 1992 and 1999, and it is definitely linked to the pumping out of groundwater and the lowering of the water levels.

Another problem resulting from pulling out too much groundwater is saltwater encroachment, especially along the coastlines. In places such as New York City, potable water supplies are pumped out of the groundwater. But the city needs a great deal of water for its huge population. As the groundwater is pulled out of the ground, the Atlantic Ocean's saltwater is also pulled toward the freshwater, replacing the amount lost. If the amount of pumping is too great or groundwater levels are too low, the saltwater can eventually encroach on the freshwater, causing the pump to extract briny water instead of fresh.

FLOODING WATERS

What was the **number one cause of natural disasters** in the United States during the 20th century?

According to the United States Geological Survey, the number one natural disaster during the 20th century was flooding, especially in terms of lost life and property damage.

An insurance adjuster takes a canoe ride along the streets of Russells Point, Ohio, in July 2003. The area flooded badly after storms dumped more than 15 inches of rain. *AP/Wide World Photos.*

How and where do most floods occur?

Floods are caused by a multitude of naturally occurring and human-induced factors. In general, all floods can be defined as the accumulation of too much water in a short amount of time over a specific area. Most floods occur as water reaches high levels, often overflowing banks along streams, lake shores, or along an ocean coast. Sometimes large tracts of land are submerged, other times not. Most floods occur when there is a prolonged precipitation event, storm surge from a hurricane, or a dam collapse. Damage is greatest where populations are high. Because the majority of the world's population lives along coastlines or near floodplains, flooding affects hundreds of millions of people each year.

Where do floods occur? They occur in almost every corner of the world, from arid deserts to rainforests, and from high to low elevations. The only exceptions are parts of the north and south polar regions.

What is a 100-year flood?

A 100-year flood is merely based on statistics. To a geologist or hydrologist, this means that a major flood occurs along a certain part of a floodplain only once every one hundred years. Or statistically speaking, flooding waters will submerge that area of the floodplain once every hundred years on the average. Based on this information, geologists develop maps to be used by town planners and developers based on 100- (and often 50-) year floods in a certain region.

Why do flash floods cause so much harm?

Flash flooding is considered to be the number one reason for deaths associated with thunderstorms, especially when they occur at night or when people are trapped in automobiles. In most scenarios, a vehicle can stall when a driver decides to drive though a flooded or washed-out road—an often fatal mistake. There is a good reason why people are warned not to drive through flooded streets: even a small stream of water covering the road can wash away the driver and the vehicle. Literally thousands of gallons of water pressing against your car or your body can sweep you away, and this is the main reason most lives are lost in a flood. This is why the National Weather Service recent slogan is "Turn Around, Don't Drown."

But just like everything else in nature, it is not that easy to estimate 100-year floods. In fact, there may be several 100-year floods in a single year. This is why people who feel safe living in the 100-year floodplain should not be complacent when it comes to a flood possibly affecting their property.

Why do **floods** cause so much **destruction**?

Floods are very unpredictable and can occur at any time of the year, in any part of the country, and at any time of the day or night. But what most people don't realize is the power of floodwaters: Flood currents have tremendous destructive potential. The lateral forces alone can cause massive amounts of erosion, undermining bridges and foundations, and cracking or collapsing buildings. Floodwaters also carry a great deal of sediment and debris, coating the inside and outside of a building, which is usually evident as the watery slurry recedes.

What are **flash floods**?

Flash floods are floods that occur extremely quickly—usually within several seconds or hours. They cause streams and rivers to rapidly rise and wash over the land, destroying almost everything in their path. A flood's destructiveness is based on several factors, including rainfall intensity, duration, surface conditions, and slope of the area. Urban areas are the most susceptible to flash floods, since a high percentage of the surface area has been made impervious by streets, roofs, and parking lots, while gutter systems help carry the water even faster in times of intense rainfall. Mountain regions are also prone to flash flooding, mainly because of steep slopes, narrow canyons, and less soil or vegetative resistance, allowing storm runoff to travel rapidly downhill. Deserts and arid regions are not immune to flash floods, either: Many arid regions are well known for their intense thunderstorms. Such storms can produce a great deal of water in a short amount of time, causing a rapid rise in otherwise dry channels.

What are **ice jam floods**?

Ice jams most often occur in the colder climates of the world during a thaw or in early spring. As a totally or partially frozen river rises in response to spring runoff or an early intense storm, the ice is moved and broken apart. As the ice begins to flow, it becomes jammed either by natural obstructions (such as a shallow part of a river) or a human-made structure (such as a bridge pier). If the ice jam is large enough, it will cause the water to back up, causing flooding upstream. If the ice jam then fails, a wall of water can flood the downstream area as it flows out; the jam can also reform further downstream, creating more flooding problems upstream. One other problem with an ice jam is that flowing ice can literally chop down trees, vegetation, and structures along the banks of a flooded river, the thick ice acting like a knife cutting through butter.

What are **storm surge floods**?

Storm surge flooding occurs when a hurricane or an intense low-pressure system approaches a coast. As the hurricane moves, it sucks up a dome of water at its low-pressure region, which is pushed ahead of the storm. Along with this, hurricane winds push the water toward shore. As the excess water from the surge and winds reaches the shoreline, it has nowhere else to go and spills onto the coastline with waves often greater than 20 feet (66 meters) high. The worst-case scenarios occur when the storm surge happens at the same time as high tide, causing an even higher destructive surge. Overall, the storm surge is thought of as the most dangerous part of a hurricane, causing about 9 out of 10 fatalities during such storms. (For more about storm surges, see "Geology and the Oceans").

Have any **floods** occurred because of a **dam break**?

Yes, many floods have occurred because of a dam or levee break, even though these structures are built with flood protection in mind. However, they are designed based on past performance of flooding in a region; thus, an intense storm dropping many inches of rain per hour may produce runoff too great for the dam to hold. If the dam or levee is overtopped and fails (or is washed out), the water released can become a flash flood. Most of the damage from such failures is not only caused by flooding, but also from the power of the sudden water release.

What were some of the **most significant floods** to date in the **United States**?

Significant floods are usually judged based on the number of people killed and/or the cost of damages. But comparing past storms to our more recent ones is sometimes not fair. In the early 20th century, there were not as many people affected by floods, and the price of materials was not as high as it is today. In fact, with an increase in population and the current economic standing, damages for the most intense floods now reach into the billions of dollars.

The following lists just a few significant floods to date in the United States (as compiled by the United States Geological Survey):

Location	Date	Description	Fatalities/Costs
Ohio, statewide	March–April 1913	Excessive regional rains caused regional flooding	467 deaths/$143 million
Galveston, Texas	September 1900	Storm surge flood	6,000 deaths/costs unknown
Northeastern United States	September 1938	Storm surge flood	494 deaths/$306 million (estimated)
Gulf Coast, Mississippi, and Louisiana	August 1969	Hurricane Camille, one of the most deadly and damaging hurricanes in the history of the United States, created a storm surge flood	259 deaths/ $1.4 billion
Buffalo Creek, West Virginia	February 2, 1972	Torrential rainstorm caused a dam to burst	125 deaths/ $60 million
Willow Creek, Oregon	June 14, 1903	Flash flood destroyed the entire city of	225 deaths
Rapid City, South Dakota	June 9–10, 1972	The city experienced 15 inches of rain in 5 hours, causing a flash flood	237 deaths/ $160 million

ICE ENVIRONMENTS

ICY FEATURES

What does the term **periglacial** mean?

Periglacial refers to cold climate regions and the associated non-glacial processes that help create specific landforms. Periglacial regions are characterized by intense frost action; the resulting processes usually involve how water freezes within the ground over time.

What are some common **periglacial features**?

There are numerous geologic features associated with periglacial environments. For example, pingos are large, perennial ice mounds that commonly occur as a circle or oval. They range from about 10 feet (3 meters) or less to over 200 feet (70 meters) in height; they can be from 100 feet (30 meters) to 1,969 feet (600 meters) in diameter. Scientists believe pingos form from heaving, as the pressure from groundwater below or within the upper, icy layer pushes the ice to the surface. Another periglacial feature is thermokarst. In these areas, depressions are formed as ground ice thaws, often creating a hummocky-like terrain.

How does **patterned ground** form in **periglacial regions**?

Patterned ground, or well-organized patterns on the surface caused by freeze-thaw cycles, is also a feature of periglacial environments. They commonly occur in arctic and antarctic regions, such as northern Canada, Siberia, Antarctica, and Alaska.

Although patterned ground can also form in other ways in various climates, such features in periglacial regions form in response to the freeze-thaw cycle of water within the soil. Repeated over the seasons—sometimes for hundreds of years—this process sorts the stones from the soil, leading to patterns such as stone islands, stripes, and polygons. Patterned ground is either sorted (patterns with rocks bordering them) or nonsorted (patterns without accompanying stones).

What is **permafrost**?

In general, permafrost is permanently frozen ground. Usually found in the very northern and southern regions of the world (close to the frozen poles), permafrost is defined on the basis of temperature: soil or rock that remains below 32°F (0°C) throughout the year and, some scientists add, is continuously that cold for two years or longer. Although this is the technical definition, there are many regions considered to be permafrost in which the uppermost layer warms to above 32°F (0°C) for part of the year, usually during the summer.

Although climate is the main factor in the formation of permafrost, the characteristics of the frozen layer are dependent on temperatures at the ground surface. In turn, this depends on several other environmental factors such as vegetation type and density, snow cover, drainage, and soil type. Continuous permafrost occurs throughout an entire region in wide, evenly spread, thick layers of perennially frozen ground—interrupted only by rivers or deep, broad lakes. Discontinuous permafrost is relatively thin, occurring in specific locations with wide unfrozen gaps.

Where is **permafrost found** around the world?

Permafrost encompasses an estimated 20 to 25 percent of the world's land surface. It occurs in more than 82 percent of Alaska, 50 percent of Russia and Canada, 20 percent of China, and probably all of Antarctica. (In some high altitude regions not associated with the polar regions of the world, pockets of sporadic permafrost exist.) In addition, permafrost thickness varies around the world. For example, permafrost in northern Siberia is about 5,250 feet (1,600 meters) thick; in northern Alaska, it is about 2,100 feet (650 meters) thick.

How do human actions affect permafrost and ground ice and vice versa?

Permafrost regions are vulnerable to environmental damage. In the summer, the thin top layer of ice unfreezes in most permafrost areas with the frozen sublayer keeping the water on the surface. This temporarily unfrozen layer is where fragile plants quickly grow during the short season; it is also extremely vulnerable. For example, the tracks from a passing vehicle or even a pathway pounded by many footsteps tears up the insulating vegetation, causing the soil to thaw into scars that may remain for hundreds of years.

In turn, permafrost and ground ice have significant affects on the human development of northern regions. Energy and mining industries need to know how the icy ground will respond to digging. It is also important to know the extent of permafrost and ground ice in areas used for residential and commercial developments, and infrastructures such as roads, railways, pipelines, airfields, and utilities. Because the thawing of ice-rich soils can result in weakening or settling of the soil, engineers and geologists must understand ground conditions before, during, and after they build or dig into such an environment.

What is **ground ice**?

Ground ice occurs when almost all soil moisture freezes at temperatures below 32°F (0°C). It occurs in two main forms: as structure-forming ice (the ice bonding to the surrounding sediments) and as large chunks of more or less pure ice. Ground ice can influence local topography (because of its size and where it forms), geomorphic processes (such as melting water soaking the ground during a warming trend), vegetation, and the response of the landscape to environmental changes (natural or human-made, such as the building of roadways). In permafrost regions, ground ice is one of the most important features of the terrain. In fact, ground ice's major influence is the presence of excess ice—when melted, it is greater than the pore volume (spaces between the soil particles and rock grains) of the surrounding sediments.

What causes frost heave in the soil?

Frost heave is the movement of stones pushed to the surface during the winter months in the colder northern regions of the world. In the winter, due to differences in pressure, the subsurface soil water moves up from below, then freezes near the surface. As it freezes, it forms ice crystals. The amount of ice crystals (and frost heave) often depends on the type of soil, especially the soil pore spaces. For example, clays with smaller pore spaces exert a greater frost heave than coarser-grained soils that have more pore space for the ice crystals to expand.

241

Sunset over tundra near the town of Naryan Mar, Russia. Russian and foreign companies are exploring for crude beneath the conifers and frozen lakes of this isolated, arctic territory, one of the country's most promising frontiers for oil. *AP/Wide World Photos.*

The ice crystals soon grow and coalesce into lenses. As the lenses continue to grow, they displace the soil, heaving the ground upward. This pushes stones (even plant bulbs) upward to the surface. This explains why we find so many rocks in our northern gardens in the spring, and why stone sidewalks, fences, and other structures often need to be mended after a cold winter.

What is an **arctic environment**?

There are many ways to define the arctic environment: Geographically, it is defined as the land area north of the tree line (or the southern-most limit of the tundra). For example, in the Western Hemisphere's northern regions, this includes an area from the Aleutian Islands in the west to Greenland and Labrador in the east. Geologically, the term arctic includes high mountains, plains, exposed bedrock, and lowlands, most of it with little or no soil development. Others characterize arctic zones by the persistence of cold (long winters and short cool summers), permafrost, large seasonal differences in sunlight amounts, and a lack of extensive plant life.

What is an **alpine environment**?

Alpine environments are those areas located above the timberline and are mainly associated with high mountains; some people define it as mountainous areas with trees less than 8 feet (2.4 meters) tall. It is distinguished by certain alpine plants, especially

An example of alpine landscape, this lake is in the Loetschental valley in southwestern Switzerland. Alpine environments are located above timberlines and are mainly associated with high mountains. *AP/Wide World Photos.*

those that have adapted to the harsh effects of wind, cold temperatures, short growing seasons, dry conditions, and ultraviolet radiation. Because of the lack of extensive vegetation and harsh conditions, alpine geology deals mostly with weathering of exposed rock and glacier formation in the alpine mountains.

There are numerous alpine areas around the world that are prime places to study alpine geology. For example, the White Mountain National Forest in New Hampshire contains over 8 square miles (21 square kilometers) of alpine zones—the largest alpine area east of the Rocky Mountains. In Tibet, the alpine zone takes in all areas in which the average July temperature (the warmest month of the year) is not greater than 50°F (10°C). That includes most of the Tibetan plateau, except the southern river valleys of the Indus, Sutlej, and Tsangpo, including the Lhasa Valley.

What is **rime ice**?

Two types of rime ice form in cold regions: The milky-white, brittle ice that forms when super-cooled droplets collide with a freezing surface is called hard rime. The droplets freeze so quickly that they can't splash or spread. Instead, they remain spherical, producing a granular layer of ice with trapped air in between the grains. Hard rime is less dense than glaze ice and does not cling as much, and it generally inflicts only minor damage to vegetation and structures.

243

Opaque, soft rime ice is deposited on cold surfaces by super-cooled cloud droplets, not by droplets falling on an object. Often seen in photos of mountaintop observatories in the winter, soft rime (also called hoar frost) forms where cold clouds move across the landscape, with the ice deposits often pointing in the direction of the wind.

GLACIERS

What is a **glacier**?

Simply put, a glacier is a large body of ice. It is made up of snow that accumulates over many years, eventually compressing into a large, thickened mass of ice that remains in one location over a long time—even throughout the summers. But to be a true glacier, the ice also has to move, as indicated by the presence of an ice flowline (the theoretical path a piece of ice would take from its origin to the edge of the glacier), crevasses (fissures in the ice), and recent geologic evidence. Some glaciers are huge, growing to over 100 miles (161 kilometers) long, while smaller ones can be the size of a football field.

What are the various **types of glaciers**?

Glaciers come in all sizes, shapes, and properties. Geologists categorize them into various types, including those listed as follows:

Mountain glaciers—Most commonly thought of when the word "glacier" is mentioned, these glaciers develop high in mountainous regions of the world. The largest are found in the Andes (South America), the Himalayas (Nepal border), the Canadian Arctic, Alaska, and Antarctica.

Hanging glaciers—Hanging glaciers (or ice aprons) seem to defy gravity, literally hanging off steep mountain slopes. Hanging glaciers cling to steep mountainsides and are usually wider than they are long. This type of glacier is common in the Alps (especially in the higher Alps of Switzerland), and are often associated with avalanches.

Valley glaciers—Valley glaciers were very prevalent during the ice ages (for more about the ice ages, see below). These glaciers usually originate in ice fields or on mountains and appear like long tongues of ice flowing down valleys. If a valley glacier flows into the sea, it is then called a tidewater glacier, many of which calve (drop off huge chunks from the glacier's end), creating small icebergs. And if a valley glacier flows out onto a relatively flat plain forming curving, bulb-like icy lobes, it is called a piedmont glacier. One of the largest piedmont glaciers is in Alaska: the Malaspina Glacier of Alaska, covering over 1,931 square miles (5,000 square kilometers) across the local coastal plain.

Corrie or cirque glaciers—Corrie (also called cirque or cwms) glaciers are larger than niche glaciers (see below) and smaller than valley glaciers, although many valley glaciers have corries as their starting points. These glaciers occupy bedrock hollows in high mountainous regions. Once the ice accumulates to a certain thickness, erosion scours the hollow out even more to form smooth amphitheaters on the valley walls, holding in the corrie. A characteristic feature of corrie glaciers are arêtes (ridges), knife-like glacial ridges that form when two or more corries erode back against each other. The Matterhorn, one of the most climbed peaks

Climbers negotiate a three ladder bridge over the Camp 1 crevasse at 19,500 feet on Mt. Everest in Nepal. *AP/Wide World Photos.*

in the Swiss Alps, has three corries eroded into each other, creating arêtes similar to the spokes of a wheel. This formed the mountain's famous pyramid shape, known as a horn or pyramidal peak. Other good examples of corrie glaciers can be found in Colorado's Rocky Mountains.

Niche glaciers—Niche glaciers look like large snow fields. Actually, they are very small snow areas that also occupy gullies and hollows on a mountain's north-facing slopes (in the northern hemisphere). They are not as extensive as the typical glacier, but glacial ice does accumulate over the seasons, and they often have crevasses along the surface. Niche glaciers are often found between larger corrie glaciers, and if conditions are right, they can eventually grow into a corrie glacier.

Where are **glaciers located**?

Glaciers are present on every continent except Australia. They are found in areas of high or low precipitation and in various temperature regimes. Glaciers are most commonly found above the snow line, where large amounts of snow in winter and cool temperatures in summer occur. This allows more snow to accumulate than is melted in any given year. Glaciers need snow to survive and thrive. For example, although very cold, Siberia has almost no glaciers due to its relative lack of snowfall.

The best places for glaciers are in mountains or the polar regions, areas in which there is historically a great deal of snow. In equatorial regions, glaciers are found at high altitudes (often called temperate glaciers); they are present at lower altitudes at the polar ice caps, because of differences in the snow line altitude. For example, in Africa, the snow line starts at 16,732 feet (5,100 meters), while in Antarctica it is at sea level.

How much of the **Earth's surface** is covered by **glacial ice**?

Currently, approximately 10 percent of the Earth's total land area is covered by glaciers, with the majority located in the polar regions of the world, such as Greenland and Antarctica. That translates to almost 5.8 million square miles (15 million square kilometers) of our planet covered by glaciers, an area almost equivalent to the entire continent of South America. The following chart from the National Snow and Ice Data Center in Boulder, Colorado, breaks down this coverage (note: because this list does not include small glaciated areas or smaller polar islands, the numbers do not add up to 5.8 square miles [15 million square kilometers]):

Approximate Location	(miles²/kilometers²)
Antarctica (without ice rises and ice shelves)	4,619,712/11,965,000
Greenland	688,806/1,784,000
Canada	77,220/200,000
Central Asia	42,085/109,000
Russia	31,660/82,000
United States (including Alaska)	28,958/75,000
China and Tibet	12,741/33,000
South America	9,653/25,000
Iceland	4,348/11,260
Scandinavia	1,123/2,909
Alps	1,119/2,900
New Zealand	447/1,159
Mexico	4/11
Africa	3.9/10
Indonesia	2.9/7.5

How does **glacial ice form**?

Glaciers form in areas in which snow remains all year and enough snow accumulates over time to eventually transform into ice. Year after year, layers of new snow cover and compress the past layers. This causes the buried snow to recrystallize, forming grains similar in shape and size to sugar. As the pressure from accumulating layers increases, the buried snow continues to compact and increase in density. The grains grow larger and the air pockets within the snow get smaller. It takes about two winters for compressed snow to turn into the intermediate state between snow and glacier ice, and several more years to transform into glacier ice. At this point, ice crystals are so compressed that any air pockets between them are extremely small. After a hundred years or more, the glacier ice crystals can reach several inches in length.

What are the **three major areas** of a **glacier**?

The following lists the three major areas of a glacier—all of which keep the ice in balance:

Accumulation area—The accumulation area is the upper section of a glacier, where new snow accumulation exceeds the amount of melting.

Ablation area—The ablation area is the lower section where melting or evaporation of ice and snow exceeds the amount of snowfall.

Equilibrium line—The boundary that separates the accumulation and ablation areas is called the equilibrium line.

How do geologists **measure the health** of a **glacier**?

Geologists determine the health of a glacier by measuring its mass balance, or the net gain or loss of snow and ice in a given year, usually expressed in terms of water gain or loss. Each year a glacier may have a positive mass balance, with a net gain of new snow; or it may have a negative mass balance, with a net loss of snow and ice; or the glacier may remain in equilibrium.

To find the mass balance, geologists measure the area of each glacier and observe the amount of accumulation and ablation relative to preset stakes in the ice. After taking density measurements, the amount of water added or lost can be calculated. The equilibrium line, where the gain in mass equals the loss, is usually visible on a glacier as the boundary between the current winter's snow and the older snow or ice surface.

The altitude of the equilibrium line changes every year in response to the glacier's mass balance. If the mass balance is negative for a number of years, the glacier may retreat—meaning the snout (terminus) is melting faster than the ice is moving downwards. A positive mass balance over a number of years may mean the glacier is advancing in response to the increase in ice thickness.

What is **firn**?

Firn is old snow in the intermediate state between snow and glacial ice that is recrystallized into a denser substance by pressure from the overlying snow layers. Under this pressure, the individual crystals glide and readjust the spaces in between. Bonding occurs where the crystals make contact and the air in between is squeezed to the surface or into bubbles. Firn is relatively light in weight. To compare, the density (or the ratio of the mass of an object to its volume) of fresh water is 1.0; snow averages about 0.1; glacial ice about 0.89; and firn about 0.55, or about half the density of fresh water.

An aerial view shows an avalanche crashing through the forest near the Swiss ski resort of Evolene, located in a valley of the southern Swiss Alps, on February 22, 1999. Two other avalanches hit the ski resort, killing two people, and leaving several others missing. *AP/Wide World Photos.*

What are **ice falls**?

Ice falls are jumbled slopes of glacial ice that fall when a glacier moves over a sharp drop in ground surface—often at overhangs. For example, as ice hangs above a valley, it becomes virtually stretched as it moves down the slope. This causes the ice to fracture into huge, unstable blocks that frequently break away with explosive cracking and crashing noises. The ice tumbles to the ground, shattering as it strikes the surface and becoming a hazard for those crossing it or standing below the glacier.

Are there different **types of crevasses**?

Yes, there are several types of crevasses, the large fissures that form in glacial ice. These form in areas where the ice speeds change, such as at valley bends and icefalls, with the surface patterns revealing a great deal about the glacier's flow.

In general, there are two types of crevasses. Marginal crevasses are the rows of parallel fissures along the edge of the ice. They form as friction from the surrounding valley walls slows or stops glacial movement along the ice edges. Transverse crevasses extend the width of a glacier and are found where there is any active movement. Still another is a bergschrund crevasse, which separates flowing ice from stagnant ice at the top of a glacier, or the area where the head of the glacial ice separates from the rock.

What is the world's thickest temperate glacier?

To date, the world's thickest—and perhaps deepest—temperate glacier (glaciers closer to the mid-latitudes and at high altitudes) is Alaska's Taku Glacier. It has been discovered to be an amazing 5,400 feet (1,645 meters) thick, about four times thicker than previously thought. Currently, the only glaciers in the world known to be thicker than Taku are the polar glaciers in Greenland and Antarctica. Taku is the largest glacier, draining the Juneau Ice Field in southeast Alaska, with more than half its length thought to rest below sea level.

What are some **dangers from glaciers**?

Glaciers can certainly be dangerous, particularly those that lie close to population centers. The following lists some of the greatest dangers from glaciers:

Avalanches—Avalanches are often glacial hazards, posing an immediate and deadly threat to anyone below. For example, numerous avalanches have occurred over the centuries in the Swiss Alps. In 1965, a huge mass of ice from the Allalingletscher glacier broke off without warning. A camp being used for the construction of a new hydro-electric plant nearby was buried, killing 88 workers.

Flooding—Floods can also be a danger, as lakes form on top of glaciers during the warm season. In 1941, a glacial lake suddenly flowed out, flooding the town of Huaraz, Peru, and killing 6,000 people. A glacial outburst flood—a sudden burst of flooding water from a glacier—is also a danger. The water may originate from sub-glacial lakes, cavities in the glacier itself, or from lakes created when the glacier dams a side valley. For example, such outbursts commonly occur in Iceland, where they are called jokulhlaups.

Icebergs—If a glacier's snout (terminus) reaches the sea, ice chunks can break off (calve), becoming icebergs. These are definitely a danger in busy shipping lanes, particularly the ones along the coasts of Newfoundland and Greenland. For example, in April 1912, the luxury liner *Titanic* struck an iceberg in the North Atlantic and sank, costing the lives of 1,503 passengers.

Crevasses—Crevasses make travel across a glacier extremely dangerous. Snow often covers these deep cracks, rendering them invisible until a traveler falls in. For example, on Mount Hood's Bergschrund glacier in Oregon, a crevasse 800 feet (244 meters) below the 11,237-foot (3,425-meter) summit has been the site of several hiker fatalities, especially as the gap opens wider in the summer.

Why do **glaciers move**?

Glaciers move because of pressure: When compressed glacial ice reaches a critical threshold thickness of about 59 feet (18 meters), the heavy mass starts to deform and move. Gravity then causes the glacier to flow down mountain valleys, sometimes fanning out across plains, or even out to sea.

Glaciers are essentially very slow-moving rivers of ice. Like rivers, glaciers can change their flow rates. In general, the flow of glacial ice is faster down the center than at the margins, and quicker at the top than at the bottom. This is because of the friction created as the ice moves along the ground. In addition, a glacier's speed depends on the thickness of the ice and on the angle of its surface slope. Speeds can also change whenever the geometry changes. For example, large amounts of seasonal snow can fall, creating bulges of thicker ice that flow many times faster than the glacier's normal velocity.

What are **glacial advances and retreats**?

Glacial advance is the movement of the terminus (snout) forward, or downslope. This occurs when the snow accumulation rate and ice flow are greater than ablation (loss) due to melting or evaporation. During an advance, the glacial snout has a convex shape. Glacial retreat is the movement of the terminus backward, or upslope. This occurs when the glacier is ablating at a rate faster than its movement downslope. A retreating glacier usually has a concave shape.

Glaciers may periodically advance or retreat depending on their mass balance. It is important to note that this movement is only in reference to the position of the terminus of the glacier. Even during a retreat, the entire glacier is still moving downslope like a conveyor belt, with ice continually deforming and flowing.

Advances and retreats normally occur very slowly, and thus are noticeable only over long periods of time. But sometimes glaciers may surge, moving forward several feet per day over a period of weeks or months. For example, in 1986, Alaska's Hubbard Glacier began to surge across the mouth of the Russell Fiord at a rate of 33 feet (10 meters) per day; it took the glacier only two months to dam the nearby fjord and create a lake. Glacial retreats may also happen rapidly, with the movement noticeable over a few months or years. For example, Glacier Bay, Alaska, contains both tidewater and non-tidewater glaciers, many of which have rapidly retreated at various times since the late 1700s.

GLACIAL FEATURES

Why are **crevasses and moraines** important to geologists studying glaciers?

Created by the movement of the ice, crevasses and moraines are used by geologists to determine whether a mass of ice is truly a glacier. The forces created when a glacier moves cause it to deform, creating large, open crevasses. These form in areas where the ice speeds change, such as at valley bends and icefalls.

Moraines are another indication of a glacier. The tremendous weight grinds up large amounts of soil and rock where ice is in contact with the ground. As the glacier moves, it pushes or carries this debris, creating moraines in surrounding areas.

What are the **types of glacial moraines**?

A glacier can have lateral, medial, ground, and terminal moraines, each one based on its location on the ice. A glacier picks up or pushes material, forming the moraines; as the glacier retreats, the ice melts away from the moraine. If the glacial meltwater does not carry the material away, it leaves a tell-tale long, narrow ridge behind. Such moraines are very evident in the Finger Lakes region of New York state, which are left-overs from the last Ice Age. The following lists the major moraine types:

Lateral moraine—The lateral moraine is visible as a dark band at the side margins of a glacier. It consists of piles of loose, unsorted rocks pushed there by the moving ice or dumped from the glacier's rounded surface.

Medial moraine—Medial moraines run down the middle of a glacier and appear as dark streaks of rock. They are created when two glaciers that have lateral moraines join together.

Ground moraine—Ground moraines are the material underneath a glacier.

Terminal moraine—Terminal moraines are found at the snout, or terminus, of a glacier, and consist of piles of loose, unconsolidated rock. Again, these rocks were either pushed there by the motion of the ice or dumped from the glacier's rounded surface.

What are some **erosional features** from glaciers?

Erosion occurs as the heavy mass of glacial ice moves against the land, grinding and tearing everything in its path over a period of hundreds or thousands of years. The following are some of the erosional features created by glaciers:

Glaciated valleys—The most readily visible evidence of glacial erosion are glaciated valleys. Found all over the world, they are trough-shaped and often have steep vertical cliffs where the ice has removed the sides of mountains. One of the best examples of this phenomenon can be found in Yosemite National Park, where vertical walls surround deep valleys.

Hanging valley—Hanging valleys are most often associated with valley glaciers, joining the main valley along its sides. They are the result of different rates of erosion between the main valley and the side valleys. In particular, side valley erosion is slower than the main valley. Over time, the main valley

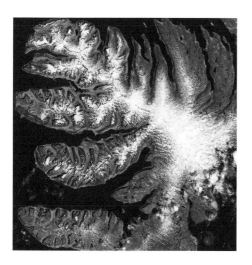
A photo of the West Fjords, a series of peninsulas in northwestern Iceland. They represent less than one-eighth of the country's land area, but their jagged perimeter accounts for more than half of the country's total coastline. *AP/Wide World Photos.*

deepens faster, leaving the side valley floors high above, or hanging on the edge, often creating high waterfalls. They are typically found in the Alps in Europe and the Southern Alps of New Zealand. Ice Age ice sheet-associated hanging valleys are found in the gorges around Ithaca, New York, part of the Finger Lakes region.

Fjords—Fjords are long, narrow coastal valleys that were carved out by glaciers, with steep sides and rounded bottoms. In this case, erosion occurred on land that was below sea level. After the glaciers retreated, the water returned and covered the valley floor. The fjords of Norway are a well known example of this phenomenon.

Cirques—Cirques are another landform created by glacial erosion. Rounded hollows shaped like shallow bowls, they are created when glaciers erode backwards into the mountainside. When the back walls of two cirque (or corrie) glaciers meet on a mountain, a jagged, narrow ridge called an arête is the result.

Horn—A horn is a mountain with several cirque glaciers that has been eroded until all that is left is a steep, pointed peak with arêtes leading up to the top. The Matterhorn in Switzerland is an example of a horn.

Roche moutonnée—A small, asymmetrically shaped hill formed by glacial erosion is called a roche moutonnée. Characteristically, the upper sides of the hill are rounded and smoothed, while the lower sides have been broken and roughened by the glacier.

What are **striations**?

Striations are scratches etched into the underlying rock on the bed of a glacier. These scratches are caused by coarse rock particles gouging into the bedrock under the glacier's extreme pressure and movement. Rocks and stones from under or around a glacier may often exhibit striations. Sometimes, when fine-grained debris is present, the underlying bedrock smoothes to a lustrous finish called glacial polish.

What are some **common glacial deposits**?

The following are some examples of glacial deposits:

Are glaciers economically important?

Yes, glaciers are economically important for a number of reasons. In some regions of the world, the water from melting glaciers is a primary source of drinking water. In fact, scientists are attempting to create artificial glaciers in Japan, a region that lacks glaciers in spite of a great deal of snow in certain areas. The hope is to create a source of drinking water to combat Japan's frequent droughts.

The water from glaciers is also often used to irrigate crops. For hundreds of years, farmers in Switzerland's Rhone Valley have diverted meltwater from glaciers to their fields. There is even an effort to dam glacial meltwater to generate hydroelectric power in New Zealand, Canada, Norway, and the European Alps. And, of course, there is the tourist impact: Many glacial areas draw huge numbers of visitors annually to see and experience these giant rivers of flowing ice.

Erratics—As a glacier melts away or retreats, it leaves behind large rocks that are not "normal" for the local area. Called erratics, these boulders are picked up and moved from their original location by the glacier. As the ice retreats, the erratics are left behind.

Kames—Kames are small, steep-sided mounds of gravel and soil adjacent to a glacier. They form as streams flowing from a glacier carry rock and soil debris, subsequently depositing this material over many years.

Kettle lake—If a piece of ice breaks off the glacier and is buried by moraine deposits or glacial till, the ice eventually melts, leaving behind a small depression filled with water called a kettle lake. These normally range from small- to medium-sized ponds, but are still referred to as lakes.

What is the difference between **glacial drift, till, and tillite**?

Glacial drift is the name given to loose and unsorted rock debris from glacial action. This material may be deposited by meltwater streams, formed at the ice edges, or dropped in place by melting ice.

Till is considered to be sediment, the fragmented debris deposited directly by a glacier, usually at the periphery. Also called glacial till, this material is an unsorted, unstratified mixture of sand, clay, gravel, and boulders (also called "glacial rubble"). It has not been reworked by meltwater, and the resulting soil can be excellent for farming. Finally, till pieces cemented into rock over time—usually by excessive pressure—are called tillite.

253

An image of the northern ice field of Mount Kilimanjaro in Tanzania, whose glaciers are rapidly retreating. Recent studies show that Kilimanjaro has lost 80 percent of its ice fields in the past century. *AP/Wide World Photos.*

What is **glacial flour**?

Glacial flour is a fine-grained sediment resulting from rock abrasion at the glacier bed. Carried by glacial rivers, it turns the waters of a lake brown, gray, or blue, depending on the original rock type.

What are **drumlins**?

Drumlins are sedimentary formations shaped like long teardrops and composed of glacial till. Hundreds to thousands can often be found in a small area. Geologists believe these phenomena were created subglacially as ice sheets moved across the land during the numerous ice ages. No one knows for sure, but the features may have formed by erosion or deposition of sediment by meltwater, directly as the glaciers scraped up sediment from the ground, or a combination of these two processes. Because they were created by glacial movement, drumlins associated with a particular glacier all face the same direction and run parallel to the original flow.

Why are **glaciers** important to **climate studies**?

Glaciers are important to climate studies in a number of ways. For instance, by their very nature, glaciers are long-lived, meaning they can be anywhere from several thousands to millions of years old. That means the masses of ice have been keeping a continuous record of the climate for all that time.

Because of this extensive record, geologists are studying the past climate of our planet by drilling into and extracting ice cores from glaciers around the world. For example, cores have been extracted from glaciers in Asia, Antarctica, Europe, Greenland, Peru, and Canada. Once the cores are in the laboratory, they are carefully analyzed. One of the most important pieces of information comes from tiny bubbles contained in the glacial ice, in which samples of the atmosphere from long ago are preserved. After analyzing the cores, geologists can determine past atmospheric conditions, types of vegetation, and temperature variations from different parts of the world over thousands and millions of years. The climate of the past can be reconstructed, including why and how the climate has changed.

Another area in which glaciers are important to climate studies has to do with global warming. Glaciers, for all their impressive size and weight, are very sensitive to temperature differences that occur during periods of climate change. Geologists are studying glaciers to determine how human activity affects climate and how much warming naturally occurs between ice ages. They have found that the vast majority of the world's glaciers have been retreating extremely rapidly since the early 20th century. Some glaciers, including ice caps and even an ice shelf, have disappeared entirely, while others may be gone within decades.

Some geologists believe this worldwide glacial retreat is due to the Industrial Revolution (starting around 1760). Human activity has been introducing carbon dioxide and other greenhouse gases into the atmosphere since that time. In the last century, the rate of greenhouse gases emissions has risen even more dramatically—due in part to power requirements and transportation needs—increasing global temperatures. Geologists are continuing to study glaciers to determine whether human activities are the entire culprit in this glacial retreat, whether it is part of a natural cycle, or a combination of both.

ICE SHEETS

What is an **ice sheet**?

An ice sheet is an extremely large, continental mass of glacial ice covering the surrounding terrain for more than 12 million acres (50,000 square kilometers). Ice sheets comprise almost 99 percent of all the glacial ice on Earth; the largest and most well-known ice sheets cover Greenland and Antarctica. In general, ice sheets completely cover the underlying land and are composed of ice domes, ice streams, and outlet glaciers.

What are the **two types of ice sheets**?

The two types of ice sheets are terrestrial- (or land-) and marine-based. Terrestrial-based ice sheets imply that most of the base lies above sea level. For example, the East

Antarctic Ice Sheet is terrestrial-based; without the weight of this ice, the underlying land would rise, with East Antarctica forming a single large landmass above sea level. Marine-based ice sheets imply that most of the ice mass lies below sea level. For example, the West Antarctic Ice Sheet is marine-based, with some spots at least 1.24 miles (2 kilometers) below sea level. If this ice sheet was removed, much of the bedrock would remain below sea level, with large islands and seas forming in its place.

What are the world's **largest ice sheets**?

During the ice ages, ice sheets grew to enormous sizes across North America, northern Europe, and Asia. Currently, ice sheets are not even half as large. The largest are in Antarctica and Greenland, covering 10 percent of the Earth's land area and containing 77 percent of the world's freshwater.

The Antarctic ice sheet covers almost all of the continent, except the protruding Transantarctic Mountains. It is divided into two parts: the East Antarctic Ice Sheet, measuring about 3,844,000 square miles (10 million square kilometers) and the West Antarctic Ice Sheet at about 768,800 square miles (2 million square kilometers). In some spots, the East Antarctic Ice Sheet measures approximately 2 miles (almost 4 kilometers) thick. Currently, many scientists believe these two ice sheets are important in determining environmental changes. The East Antarctic Ice Sheet is fairly stable and responds to environmental changes slowly; in contrast, the West Antarctic Ice Sheet might be capable of rapid changes, which means a difference might be indicative of a rapid change in the global environment.

What is the **difference** between a **glacier and ice sheet**?

The obvious difference between a glacier and ice sheet is size. Another difference is ice sheets rest on land that is relatively flat compared to their thickness; glaciers are bound within channels (such as surrounding mountain passes), creating a variation in ice thickness.

What is an **ice shelf**?

Ice shelves occur when ice sheets extend over the ocean, allowing the ice to literally float on the water. These floating ice masses are attached to the coast by at least one edge (the bases of other types of glaciers are usually in contact with the ground). Where the ice sheet is no longer attached to the ground and begins to float is called the grounding line. Ice shelves measure anywhere from a few yards to thousands of feet thick.

Antarctica is a continent surrounded by ice shelves. The Ross Ice Shelf, where the largest United States station is located at McMurdo, is the largest ice shelf known and is close to the size of Texas. Another Antarctic ice shelf is the Larsen Ice Shelf. It has been

Why do some icebergs look so blue in the waters surrounding Greenland and Antarctica?

Ice is blue for much the same reason that ocean water is blue: it absorbs more of the red than blue part of the visible light spectrum. The ice near the bottom of a glacier or ice sheet is compressed by the ice above it. Along with fewer air bubbles, the compacted ice crystals realign in such a way that when light shines through the ice, all the colors except for blue are absorbed. Icebergs that appear green instead of blue may contain more suspended sediments, algae, or air bubbles.

retreating since the spring of 1998, dropping large icebergs that have changed the local ice regime. Some scientists believe this might be an indication of climate change.

What happens when **pieces of an ice shelf** break off (or calve)?

The ice shelf pieces that break off (or calve) usually form icebergs, huge, often flat-topped chunks of ice that fall into the surrounding bays, seas, and ocean. They usually are cleaner than icebergs from valley glaciers.

Ships—both commercial and cruise—traveling in northern and southern shipping lanes are particularly affected by icebergs from ice shelves. And these icebergs cause problems in other ways. For example, an iceberg measuring 25 miles (40 kilometers) wide and 50 miles (80 kilometers) long recently broke away from Antarctica's Larsen Ice Shelf and has already changed the local ice regime in the surrounding area. It is being monitored by satellites and aircraft to determine the iceberg's possible threat to southern shipping lanes.

Are **ice domes** associated with ice sheets?

Yes, ice domes are areas of accumulation that slowly move radially outward on an ice sheet. They are almost symmetrical in outline and dome-shaped in cross-section. An ice sheet might contain several ice domes, all of which are usually topographic high spots. The largest ice domes are found in Antarctica. For example, the East Antarctic ice sheet contains Siple Dome, a rise in the ice sheet near the Ross Ice Shelf; and Taylor Dome is a rise just inland of the Transantarctic Mountains (it also provides ice to outlet glaciers entering Taylor Valley and McMurdo Sound).

What are **ice caps and icefields**?

Ice caps are smaller than ice sheets, usually forming in polar and sub-polar regions at relatively flat, high elevations. Similar to ice sheets, ice caps completely cover the

257

What would happen if all the glaciers and ice sheets melted today?

If all the glaciers and ice sheet melted today, the results would be catastrophic. After all, such an amount of fresh water ice would cause sea levels to rise nearly 262 feet (80 meters), inundating coastal areas where almost 50 percent of the world's population resides.

Some scientists estimate that the sea level is currently rising an average of about 0.08 inches (2 millimeters) per year. Even with the most accurate measurements, it is not known whether the present ice sheets are shrinking or growing. About 0.31 inches (8 millimeters) of water from the entire Earth's ocean surface accumulates as snow on Greenland and Antarctica. But scientists still do not know how much water is returned to the oceans in the form of meltwater runoff and icebergs from ice sheets. This is because no one has directly measured how ice sheet volume changes over time, a difficult task in such a harsh environment.

underlying land and are composed of ice domes, lobes, and outlet glaciers. For example, Iceland possesses an ice cap. Icefields are similar to ice caps, except they are even smaller. Their flow is influenced by the area's underlying topography.

Has a **huge lake** been found **beneath the Antarctic** ice sheet?

Yes, scientists recently discovered a huge lake under the Antarctic ice sheet. Named Lake Vostok, it is the largest of more than 70 subglacial lakes that lie thousands of feet under the continental ice sheet's surface. Scientists used radar and artificially generated seismic waves to discover a vast, warm lake. It lies beneath Russia's Vostok Station, and measures 155 miles (250 kilometers) long by 25 miles (40 kilometers) wide and 1,312 feet (400 meters) deep, comparable in size and depth to Lake Ontario, one of the Great Lakes.

Although it lies under nearly 2.5 miles (4 kilometers) of solid ice, it is warm enough to remain liquid. This is a mystery since it is in one of the Earth's coldest regions. One theory is that geothermal heat from the Earth (from radioactive decay) keeps the lake from freezing. Two other theories involve the ice sheet: It either acts as a thermal blanket, protecting the lake from the cold surface temperatures—or the pressure of the ice sheet keeps the lake liquid.

Either way, scientists are interested to see if Lake Vostok could possibly hold microbial life. If so, the life would be ancient, as the samples taken so far indicate the lake has been under the icecap for between 500,000 and a million years. But the lake is not easy to study: Vostok Station is located in one of the world's most inaccessible places, near the South Geomagnetic Pole at the center of the East Antarctic Ice Sheet. The station is 11,484 feet (3.5 kilometers) above sea level and recorded the coldest temperature ever on Earth: –128.6°F (–89.2°C) on July 21, 1983.

ICE AGES

What is an **ice age**?

Geologists define an ice age as a period when more of the Earth's surface was covered by larger ice sheets than in modern times, or when cool temperatures endured for extended periods of time, allowing the polar ice to advance into lower, more temperate latitudes. The last ice age event is referred to as the "Ice Age" or the "Great Ice Age"—both usually capitalized—a time when ice covered nearly 32 percent of the land and 30 percent of the oceans.

Who **first proposed** the idea of **ice ages**?

Several scientists over the centuries came close to proposing the idea of ice ages. Scottish naturalist James Hutton (1726–1797) observed strangely shaped glacial boulders (erratics) near Geneva, Switzerland. Based on this, he published his theory in 1795—that alpine glaciers were more extensive in the past. In 1824, Jens Esmark (1763–1839) proposed that past glaciation had occurred on a much larger continental scale.

But the most persuasive argument for ice ages came in 1837, when Swiss-American geologist Louis Agassiz (1807–1873) gave his now-famous speech on past widespread ice age conditions. He proposed that nearly all of northern Europe and Britain had once been covered by ice, and he subsequently found evidence for his theory in New England.

Others eventually uncovered additional evidence. In 1839, United States geologist Timothy Conrad (1803–1877) discovered evidence of polished rocks, striations, and erratic boulders in western New York, supporting Agassiz's theory that Ice Age glaciation was worldwide. In 1842, the first attempt to explain the ice ages using an astronomical connection was made by French scientist Joseph Adhemar (1797–1862). He proposed that the ice ages were the result of the 22,000-year precession of the equinox, a natural movement of the Earth's axis that causes the seasons to switch over thousands of years.

What **causes ice ages**?

No one knows why ice ages occur, but there are several theories. One possibility is that the sun's energy varies in intensity over time. Each time there is a decrease in activity, an ice age may occur as the Earth cools. Another possibility is an increase in dust in the atmosphere, either from volcanoes or a large meteorite impact. The debris from either event would reflect more of the sun's light into space (albedo), cooling down the atmosphere and causing more snow and ice to form. This would also further increase the world's albedo, as even more sunlight would reflect off the ice and snow. However, this theory has a problem, as many of the other theories do, in that it doesn't explain what causes the ice sheets to retreat.

What were the **major glacial ice ages** over time?

Evidence from the geologic record shows that ice ages have occurred relatively few times in Earth's history, with the first known large-scale ice age taking place approximately 2.3 billion years ago. (In the last about 670 million years, ice ages have occurred less than 1 percent of the time.) Geologists believe there have been five major ice ages over geologic time. The following lists the occurrences:

- 1.7 to 2.3 billion years ago (Huronian Era, Precambrian)
- 670 million years ago (Proterozoic Era, Precambrian)
- 420 million years ago (Paleozoic Era, between the Ordovician and Silurian Periods)
- 290 million years ago (Paleozoic Era, between late Carboniferous and early Permian Periods)
- 1.7 million years ago (Cenozoic Era, Quaternary Period, Pleistocene epoch)

How did the **last Ice Age** begin and end?

Approximately 1.7 million years ago (the beginning of the Quaternary Period, Pleistocene epoch) geologists believe the plains of North America cooled. As a result, large ice sheets began to advance south from the Hudson Bay area of Canada and eastward from the Rocky Mountains. These ice sheets advanced and retreated many times toward the end of the Pleistocene epoch in intervals lasting from 10,000 to 100,000 years. This most recent ice age ended about 10,000 years ago, when the ice retreated to its present polar positions. Currently, the Earth is nearing the end of an interglacial (warmer) period, meaning that another ice age might be due in a few thousand years.

What are **major and minor ice age periods**?

Of course, not all scientists agree on how to divide the periodic ice ages, with some calling for a more strict division between the times and temperatures. For example, some scientists believe a major ice age period should be defined as lasting about 100,000 years, with a 9°F (5°C) decrease in temperature between glacial and interglacial periods; a minor ice age period, lasting about 12,000 years with a 5°F (2.8°C) decrease in temperature; and a smaller ice age period, lasting about 1,000 years with a 3°F (1.7°C) decrease in temperature.

What were the **glacial and interglacial periods** in the last Ice Age?

In the last Ice Age (and in all ice ages), there were cycles of glacial (when ice covered the land) and interglacial (relatively warmer temperatures) times. Corresponding with these times, the glaciers advanced or retreated. Scientists believe the last Ice Age—also called the Pleistocene Ice Age—had eight cycles. The following lists these stages for North America (stage names for northern and central Europe differ). All dates are approximate:

Approximate Years Ago	North American Stage
75,000–10,000	Wisconsin*
120,000–75,000	Sangomonian (interglacial)
170,000–120,000	Illinoian
230,000–170,000	Yarmouth (interglacial)
480,000–230,000	Kansan
600,000–480,000	Aftonian (interglacial)
800,000–600,000	Nebraskan
1,600,000–800,000	Pre-Nebraskan

* Note: During the Wisconsin glacial stage, an interstadial period occurred—a time not warm or prolonged enough to be called an interglacial period.

What is the **Channeled Scablands**?

The Channeled Scablands is a region in eastern Washington state carved by the sudden release of water from glacial Lake Missoula, an ancient, ice-dammed lake that was created as the last Ice Age ice sheet retreated. Lake Missoula covered a large area of western Montana. A glacial dam in northern Idaho repeatedly failed over time, releasing the lake water in huge discharges. The resulting floods were the greatest documented in the geologic record, spilling across the northern rim of the plateau and eroding great flood channels, waterfalls, and sundry other river features. The big difference between these and regular river features was size. For example, the Dry Falls is a huge waterfall 3.4 miles (5.5 kilometers) wide and 394 feet (120 meters) high, with giant ripple marks measuring 16 feet (5 meters) high and spaced 328 feet (100 meters) apart.

What was **Lake Bonneville**?

Lake Bonneville was another ancient, huge glacial lake that covered northwest Utah during the end of the last Ice Age. After the last Ice Age ice sheet retreated, the lake shrank, leaving the Great Salt Lake as a remnant.

What are the **Finger Lakes**?

The Finger Lakes are long troughs of lakes located in central New York state; they were all formed by glacial action from the last Ice Age. They include eleven long, narrow, roughly parallel lakes oriented north-south. The southern ends of the lakes all have high walls cut by steep gorges. Seneca and Cayuga Lakes are among the deepest in North America and have bottoms below sea level. Cayuga Lake, the longest and the second deepest of the Finger Lakes, measures 38.1 miles (61 kilometers) long and 435 feet (131 meters) at its deepest spot (with 53 feet [16 meters] below sea level). The carved rock of the lake—now filled with close to 1,000 feet (305 meters) of glacial sediment—is actually more than twice as deep.

These lakes formed over the past 2 million years, originating as a series of north-flowing rivers. As the ice sheet advanced from the Hudson Bay area, it gouged out the old stream valleys. With the many advances and retreats over hundreds of thousands of years, ice deepened the valleys into troughs. Around 19,000 years ago, the climate warmed, and the glacier began to retreat, adding meltwater to the lakes.

How much of the **Earth was covered in ice** during the various ice ages?

Because of extensive erosion, it is difficult to determine the extent of ice sheets during the various ice ages. But scientists do know some things about the last ice age—the Pleistocene Ice Age—a time when up to 10 percent of the Earth was covered (although not simultaneously) by often miles-high ice. At their greatest extent, the Northern Hemisphere glaciers and ice sheets covered most of Canada, all of New England, much of the upper Midwest, large areas of Alaska, most of Greenland, Iceland, Svalbard, and other arctic islands, Scandinavia, much of Great Britain and Ireland, and the northwestern part of the former Soviet Union. In the Southern Hemisphere, glaciers were much smaller, with the main effects being cooler and much drier weather.

During the last stage of the Ice Age (the Wisconsin stage in the United States), ice sheets covered parts of Eurasia and much of North America, extending as far south as Pennsylvania. As the climate warmed up, scientists estimate that sea level rose about 410 feet (125 meters), an average rate of an inch (2.5 centimeters) per year for roughly 5,000 years. Interestingly, although most of the huge northern ice sheets melted, the Antarctic ice sheet decreased by only 10 percent.

What was the **"Little Ice Age"**?

The "Little Ice Age" was an interval of relative cold, beginning about 1450 and lasting until about 1890 (the coldest periods of 1450 and 1700 are often divided into the two Little Ice Ages). It occurred during the current warm, interglacial period, but is not considered a full glacial episode, since the high latitudes of the Northern Hemisphere landmasses remained largely free of permanent ice cover.

Even so, much of the world experienced cooler temperatures during this time of at least 2°F (1°C) lower worldwide average surface temperatures. It was a time of renewed glacial advance in Europe, Asia, and North America, with sea ice causing havoc in the colonies of Greenland and Iceland. In England, the Thames River froze; in France, bishops tried to halt glacial advances with prayer. Several historians also believe the low temperatures caused social conflict and poor food production. Thus, this may have been partially responsible for war and hunger during that time.

Just like the major ice ages, no one really knows what caused the Little Ice Age. Some scientists attribute the cooling down to volcanic eruptions, variations in the sun's energy output, changes in the ocean circulation, changes in the Earth's orbit, the wobbling of the Earth's axis, or even our planet's passage through clouds of interstellar dust.

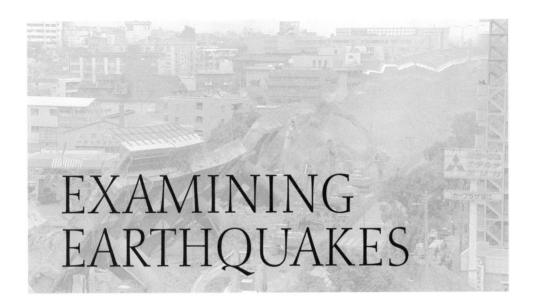

EXAMINING EARTHQUAKES

EARTHQUAKE HISTORY

Why are **earthquakes important**?

Records from our past history show that earthquakes have been responsible for death and injury to countless human beings. In addition, they have destroyed cities, both large and small, changed the existing geography, and been responsible for the fall of many civilizations. But earthquakes are not just past events that only happen in distant places. In recent years, the power of earthquakes has destroyed countless lives and wrecked infrastructures, sometimes even in places not historically associated with the phenomena.

When did the **first earthquake** occur?

Although it is scientific speculation, the first earthquake probably occurred sometime after the Earth's crust had cooled enough to become solid some 4.55 billion years ago. Although modern writing goes back 5,500 years, ancient records of earthquakes are scarce. For example, even though certain places in China experience frequent earthquakes, records of these events date back only to about 1177 B.C.E.; and one of the first European records of an earthquake only dates back to about 580 B.C.E..

What were some **ancient explanations** for earthquakes?

Humans in each culture used their religious beliefs and unique circumstances to try to explain the earthquake phenomenon. For example, some believed earthquakes were caused by dancing gods or giant burrowing moles; in Japan, earthquakes were interpreted as being the result of catfish thrashing about inside the Earth. Still others felt that winds rushing through caves in the center of the Earth were to blame. Another popular theory was that the Earth was carried on the back of some giant creature—like a frog or tortoise—whose movements caused earthquakes.

265

Starting around the 7th century B.C.E., the Greeks became keenly interested in earthquakes, the Mediterranean region being very seismically active. Philosopher Thales of Milet suggested around 600 B.C.E. that earthquakes were caused by powerful waves hitting the shorelines, but his theory could not explain quakes that occurred far inland. Others believed earthquakes occurred because of hot and dry weather, large landslides in underground caverns, or the release of highly compressed gases from deep within the Earth. The Greeks also believed that Poseidon, god of the sea, shook the Earth whenever he was displeased with people.

By the 4th century B.C.E., Greek philosopher and scientist Aristotle (384–322 B.C.E.), proposed his own theory of earthquakes that would be accepted for many centuries afterward: He believed warm, still air moved down into large underground caverns inside Earth, where it was superheated by fires. When the pressure built up sufficiently, the gases erupted out of the ground, resulting in earthquakes and volcanic eruptions.

Throughout history, others have also speculated on the cause of earthquakes, including the famous scientist and inventor Leonardo da Vinci (1452–1519). He believed earthquakes were caused by escaping compressed air from subsiding caves. Up until the mid-1750s, many explanations for earthquakes seemed to be limited to religious, supernatural, or cultural explanations. For example, the Lisbon, Portugal, earthquake of 1755 that killed approximately 70,000 people was interpreted to be God's punishment for Lisbon's sins. Survivors were consequently executed during the Inquisition for their supposed transgressions.

When did the **modern study of earthquakes** begin?

The start of modern earthquake study occurred toward the mid-1700s. Much of the work was precipitated by studies of the devastating quake and tsunami that struck Lisbon, Portugal, in 1755. One of the first scientific works was by English clergyman John Mitchell (1724–1793), who studied the timing and severity of quakes. Independently, Swiss scientist Elie Bertrand (1713–1797) worked on the reasons behind earthquakes and even compiled a history of earthquakes back to the year 563.

Alexander Von Humboldt's (1769–1859) work represents the change from old to modern earthquake studies. He was the first to understand the connection between earthquakes and volcanoes. In the 1850s, Robert Mallet (1810–1881) figured out how to measure the velocity of seismic waves by using explosions of gunpowder. Based on this, he believed that changes in seismic velocity were indications of various properties in the Earth. (This same method is still used today, especially in the search for oil.) Others followed, including scientists who estimated the depth of an earthquake, cataloged earthquakes, and determined that earthquakes usually center around a fault line.

Who is considered the **"father of seismology"**?

Most scientists agree that geologist John Mitchell (1724–1793), an English clergyman at Cambridge University, should be given the title "father of seismology." In 1760, Mitchell was the first to theorize that earthquakes were generated when water met subterranean fires, a force that generated waves in the Earth's crust. In his paper *Conjectures Concerning the Cause, and Observations upon the Phenomena of Earthquakes* he postulated that two waves from this activity arrived at different times. Mitchell also believed the center of an earthquake (now called the *epicenter*) and speed could be determined by measuring the arrival times of the two waves at different locations, a method that (although modified) is still used today. Although his mechanism for earthquakes was incorrect, the idea of earthquake waves propagating through the Earth was true.

Who is considered the **"father of modern seismology"**?

Most scientists agree that British geologist and astronomer John Milne (or "Earthquake Milne," as he has been called) deserves to be called the "father of modern seismology." Milne (1850–1913) was the inventor of the horizontal pendulum seismograph (1880), a device for measuring ground shaking during an earthquake. He also spent twenty years working in Japan, along with British scientists Sir James Alfred Ewing (1855–1935) and Thomas Gray, establishing the world's first earthquake laboratory.

EARTHQUAKES DEFINED

What is an **earthquake**?

In general, an earthquake is a vibration measured or felt at the Earth's surface and produced by seismic, or shock, waves. These waves are created as rocks rupture under stress or friction at or below the surface, causing energy to be released. The vast majority of quakes occur along faults, or cracks, in the Earth's crust or upper mantle, ranging from the surface to about 500 miles (800 kilometers) deep.

Earthquakes can be generated by a number of sources. Most are the result of natural tectonic processes, usually caused by the interaction between two lithospheric plates. (For more information on lithospheric plates and tectonics, see "The Earth's Layers.") Other quakes can be generated by volcanoes as magma is injected into the Earth's crust. For example, many of the quakes on the island of Hawaii are volcanic earthquakes. Still others are artificially generated by nuclear explosions from military tests.

What are **foreshocks, mainshocks, and aftershocks**?

Earthquakes come in clusters, usually in three forms: foreshocks, mainshocks, and aftershocks. Foreshocks are quakes that occur before a larger one in the same location; one quarter of all mainshocks happen within an hour of their foreshock. Mainshocks and aftershocks are more well known. An earthquake mainshock is easy to understand: It is the main, and usually the highest magnitude, earthquake generated. Aftershocks

Do underwater landslides generate earthquakes?

Yes, marine scientists know that underwater landslides can generate earthquakes. For example, there are over 15 giant landslides surrounding the Hawaiian Islands. These are among the largest slides known to exist on the Earth and are thought to have occurred within the past 4 million years, with the most recent being only about 100,000 years old. As slides moved, huge chunks of the island's landmass slid into the ocean, earthquakes were generated, and large waves (called tsunami) sometimes crashed into the coastline. Currently, some scientists have evidence that large blocks of Hawaii are beginning to slide again. If such an underwater slide were to occur, there would be the potential for enormous loss of life, property, and resources.

are smaller quakes that occur in the same general geographic area for days—and even years—after the larger, mainshock event (although some aftershocks can be comparable in scale to the mainshock earthquake). They can occur not only along the fault that precipitated the earthquake, but also along nearby faults as well. In general, larger quakes have more aftershocks. In addition, aftershocks most often occur after shallow quakes and appear to be minor readjustments along the fault that caused the major event.

How many **earthquakes occur** each year?

According to the United States Geological Survey, several million earthquakes occur around the world each year. This high number is only an estimate, because much of the world's surface—mostly remote, unpopulated areas—has no instruments to measure earthquakes, so many quakes go undetected. The following table breaks down the average annual number of earthquakes since 1900, based on Richter scale magnitude (for more information about the Richter scale, see below):

Description	Magnitude	Average Annually
great	8 and higher	1
major	7–7.9	18
strong	6–6.9	120
moderate	5–5.9	800
light	4–4.9	6,200 (estimated)
minor	3–3.9	49,000 (estimated)
very minor	<3.0	magnitude 2–3: about 1,000 per day magnitude 1–2: about 8,000 per day

Source: United States Geological Survey

Los Angeles City Fire Department chief looks over the remains of an unoccupied house. The house, partially knocked off its foundation in the January 17, 1994, Northridge earthquake, fell completely off its foundation in the strong aftershock, which registered 5.3 on the Richter Scale. *AP/Wide World Photos.*

What are **faults**?

Faults are fractures (cracks) in the crust of the Earth. They are places in which great masses of rock are moving, such as at the boundary between two crustal plates or even between pieces of a crustal plate. Faults may or may not be visible on the ground surface; many are hidden deep within the Earth's crust.

How are **earthquakes** generated along **faults**?

Most faults are the result of movement within the Earth's crust. When the surfaces on each side of a fault become stuck together (normally due to friction), pressure and strain begin to build. Eventually, enough energy builds up, the friction is overcome, the ground moves on one or both sides of the fault.

If the amount of friction holding the two surfaces together is small, the ground movement will also be small. This process is called "creep" and produces small earthquakes known as tremors. On the other hand, if the amount of friction is large, with the surfaces firmly locked together, then the energy builds up precipitously, eventually resulting in one large, sudden movement along the fault. This single displacement produces a great earthquake, usually with accompanying serious damage and destruction.

269

What are the **types of faults**?

There are three major types of faults associated with earthquakes. The following lists the type and associated fault zone:

Dip-slip faults—Dip-slip faults move mostly vertically, with no sideways component. They are divided into "normal" and "thrust" (reverse). Normal dip-slip faults have a block that slips down in relation to the other, down the inclined fault plane, usually caused by tension as the rocks pull apart. Some examples of earthquake movement along such a fault include the 1959 Yellowstone; 1954 Dixie Valley, Nevada; and 1993 Klamath Falls, California, earthquakes. Thrust (or reverse) faults occur as one block moves up the fault plane in relation to the other. Usually this is caused by compression. Examples include the 1700 Eureka-Seattle; 1964 Alaskan; 1960 Chile (one of the largest ever recorded); 1988 Armenia; and the 1994 Northridge, California, quakes.

Strike-slip faults—Strike-slip (or lateral) faults move predominantly in a horizontal direction. California's San Andreas fault is one example of a strike-slip fault—one that has caused numerous earthquakes over the centuries, including large ones in 1857 and 1906. Other examples of earthquakes caused by this type of fault include the 1992 Big Bear and Landers earthquakes, both of which were also in California.

Oblique-slip faults—Oblique-slip faults are those in which the movements contain both horizontal and vertical components. They include the 1989 Loma Prieta, California, and 1995 Kobe, Japan, quakes.

What are a **graben and horst**?

A graben is a movement associated with faulting: When a block of land moves vertically downward between two faults, it creates a graben. At the surface, this is often seen as a rift valley, such as those found in Africa. A horst is when a block is left standing and the rock masses on each side slide down. If this is seen at the surface, it is called a block mountain.

What is the **San Andreas fault**?

The San Andreas fault is often thought of as one long, strike-slip fault in the Earth's surface, extending for at least 807 miles (1,300 kilometers) in western North America. In reality, it is also a swarm of right-lateral faults that are mostly parallel to one another. The faulting is caused by two lithospheric plate boundaries butted up against one another and moving in opposite directions. The Pacific plate moves northwest in relation to the North American plate, which moves to the southeast at about 1 to 1.5 inches (2.5 to 4 centimeters) per year. It is estimated that in the last two million years it has moved about 10 miles (16 kilometers).

A view of the destruction resulting from the April 18, 1906, earthquake in San Francisco, California. Following the powerful quake, fire spread throughout the city reducing much of the remaining structures to rubble. *AP/Wide World Photos.*

What is **elastic rebound**?

Elastic rebound is a mechanism used by geologists to explain earthquakes and surface ruptures (also called surface faulting) seen when a deep fault's movement breaks through to the Earth's surface. It was first proposed in 1907 by American seismologist and glaciologist Henri Victor Reid (1859–1944). Reid based his ideas on the great 1906 San Francisco earthquake, observing that fences, streams, and roads had been deformed for years preceding the quake. Afterward, they had mostly returned to their normal state, but were initially offset by up to 21 feet (6.4 meters).

Based on these observations, Reid theorized that the ground had become locked together at the fault due to friction, while on each side of the fault the ground had continued moving slowly in opposite directions, leading to the elastic deformation. When the buildup of energy was enough to overcome the friction at the fault, the ground suddenly moved. Although the ground snapped back to its original state, there was an offset due to years of ground motion on either side.

Reid's ideas, although modified, are still used today. Scientists can now explain the role of tectonic plates in elastic rebound. As the plates move relative to each other, energy builds up along the fault plane rocks. Because these planes (edges) are not smooth, the rock "locks" the movement in place, storing great amounts of energy. When the stresses in the rock exceed the ability of the rock to stay locked, a rupture occurs.

271

What is the **focus (or hypocenter)** of an earthquake?

The focus, or hypocenter, of an earthquake is the precise spot where the ground initially moves. This is normally found deep beneath the surface, often miles into the crust. At this point, the built-up energy is converted into waves that move through the interior of the Earth toward the surface.

What is the **epicenter** of an earthquake?

The epicenter of an earthquake is a more widely known concept: It is the location on the Earth's surface directly above the focus. Because it is so close to the earthquake focus, the waves arrive at this point with the highest energy. In turn, the resulting surface waves have the most energy and potential for destruction and death.

What **type of seismic waves** are generated by an earthquake?

An earthquake generates two types of seismic waves: body (or interior) and surface waves. Body waves are generated at the earthquake focus as the ground shifts and spread out in all directions in the Earth's interior. They are further broken down into P- (primary) and S- (secondary) waves. (For more information about P- and S-waves, see "The Earth's Layers.") The fastest waves are the P-waves, which move away from the earthquake's focus at speeds up to 4 miles (7 kilometers) per second and are first to register on a seismograph. Similar to sound waves, P-waves compress and expand materials in the direction of travel. When they reach the Earth's surface, the movement produced is in an up and down (vertical) direction. If the P-waves are powerful enough—such as in a high-magnitude quake—they will compress the air, creating a characteristic train-like roar.

S-waves travel slower through the Earth than P-waves and are recorded second on a seismograph. They move through material in a side-to-side (shearing) motion, similar to an ocean wave rippling through the Earth. Upon reaching the Earth's surface, the movements produced are in up, down, and sideways (horizontal and vertical) directions. Because of this multitude of motions, S-waves are some of the most destructive to humans and infrastructures.

What **types of surface waves** are generated by earthquakes?

When seismic waves reach the surface, they generate two major types of surface waves: Love and Rayleigh. In the late 1800s, English mathematician and physicist Augustus Edward Hough Love (1863–1940) discovered the Love wave; in 1884, English physicist Lord John William Strutt Rayleigh (1842–1919) detected the Rayleigh wave.

Love waves are similar to the motion of a snake as it travels along the ground: The waves shake the ground in a side-to-side motion perpendicular to the wave's direction of travel. These are registered as horizontal movements on a seismograph. Rayleigh waves are similar to ocean waves rippling along the ground: The waves shake the ground in a side to side, and up-and-down movement and register as vertical and horizontal movements on a seismograph.

MEASURING EARTHQUAKES

Who invented the **first earthquake detection device**?

Zhang (Chang) Heng was the first person in recorded history to invent an earthquake detection device. Around 132 C.E., in China, he suspended a pendulum inside a pot with linkages to the outside. Dragon ornaments were attached around the circumference. An earthquake would shake the pot, while the pendulum would remain steady. This action caused the release of a ball from one of the dragon's mouths; the ball would fall into the mouth of a frog statue beneath. The direction of the earthquake's center was indicated by which ball was released. Use of this device allowed Zhang Heng to determine the direction and occurrence of an earthquake northwest of Loyang, China, some 400 miles away, long before horse-bound messengers brought the news.

What is a **seismoscope**?

Zhang Heng's instrument could technically be called a seismoscope—an instrument that indicates the shaking of an earthquake—but it made no record of ground oscillations (amplitude) or timing. The history of seismoscope invention is complex, with many people contributing to the design over time. Some scientists give Italian naturalist and clockmaker Ascanio Filomarino (1749–1799) credit for inventing the first true seismoscope in 1795. The instrument had a heavy suspended pendulum, bells, and a clock, and was based on Zhang Heng's 1,700-year-old principles. During an earthquake, the pendulum would remain stationary, while the rest of the apparatus would shake, ringing the bells and starting a clock. Sadly, Filomarino never improved his design: He was killed by an angry mob on Mt. Vesuvius who objected to his experiments. They subsequently destroyed the seismoscope and burned Filomarino's workshop to the ground.

What are **seismometers, seismographs, and seismograms**?

Seismometers are also called seismographs (or accelerographs). They measure seismic waves and are capable of creating a continuously timed, graphical record of ground motion. Early seismographs recorded their data on smoked glass or paper, tape, film, or rotating drums; modern seismographs record data digitally. Such recordings are called seismograms (or traces).

A great deal of information can be gleaned from analysis of a seismogram. For example, geologists can determine how much energy was released at the earthquake's

A seismologist shows the seismographic activity of an earthquake measuring 6.5 on the open-ended Richter scale that hit the island of Taiwan in June 2003 near Taipei. The tremor's epicenter was in the Pacific Ocean, about 22 miles (35.5 kilometers) off the coast. *AP/Wide World Photos.*

focus; the exact location of the earthquake can be determined by comparing the different arrival times of the waves on seismographs throughout the world; and, finally, geologists can tell the strength of the quake and its duration.

Who invented the **first seismograph**?

There are several contenders for the invention of the first seismograph, depending on a person's definition. In 1855, Italian seismologist Luigi Palmieri (1807–1896) invented the first electric seismograph which was thought to be the first seismic instrument capable of routinely detecting earthquakes imperceptible to human beings. Also the director of an observatory near Mt. Vesuvius, Palmieri realized that an instrument capable of measuring small ground tremors would aid in predicting impending volcanic eruptions. The first part of his instrument consisted of pendulums, tubes of mercury, and springs—all to detect the earthquake. The second section had a clock and a mechanism for recording the vibrations on a strip of paper—thought to be the first seismogram. Palmieri discovered that small foreshocks could be precursors to larger earthquakes and that volcanic eruptions were accompanied by tremors. His original instrument is still in use today.

Although highly debated, Italian scientist Filippo Cecchi is often given credit for inventing the first true seismograph, an instrument he built around 1875. Using two simple pendulums to measure horizontal movements, one swinging north to south, the other east to west, Cecchi recorded the relative motion of the pendulums and the Earth as a function of time—the first seismograph to do so.

Who invented the **modern seismograph**?

Most scientists give credit to British geologist John Milne (1850–1913) for the invention of the modern seismograph. Milne had grown dissatisfied using written reports as a means to collect data about the earthquakes of Japan, where he was a professor at the Imperial College of Engineering. During the 1890s, Milne—along with help from his colleagues James Ewing, Thomas Gray, and F. Omori at the same college—developed a device that would both detect and record the vibrations associated with earthquakes.

The seismograph had three pendulums supported in frames that were oriented so vertical and horizontal motions of an earthquake could be measured. The frames were anchored to the ground and moved during an earthquake, while the pendulums stayed relatively motionless. A revolving drum of smoked paper moved with the frame; a stylus on each pendulum would then trace the motion.

In 1893, Milne modified the seismograph to allow the data to be recorded on revolving photographic film. Milne's horizontal pendulum seismograph would be further improved after World War II, including the development of the Press-Ewing seismograph (after Milne's colleague James Ewing)—a way of recording long-period waves. This instrument—still used today—is similar to Milne's, but in this case the pivot supporting the pendulum has been replaced by an elastic wire to cut down on friction. Other seismographs are also in use around the world, each modified to record certain aspects of seismic information.

Why are **seismographs important** to the study of earthquakes?

With the advent of the modern seismograph, the vibrations (seismic waves) associated with earthquakes could finally be measured reliably and accurately and recorded for later study and analysis. As a result of this breakthrough, geologists can now determine an earthquake's duration, magnitude, and the exact location of the focus and epicenter. Details about the fault can also be determined—including orientation, extent, magnitude, and direction—along with knowledge of the Earth's interior. (For more about using earthquakes to study the Earth's interior, see "The Earth's Layers.")

Who developed the **first earthquake maps**?

Robert Mallet (1810–1881) was an Irish engineer who made two outstanding contributions to the then newly emerging field of seismology: In 1857, after an earthquake struck Naples, Italy, Mallet examined the area and generated maps (known as isoseismal maps), revealing contours of damage and intensity. He also gathered more than 20 years worth of historical earthquake data to generate a world map. This map—one of the first of its kind—revealed that earthquakes happened in clusters and in specific localized areas around the planet.

What is earthquake **magnitude**?

Magnitude is an objective measurement of the energy released by an earthquake. It is determined by the amplitude of the seismic waves at a specific distance from the epicenter. The data is recorded by a seismograph; it can be measured and calculated from anywhere in the world. Thus, it is not dependent on local observation.

What is the **Modified Mercalli Intensity Scale**?

Along with magnitude, earthquakes can be measured by their intensity, especially in remote places in which no seismic instruments are available to record quake magnitude. Intensity is a subjective measurement of the strength of an earthquake and is based on the effects on local population and structures. One of the most common scales used to measure earthquake intensity is the Modified Mercalli Intensity Scale. It was first developed in 1902 by Italian seismologist Giuseppe Mercalli (1850–1914) and has subsequently been modified. Below is a table explaining the Modified Mercalli Intensity Scale.

Number	Comments
I	Not felt except by a very few under especially favorable circumstances.
II	Felt only by a few persons at rest, especially on upper floors of buildings. Delicately suspended objects may swing.
III	Felt quite noticeably indoors, especially on upper floors, but many people do not recognize it as an earthquake. Standing motor cars may rock slightly. Vibration is like a passing truck.
IV	During the day, felt indoors by many; outdoors by few. At night are some awakened. Dishes, windows, doors disturbed; walls make creaking sound. Sensation like heavy truck striking building. Standing motor cars noticeably rocked.
V	Felt by nearly everyone; many awakened. Some dishes, windows, etc. broken; a few instances of cracked plaster; unstable objects overturned. Disturbances of trees, poles, and other tall objects sometimes noticed. Pendulum clocks may stop.
VI	Felt by all; many frightened and run outdoors. Some heavy furniture moved; a few instances of fallen plaster or damaged chimneys. Damage slight.
VII	Everybody runs outdoors. Damage negligible in buildings of good design and construction; slight to moderate in well-built ordinary structures; considerable in poorly built or badly designed structures; some chimneys broken. Noticed by persons driving motor cars.

Number	Comments
VIII	Damage slight in specially designed structures; considerable in ordinary substantial buildings, with partial collapse; great in poorly built structures. Panel walls thrown out of frame structures. Chimneys, factory stacks, columns, monuments and walls may fall. Heavy furniture overturned. Sand and mud ejected in small amounts. Changes in well-water levels. Persons driving motor cars disturbed.
IX	Damage considerable even in specially designed structures; well-designed frame structures thrown out of plumb; great damage in substantial buildings, with partial collapse. Buildings shifted off foundations. Ground cracked conspicuously. Underground pipes broken.
X	Some well-built wooden structures destroyed; most masonry and frame structures along with foundations destroyed; ground badly cracked. Rails bent. Landslides considerable from river banks and steep slopes. Shifted sand and mud. Water splashed over banks.
XI	Few, if any, masonry structures remain standing. Bridges destroyed. Broad fissures in ground. Underground pipelines knocked completely out of service. Earth slumps and land slips in soft ground. Rails bent greatly.
XII	Damage total. Waves seen on ground surfaces. Lines of sight and level distorted. Objects thrown upward into the air.

What is the **Richter scale**?

The Richter scale is an objective means to quantify the magnitude (energy) released by an earthquake. In 1935, United States seismologist Charles Francis Richter (1900–1985) and German-born seismologist Beno Gutenberg (1889–1960) borrowed the magnitude scale from astronomers, who use magnitude to describe the brightness of stars. They defined magnitude as how fast the ground moved as measured on a particular seismograph at a specific distance from the earthquake.

The scale is a mathematical construct, not a physical scale like a ruler, and it is logarithmic, not linear. The numbers represent the maximum amplitude of seismic waves that occur 62 miles (100 kilometers) from the epicenter of an earthquake. Since seismographs are never located at this precise distance, the actual measurement takes into account the difference in time between the arrival of the P- and S-waves. An increase in each whole number on the scale represents a ten-time increase in power. For example, an earthquake with a magnitude of 8 on the Richter scale is 10 times the amplitude of a magnitude 7 earthquake, and the energy released is 31 times greater.

What **new way** do scientists **measure earthquakes**?

Scientists realize the Richter scale does not address all facets of earthquakes. Therefore, the various earthquake "dimensions" are measured with different magnitude scales. Each scale measures how fast the ground moves at a different distance and in various

American seismologist Charles Francis Richter, who developed the first widely used seismic magnitude scale, studies earthquake tremors in his laboratory in Pasadena, California, in 1963. Dr. Richter developed his scale in 1935. *AP/Wide World Photos*.

frequency bands. Each has its uses, but in reality they are limited because they measure only part of the ground motion.

To address this gap in data, scientists are developing a new scale—called "moment magnitude"—that eliminates many of the previous limitations. Moment is a physical quantity related to the total energy released in the earthquake. Scientists can also estimate moment magnitude by examining a fault's geometry in the field or by analyzing a seismogram recording. This is a great boon for the study of past and future earthquakes: We can use the method to measure old earthquakes and compare them to instrumentally recorded events, allowing scientists to know just a little more about how quakes occur.

But you will probably not hear anyone use "moment magnitude" in the media when talking about an earthquake. Because the units are harder to understand, moment magnitudes are converted to Richter scale magnitudes for the general public.

Has anyone successfully **predicted an earthquake**?

Although there is disagreement as to the definition of "prediction," there is one record of a successful prediction of an earthquake in southern Liaoning Province, China, by scientists at the State Seismological Bureau in Beijing.

In early 1974, this Manchurian province experienced numerous tremors, magnetic field changes, and uplifting. Based on these phenomena, seismologists made a prediction that a moderate to strong earthquake would occur in the region within two years. Another series of tremors in late 1974 allowed them to refine their predictions to a magnitude 5.5 to 6 (Richter Scale) earthquake occurring around Yingkou, Manchuria, in the first half of 1975. When a large group of tremors and a magnitude 4.8 quake was followed by silence in early 1975, the order was given to evacuate. More than 3 million people left their homes in southern Liaoning province on the afternoon of February 4, 1975. At 7:36 p.m., the predicted large earthquake struck, damaging roads and bridges and destroying most of the build-

Can animals detect earthquakes?

People have long claimed that animals—from cats and dogs to horses and snakes—can predict or detect a coming earthquake. No one quite knows how, but the animals may hear sounds and feel vibrations associated with deep Earth movements, or even detect electromagnetic changes better than humans. While most scientists believe changes in animal behavior prior to a quake are not consistent enough, there is still not enough solid data to say whether it is true or not.

Still many people believe changes in animal behavior are connected to quakes, an idea that has been observed and documented in different parts of the world for centuries, especially in places where earthquakes are prevalent. For example, the first known record of an animal-earthquake connection occurred in earthquake-prone Greece in 373 B.C.E. The report stated that dogs howled, and rats, snakes, and centipedes ran to safety several days before a destructive earthquake struck the area.

More recently, in California, a study in the south San Francisco Bay area showed that when local newspapers listed an increase in lost and found cats and dogs, the probability of a quake increased. And in Japan and China, animals are often considered an essential part of the earthquake warning system, along with the usual high-tech seismic instruments. In one incident in 1975, strange behavior in animals caused officials in Haicheng, China, to issue an earthquake warning, ordering 90,000 residents to evacuate the city. In a matter of hours, a huge, magnitude 7.3 quake destroyed 90 percent of the city's buildings, with little loss of human (and animal) life.

ings of Yingkou and Haicheng. Because of the early prediction and evacuation, only some 300 people died.

There have been other such predictions. For example, during the 1980s, Greek scientists studied the changes in the natural electric currents in the ground caused by impending earthquakes. As a result, they were able to successfully predict the location and magnitude of numerous earthquakes in Greece in 1988 and 1989.

But along with a few accurate short-term earthquake predictions come mostly failures. There is still much to learn in the field of earthquake prediction, which is almost as much art as science. The number of alleged earthquake precursors is long: changes in ground strain, tilt, resistivity and elevation, foreshocks, ground water level shifting, changes in gravitational and magnetic fields, radon gas emissions, flashes of light, sounds—and even the behavior of animals. Not only must all of this data be collected, it must then be accurately interpreted.

Unfortunately, we are not at the stage where scientists can confidently and accurately predict the occurrence of short-term or long-term earthquakes. More scientists are now focusing their efforts on the long-term elimination of earthquake hazards by improving the structural safety of buildings, rather than trying to accomplish short-term quake predictions.

THE RISKS AND DAMAGE FROM EARTHQUAKES

How does an **earthquake feel**?

Depending on the severity, there are certain motions you will feel during an earthquake. In a small quake, you may hear and feel a sound similar to an 18-wheeled truck rolling past your house. In general, in a larger quake you will feel a swaying or small jerking motion, then a pause followed by a more intense rolling or jerking motion.

Is it safe to **stand under a door frame** in an earthquake?

We have all seen pictures of a door frame as the only house item left standing after an earthquake, but it is probably the same picture shown over and over. If you lived in an adobe home, such a door frame would probably be the safest place, but most modern homes are built differently. Doorways in modern homes are no stronger than the rest of the house, and the door can swing open or shut, causing injury. Earthquake authorities advise people to hide under a table or desk during an earthquake.

What factors affect the **average duration** of an earthquake?

Although most earthquakes last in terms of seconds, to someone going through the quake it seems like hours. The actual duration of an earthquake depends mainly on the physical conditions around the quake and your location.

Can faults swallow people during an earthquake?

No, faults do not swallow people during an earthquake. This only happens in books, television shows, or the movies. If a fault opened up, there would be no ground friction; without friction, there would be no earthquake. (Movement occurs along the plane of a fault, not perpendicular to it.) Most cracks in the surface form after the ground shakes, causing slope failure, ground settling, tears in roads and sidewalks, or slumping in water-saturated, loose sediments, especially along steep slopes and bodies of water.

How do the Japanese plan for major earthquakes?

It's not easy to plan for an earthquake. Japan alone receives almost 10 percent of the total seismic energy generated each year by our planet, and it has almost 1,000 earthquakes per year that can be felt.

The Japanese are no strangers to damaging quakes. On January 17, 1995, at 5:46 a.m., a major earthquake struck the Japanese city of Kobe. The quake had the energy of eight Hiroshima bombs released within 20 seconds. About 6,425 people died—about 80 percent of whom were killed by falling debris—with tens of thousands wounded; 240,932 houses were completely or partially destroyed. Although the city thought it was prepared, it wasn't. Even houses certified to withstand earthquakes folded, and major highways crumbled.

Kobe taught the Japanese a great deal about being prepared, and they have since built more earthquake-resistant buildings and infrastructures. And although the government continually prepares people with earthquake drills, they are also realistic about determining the odds if a major earthquake occurs. According to the Tokyo Metropolitan Disaster Prevention Center, all sorts of disaster scenarios are now stored on their computers. They have even predicted the number of dead and wounded if another earthquake occurs. After all, without warning, and with such a huge population, it would not be possible to accomplish such measures as an evacuation. Many of the scenarios are for Tokyo, whose last big quake was in 1923: a magnitude 7.9 Kanto earthquake that flattened the city. Studies by the Tokyo Metropolitan Government estimate that if a similar-sized earthquake were to occur again the results would be disastrous. For example, in one scenario it is estimated that on a weekday at 6 p.m. with sunny weather and only mild winds, a Kanto-type quake would leave 156,431 dead, 155,416 houses destroyed, and 12 percent of Tokyo leveled by subsequent fires.

In general, the greater the length of the fault, the bigger the earthquake, and the longer the duration of the quake. For example, the magnitude 7.8 quake that occurred January 9, 1857, at Fort Tejon, California, was generated by a 224 mile (360 kilometer) fault and lasted a record time in the United States—about 130 seconds. The famous April 18, 1906, San Francisco earthquake lasted for 110 seconds; the fault line there is about 250 miles (400 kilometers) long. But this rule is not ironclad: The Loma Prieta, California, quake on October 17, 1989, lasted 7 seconds from a fault 25 miles (40 kilometers) long; the January 17, 1994, Northridge, California, quake also lasted 7 seconds and was caused by a 9-mile (14-kilometer) fault.

Other factors are important to a quake's duration: The amount of time you shake depends on how far you are from the epicenter, the magnitude of the earthquake, and the type of rock under your feet. For example, if you are close to the epicenter you will

Heavy construction machines break up concrete from a toppled freeway in Kobe, Japan. Portions of the Kobe-Osaka freeway collapsed during the earthquake that hit this western Japanese city early on January 17, 1995. *AP/Wide World Photos.*

experience a longer shaking; and the ground will usually shake longer for higher magnitude quakes (seconds for small quakes, close to a minute for major quakes). The type of rock is important, too. If you are on sand, the shaking will last almost three times as long as if you stood on a stable bedrock such as granite.

What **cities** are at **risk for earthquakes**?

Some well-known cities located in seismically active areas include Rome, San Francisco, Istanbul, Cairo, Hong Kong, Singapore, Jakarta, Los Angeles, Beijing, Athens, Manila, Tehran, Mexico City, and Tokyo. And even with all the seismic activity, these cities continue to grow in population.

Who is at risk for earthquakes?

The Global Seismic Hazard Assessment Program (GSHAP) was launched in 1992 by the International Lithosphere Program and ended in 1999. Scientists from all over the world pooled their earthquake data and brought out a new global earthquake potential map. Data was collected from various sources, including seismic monitoring stations and geologic and historic data, with sites given a seismic hazard value based on a specific measurement called Peak Ground Acceleration (or PGA).

The resulting maps showed that 90 percent of all earthquakes occur along the plate boundaries, which is to be expected. But it also showed some surprises. For example, the hazard map showed a definite earthquake potential in the central United States in an area known as the New Madrid seismic zone right around the Mississippi River; another was in Charleston, South Carolina. (Both areas had previously experienced two of the largest earthquakes in United States history.) And although the largest earthquakes ever recorded in the Americas are the 1960 Chile and 1964 Alaska earthquakes, the largest seismic hazard value (PGA) in the Americas is in southern California, along the San Andreas fault.

Why do **earthquakes** cause so much **damage**?

There are a number of factors—direct and indirect—that cause damage from earthquakes. The following lists some of the more common:

Ground shaking—The vertical and horizontal shaking of the Earth's surface during a quake causes a great deal of damage—it's like putting structures on a wild ride at the carnival. The shifting of the ground causes cracks in buildings and infrastructures. The cracks can cause further problems by being unstable during aftershocks or another major quake.

Tsunamis—Tsunamis are huge waves formed in the oceans, often as the result of an earthquake. When these huge walls of water strike a coastline, they can cause immense damage to structures and loss of human life.

Landslides—Landslides are an indirect result of earthquakes, a slipping and shifting of the land caused by the shaking from the quake. Structures can sink, split, or crack, becoming unstable during aftershocks; other structures are shattered as they are carried with the soils downslope.

Subsidence—Subsidence, or the sinking of the land, is also an indirect result of a quake. Like landslides, it can cause structures to shatter as they subside with the local soils.

Fires, gas leaks, pipeline breaks—The shaking of the ground not only cracks structures but stresses underground and above-ground pipes, cables, and wires. This causes fires due to breaks in gas or power lines; pipeline breaks cause environmentally damaging chemical/oil spills; and water pipelines can either be compromised or broken during a quake.

What is **liquefaction**?

Liquefaction (also spelled liquifaction) is another way earthquakes can do damage. Its most common definition means to turn a gas or solid into a liquid, which is close to what happens in certain areas during an earthquake. The shaking ground causes the soil to lose strength and stiffness. The types of sediments most susceptible are clay-free deposits of sand and silts; occasionally, even gravel liquefies. Most of the time, liquefaction occurs in water-saturated soils, either by heavy rains or in a high groundwater area.

A small fishing boat, rear, is washed aground on a breakwater as a roof sits on what was once a wharf of a port on Okushiri Island, off the southwestern coast of Hokkaido in July 1993, one day after a major earthquake shook the Japanese island and a tsunami flattened the seaside area. *AP/Wide World Photos.*

In stable conditions, low water pressure is usually exerted on the soil particles. But when the quaking starts, the water pressure increases to a point where the soil particles begin to move with respect to each other. Thus, a semi-solid soil becomes like a viscous liquid. This motion can cause the ground to settle, slide, or move, causing destruction to structures sitting on top of the soils (including houses, highways, and dams) or even underground (cables, wires, and pipelines). In addition, a liquefied sand layer can shoot to the surface through cracks, forming a sandblow or sandboil and depositing a characteristic lens of sand on the ground with a volcano-like vent in the center.

Historically, liquefaction has been responsible for tremendous amounts of damage around the world. For example, in quake-prone California, the Lower San Fernando dam suffered an underwater landslide from liquefaction during the San Fernando earthquake in 1971. Fortunately, it stood, avoiding disastrous flooding of the populous below the dam. After the 1989 Loma Prieta earthquake, liquifaction caused damage to San Francisco's Marina district, an area built on landfill. Although most California cities already have stringent building requirements for quakes to cut damages, liquefaction still has potential to damage other infrastructures, such as gas or electrical lines.

What often happens when an earthquake **occurs in the ocean**?

If the earthquake causes a shifting of the ocean floor (especially around faults) the resulting displacement can generate a seismic sea wave, or tsunami. From the Japanese words *tsu* (meaning "harbor") and *nami* (meaning "waves"), a tsunami is sometimes incorrectly referred to as a "tidal wave." In actuality, it is a gravity or deep-ocean wave. Large tsunamis result from stronger earthquakes and there may be one, or a succession, of waves.

In the open ocean, these huge waves are harmless and are not even felt on ships. The crest-to-crest distance can be several miles, and the height from crest to trough on the order of a few feet. Tsunamis move very swiftly through the deep water because of their energy and a squeezing effect of the atmosphere on the ocean's surface. Tsunamis become dangerous when they reach a shoreline: As the wave velocity slows toward the more shallow water, the wave height increases, with most crest heights reaching between 98 to 164 feet (30 to 50 meters). A major earthquake can generate a tsunami wave up to 200 feet (61 meters) high, traveling at speeds up to 150 miles (241 kilometers) per hour. When these large waves hit the coast, they can wipe out entire villages and towns.

Why are **near-shore earthquakes** so dangerous?

Earthquakes occurring close to shore may cause offshore landslides. These landslides, in turn, may generate huge seismic waves (tsunami) that strike before any chance of advanced warning. In July 1998, a magnitude 7.0 earthquake occurred just off the northwest coast of Papua New Guinea, triggering an offshore landslide. This created a huge tsunami that struck within 10 minutes after the earthquake, destroying villages and killing more than 3,000 people, making it one of the most deadly tsunami of the 20th century.

FAMOUS FAULTS AND EARTHQUAKES

What was the **strongest earthquake** ever recorded?

Scientists did not measure quakes until just over a hundred years ago, so it's difficult to know the strongest earthquake ever recorded by humans. To date, the largest earthquake recorded was on May 22, 1960, in Chile at magnitude 9.5. The largest to date in the United States occurred on March 28, 1964 (nicknamed the Good Friday Earthquake) in Prince William Sound, Alaska. It was a magnitude 9.2 quake that killed 131 people.

What earthquake had the **highest death toll** ever recorded?

The highest death toll ever recorded from an earthquake was on January 23, 1556 in Shensi, China. According to ancient records, about 830,000 people died in the magnitude 8 quake, although historians can't separate how many died from the quake or from other consequences of the shaking.

285

What are some **famous ancient earthquakes**?

The following represent some of the most famous ancient earthquakes:

Location	Date	Approximate number of deaths
Antioch, Asia Minor	115, 458, 526 C.E. (around 500 severe earthquakes in that time)	unknown
Iran, Damghan	856	200,000
Ardabil, Iran	893	150,000
Aleppo, Syria	1138	230,000
Chihli, China	1290	100,000
Asia Minor, Silicia	1268	60,000
Shensi (Shansi), China	1556	830,000
Caucasia, Shemakha	1667	80,000
Calcutta, India	1737	300,000*
Lisbon, Portugal	1755	70,000

* Note: Recent studies indicate deaths were due to a cyclone, not a quake.

What are the **ten largest earthquakes** in the **United States** by magnitude?

The ten largest earthquakes in the United States are as follows:

Location	Date	Magnitude
Prince William Sound, Alaska	1964	9.2
Andreanof Islands, Alaska	1957	9.1
Rat Islands, Alaska	1965	8.7
East of Shumagin Islands, Alaska	1938	8.2
New Madrid, Missouri	1811	8.1
Yakutat Bay, Alaska	1899	8.0
Andreanof Islands, Alaska	1986	8.0
New Madrid, Missouri	1812	8.0
near Cape Yakataga, Alaska	1899	7.9
Fort Tejon, California	1857	7.9

What are some earthquake caused **death tolls above 5,000** since 1900?

The United States Geological Survey (an agency under the Department of the Interior) has used several sources to compile a list of some of the largest death tolls since the turn of the 20th century. The following table lists the date, location of the quake, the number of deaths, and the magnitude of the earthquake responsible:

Location	Date	Deaths	Magnitude	Comments
Kangra, India	April 4, 1905	19,000	8.6	
Santiago, Chile	August 17, 1906	20,000	8.6	
Central Asia	October 21, 1907	12,000	8.1	
Messina, Italy	December 28, 1908	70,000 to 100,000	7.5	Deaths from earthquake and tsunami
Iran	January 23, 1909	5,500	7.3	
Avezzano, Italy	January 13, 1915	29,980	7.5	
Kwangtung, China (Guangdong)	February 13, 1918	10,000	7.3	
Gansu, China	December 16, 1920	200,000	8.6	Major fractures, landslides
Xining, China	September 1, 1923	200,000	8.3	
Kwanto, Japan Yokohama	May 1, 1929	143,000	8.3	Known as the Tokyo-Great Tokyo fire
Yunnan, China	March 16, 1925	5,000	7.1	Talifu almost completely destroyed
Gansu, China	December 25, 1932	70,000	7.6	
China	August 25, 1933	10,000	7.4	

287

Location	Date	Deaths	Magnitude	Comments
Bihar-Nepal, India	January 15, 1934	10,700	8.4	
Quetta, Pakistan	May 30, 1935	30,000 to 60,000	7.5	Quetta almost completely destroyed
Chillan, Chile	January 25, 1939	28,000	8.3	
Erzincan, Turkey	December 26, 1939	30,000	8.0	
San Juan, Argentina	January 15, 1944	5,000	7.8	
Fukui, Japan	June 28, 1948	5,390	7.3	
USSR (Turkmenistan, Ashgabat)	October 5, 1948	110,000	7.3	
Ambato, Ecuador	August 5, 1949	6,000	6.8	Large landslides, topographical changes
Agadir, Morocco	February 29, 1960	10,000 to 15,000	5.9	at a shallow depth just under the city
Chile	May 22, 1960	4,000 to 5,000	9.5*	Earthquake, tsunami, volcanic activity, and floods
Qazvin, Iran	September 1, 1962	12,230	7.3	
Iran	August 31, 1968	12,000 to 20,000	7.3	
Yunnan Provence, China	January 4, 1970	10,000	7.5	
Peru	May 31, 1970	66,000	7.8	Great rock slides and floods
Southern Iran	April 10, 1972	5,054	7.1	
China	May 10, 1974	20,000	6.8	
Pakistan	December 28, 1974	5,300	6.2	
China	February 4, 1975	10,000	7.4	
Guatemala	February 4, 1976	23,000	7.5	
Tangshan, China	July 27, 1976	255,000 (official, but estimated to be as high as 655,000)	8.0	
Mindanao, Philippines	August 16, 1976	8,000	7.9	
Iran	September 16, 1978	15,000	7.8	

* Recomputed magnitude.

Location	Date	Deaths	Magnitude	Comments
Michoacan, Mexico	September 19, 1985	9,500 (official, but estimated as high as 30,000)	8.1	
Turkey-USSR border	December 7, 1988	25,000	7.0	
Western Iran	June 20, 1990	40,000 to 50,000	7.7	
Near southern Coast of Western Honshu, Japan	January 16, 1995	5,502	6.9	Landslides and liquefaction
Turkey	August 17, 1999	17,118	7.4	
India	January 26, 2001	20,023	7.7	166,836 injured; 600,000 homeless

How can an earthquake with a **low magnitude** cause many **deaths**?

There have been many low-magnitude earthquakes that have caused much damage and many deaths. For example, on February 29, 1960, in Morocco, Agadir, 10,000 to 15,000 people were killed in a 5.9 magnitude quake. The reason for the huge death toll had to do with the location of the quake: It occurred at a shallow depth just under the city, literally shaking apart structures and crushing thousands of people. Another such shaking under a city occurred in Skopje, Yugoslavia, in 1963, when 1,100 died during a magnitude 6.0 quake.

VOLCANIC ERUPTIONS

VOLCANO HISTORY

When did the **first volcanoes** form?

Geologists believe that volcanoes were present as far back as 4.55 billion years ago, the era when our planet first formed its crust. Some scientists believe these early volcanoes had a major impact on the Earth by providing early atmospheric gases, including water vapor and carbon monoxide. In addition, huge volcanic eruptions and accompanying debris may have dimmed the sun's light periodically, contributing to global climate changes.

What were some **early explanations** for **volcanoes**?

Ancient cultures often explained active (and dormant) volcanoes using myths, religion, or superstition. Because of the violence of volcanic activity, most cultures believed eruptions and tremblings had to be the fault of humans. Some made offerings to a volcano, including human sacrifices, to appease gods or monsters. The following lists just a few of the many cultural beliefs surrounding volcanoes:

- The Polynesians believed volcanoes were controlled by the demi-goddess Pele, who could appear in human form as an old, ugly woman or a beautiful young girl.

- Chileans believed a giant whale lived inside volcanoes.

- The Japanese thought a giant spider lurked inside active volcanoes.

- Indonesians believed the world was held up by the snake Hontobogo, whose movements shook the ground and caused fire to erupt from the mountains.

- Indians thought there was a giant boar or mole living in each volcano.

- Ancient Greeks believed volcanic eruptions were the exhalations of the Titans— giants the god Zeus had buried beneath the mountains.

- The people of the Russian Kamchatka peninsula believed volcanoes were young men that had quarreled over a beautiful girl; they were turned into mountains by a shaman to preserve the peace.

What is the **earliest known** record of a volcanic **eruption**?

One of the earliest known records of a volcanic eruption comes from the 5th century B.C.E.: the poet Pindar's ode about the myth of Typon. In this work, he characterized the Sicilian volcano Etna as a "pillar of the sky" from which fire shot with a loud roar. Pindar also recounted how white-hot rocks crashed into the sea and land, and how the lava flowed down the volcano. Geologists believe Pindar was accurately describing the eruptions of Etna that occurred in 479 B.C.E.

What were some **early scientific explanations for volcanoes**?

The Greek philosophers where among the first to explain volcanoes using scientific methods. Plato (429–347 B.C.E.) thought volcanic eruptions were caused by air escaping from the Pyriphlegethon, the river of fire thought to flow through the inside of the Earth. On the other hand, philosopher Aristotle (384–322 B.C.E.), had a "pneumatic" theory. He believed waves breaking on the shore compressed and forced air into caves located deep within the Earth. The air came in contact with bitumen and sulfur, creating fire and resulting in smoke, lava fragments, ashes and flames erupting out of openings in a volcano. Greek geographer Strabo (c. 53 B.C.E.–c. 39 B.C.E.) published *Geography* c. 30–20 B.C.E., in which he detailed descriptions of volcanoes such as Etna and Vesuvius, accurately recognizing the then-forgotten volcanic nature of the latter.

During the Middle Ages volcanoes were again given religious and superstitious explanations; but by the Renaissance, scientific study of volcanoes resumed. In 1546, German geologist Georgius Agricola (Georg Bauer, 1494–1555) proposed that Earth's subterranean heat was localized at volcanoes; he theorized that this heat was derived from the combustion of bitumen, sulfur, or coal, and ignited by very hot vapors. In 1583, the first full scientific study was conducted on the birth and evolution of a new volcano: Monte Nuovo (New Mountain), located near Pozzuoli in Campania, Italy. The study generated heated debates between religious leaders and scientists.

When did the **modern study of volcanoes** begin?

In the mid-1700s, a new age of scientific inquiry began, leading to renewed interest in natural phenomena such as volcanoes and earthquakes. For example, in the early 1750s, French geologist Jean-Etienne Guettard (1715–1786) noted the similarity between black rocks found on Etna and Vesuvius and those used in construction in Auvergne. When he discovered the source rocks used for building in a quarry at Volvic, he was able to conclude that the rocks were the weathered remains of ancient volcanic cones. In 1752, Guettard delivered a paper to the French Academy of Science titled "Memoir on Certain Mountains in France that Have Once Been Volcanoes." He spent the next decades mapping and studying the extinct volcanoes of France.

Who conducted the **first modern study in volcanology**?

The first modern scientific study in volcanology consisted of a series of letters published in 1772 by Sir William Hamilton (1730–1803) to the president of the Royal Society in London. Hamilton was the British envoy to Naples from 1764 to 1800, and the letters contained his observations about Mt. Vesuvius in Italy. During his entire career in Naples, Hamilton made more than 200 trips up the volcano and studied nine eruptions. He also compiled a list of the volcano's past eruptions.

Who discovered how **molten rock crystallizes**?

During the 1790s, Scottish geologist Sir James Hall (1761–1832) conducted a series of experiments to determine how molten rock crystallized upon cooling. Hall obtained samples of local basalt-type rock and melted them in a forge. He named the resulting fluid magma after a Latin chemical term used at that time for a pasty substance. Hall then cooled the rock under different conditions: The magma formed a glass if it cooled quickly, and developed a crystalline structure if cooled over several hours. Hall also discovered he could change the crystals' sizes by varying the cooling time; and he experimented with the effects of enormous pressure, as well as temperature, on rock.

VOLCANOES DEFINED

What is a **volcano**?

A volcano is simply an opening in the Earth's surface in which eruptions of dust, gas, and magma occur; they form on land and on the ocean floor. The driving force behind eruptions is pressure from deep beneath the Earth's surface as hot, molten rock wells up from the mantle. The results of this activity are a number of geological features, including the buildup of debris that forms a mound or cone commonly thought of as a volcano.

What is **volcanology**?

The study of volcanoes is called volcanology (also spelled vulcanology). Volcanologists examine an entire range of subjects, such as understanding the gases emitted by volcanoes, structures of volcanoes, effects of volcanoes on current climate and paleoclimates, reduction of volcanic hazards, and monitoring active volcanoes for possible eruptions.

What is the **origin** of the word **volcano**?

The word volcano comes from the Mediterranean island of Vulcano, site of an active volcano in ancient times. Located off the northern tip of Sicily in the Tyrrhenian Sea, this volcano was thought by the Romans to be the entrance to the nether regions of Earth. It was the domain of Vulcan, god of fire, who forged armor for the gods and thunderbolts for Jupiter.

What is **magma**?

Magma is hot, liquid (molten) rock originating from deep within our planet. The energy source that turns rock into magma is still a highly debated topic in geology. Theories include heat from the Earth's natural radioactivity, compression from pressures deep within the planet, past meteorite collisions, and the original heat from when the planet formed.

What is **lava**?

Lava is magma that flows along the ground (on land or the ocean floor), eventually cooling to form volcanic rock. There are two basic types of lava, both having Hawaiian names:

Pahoehoe—A smooth lava that forms a ropy surface is called pahoehoe (pronounced pa-hoy-hoy); it comes from higher-temperature and lower-volume volcanic eruptions. Pahoehoe flows readily and forms a skin. It normally travels at approximately 3 feet (1 meter) per minute in a creeping motion, but under the right conditions (high rate of eruption and/or steep slopes), it can reach speeds of up to 14 miles (23 kilometers) per hour. Typically, the thickness of a pahoehoe lava flow is approximately 1 foot (0.33 meters).

Aa—Aa lava (pronounced ah-ah, the name expressive of the way you would feel if you fell or stepped on its sharp surface) is the opposite of pahoehoe: It is coarse and rough, and occurs under lower-temperature and higher-volume eruptions. Aa flows in surges, with a slow moving front (a few yards per hour) that builds up height, then quickly releases. The resulting aa flow can cover ground at the rate of a hundred yards in minutes, then return to its original thickness. The flows are very large, with widths longer than a football field, the thickness ranging from 6.5 to 16.5 feet (2 to 5 meters).

Are there **other types of lava**?

Yes, there are a few other variants of pahoehoe and aa. For example, Pele's hair is formed when very viscous lava is forced through a small opening and modified by the wind. Obsidian is natural glass and considered a lava, forming as the lava cools quickly and solidifying into a dense, usually dark, glassy material. Pumice is lava that rapidly cools, trapping gas bubbles inside. This results in a low density, light rock, one of only a few that float in water.

Why are volcanoes dangerous?

Volcanoes are dangerous for a number of reasons, and this puts a lot of people at risk because millions of human beings live very close to active volcanoes. For example, about 500 million people live dangerously close to volcanoes with the potential to spread deadly ash and lava. For example, Seattle, Washington, lies within the destructive range of Mt. Rainier, a volcano in the Cascade Range. The huge volcano El Popocatepetl is visible from Mexico City, Mexico; thus, the world's largest city with about 30 million people lies in the shadow of the volcano.

Deaths are usually the result of suffocation (by volcanic ash or gases), lava flows, erupting bombs (rocks that are still molten on the inside), or even shock waves from the eruption. Indirectly, a volcano can cause mudflows that cover villages below the mountain, killing people and causing others to abandon homes and land forever. Aircraft can be affected by the huge ash plumes sent up by an erupting volcano. And, depending on the amount of volcanic debris sent into the atmosphere, even the local or global climates can be changed after an eruption.

How does a **volcano form**?

Volcanoes are geologic features usually occurring at the boundaries between separating or colliding lithospheric plates. When plates separate, openings are created at the boundaries, allowing magma to flow to the surface and creating volcanoes. For example, there are numerous volcanoes located along the Mid-Atlantic Ridge, an area where the North American and Eurasian plates are currently pulling apart. (For more information on lithospheric plates, see "The Earth's Layers.")

When lithospheric plates collide, one plate may move beneath the other, or subduct. As the plate sinks into the hot mantle of the Earth, it becomes molten; the resulting magma eventually rises to the surface, forming volcanoes. These types of volcanoes can be found along the edges of continents, such as the spectacular Andes Mountains of South America, or rising from the ocean, such as the islands of Japan.

Volcanic islands and/or chains can also form in the middle of a plate. Most are created at "hot spots," areas in which magma bubbles to the Earth's surface (for more about hot spots, see below).

What are some **types of volcanic rocks**?

The types of volcanic rocks are often based on where they formed (under or at the surface) and by the amount of silica (silicon dioxide or SiO_2) content. For example, volcanic rocks on the Earth's surface (extrusive rocks) are basalt (45 to 54 percent silica), andesite (54 to 62 percent silica), dacite (62 to 70 percent silica), and rhyolite (70 to

78 percent silica). Rocks containing more silica also have more sodium and potassium, but less iron, calcium, and magnesium. These last three elements are darker colored. Thus, basalts (with more of these elements) are dark gray to almost black, andesites are medium gray, and dacites and rhyolites are light tan to gray. The one exception to this rule is obsidian, a glassy rhyolite that is nearly black.

Volcanic rocks that solidify below the surface (intrusive rocks) are gabbro (45 to 54 percent silica), diorite (54 to 62 percent silica), granodiorite (62 to 70 percent silica), and granite (70 to 78 percent silica).

What are the differences between **active, dormant, and extinct volcanoes**?

Volcanoes are classified as active (eruptions are occuring regularly), dormant (no eruptions at present), or extinct (no longer any chance of eruptions). However, this last classification is by no means one hundred percent accurate. On average, one "extinct" volcano erupts every five years.

What are the **types of volcanoes**?

There are several major volcano divisions based on such parameters as how the volcano formed, or its shape, size, or predominant rock type. But scientists have learned that volcanoes are not easy to classify, especially since most are a complex combination of different types, such as a composite caldera. Two classifications of volcanoes follow:

Stratovolcanoes—Stratovolcanoes are the most prevalent on the planet, totaling about 60 percent of all individual volcanoes. Alternately layered with half lava and half pyroclastic material, they are commonly referred to as composite volcanoes. Because of their cooler and more viscous lava, gases tend to build up and cause explosive eruptions. Lava from these volcanoes barely flows at all, and so it piles up in the vent to form volcanic domes, with some being comprised of just domes piled one on top of the other. Examples include Mt. St. Helens and Mt. Rainier (United States), Mt. Pinatubo (Philippines), and Mt. Fuji (Japan).

Shield volcanoes—Shield volcanoes are almost all basalt and are created by a very fluid lava. This results in volcanoes that are not steep, with 90 percent of the volcano made of lava rather than pyroclastic material. They are some of the largest volcanoes in the world, and include the Hawaiian volcanoes Kilauea and Mauna Loa.

When does a **volcano not look like a volcano**?

This may sound like a trick question, but there are three other types of volcanoes that don't look quite normal. One includes rhyolite caldera complexes, the Earth's most explosive volcanoes. But they are not typical. Because of the volcano's explosive power, it collapses in on itself rather than building up the usual volcanic structure. The collapsed depressions at the eruption point are large calderas associated with huge magma chambers. The layers of ash falls or flows that comprise the complexes often extend over thousands of square miles in all directions from the central caldera. A good example includes the Yellowstone National Park area in Wyoming.

Monogenetic fields are another type of "volcano" that doesn't look quite right. They are collections of sometimes hundreds to thousands of separate vents and flows and are the result of very low amounts and slow movement of magma. Each magma push takes its own pathway to the surface, similar to spreading a volcano's single eruption point over a large area. One famous monogenetic field is the San Francisco volcanic field in northern Arizona, a region that covers about 1,800 square miles (4,662 square kilometers).

Another strange type of "volcano" includes flood basalts, which are thick, basaltic lava flows that cover thousands of square miles and can be more than 160 feet (50 meters) thick. One of the most well-known examples is the Columbia River Basalt area, covering most of southeastern Washington state and part of Oregon; two others are the even larger Deccan Traps of northwestern India and the Siberian Traps. (Some scientists believe the volcanic activity that formed the Deccan Traps may have contributed to the demise of the dinosaurs.)

What **type of volcano** has caused the most **casualties**?

Based on past records and current observations of volcanoes, stratovolcanoes have caused the most casualties. Statistically speaking, there are more stratovolcanoes than any other type, making it more likely that populations will live around such mountains. Their slopes contain ash and lava, which can easily slide or collapse during heavy rains or earthquakes, creating dangerous mudflows (lahars), landslides, and volcanic debris avalanches. Finally, entire sides of these volcanoes often collapse. For example, in 1792, the Unzen volcano in Japan collapsed, dumping a huge chunk of land into a shallow inland sea and generating a sea wave that killed close to 15,000 people along nearby coastlines. A more recent example occurred on Mt. St. Helens in 1980, when the north flank of the volcano collapsed after a major eruption.

What are the **two major characteristics** of volcanic **eruptions**?

A volcano can experience violent eruptions or less explosive, more effusive eruptions that produce wide-ranging lava flows. In general, the eruptive characteristics of a volcano are based on the silica and water content of the magma.

What is **tephra**?

Tephra is the name given to all the material that erupts from a volcano, excluding lava. Tephra comes in all shapes and sizes, and is also referred to as pyroclastic material ("fire particles"). A pyroclast is material that is ejected during the explosive eruption of a volcano in the form of fragments; pyroclastic material that is hot enough to fuse together before it falls to the ground is called welded or volcanic tuff. Geologists classify tephra according to size. The following lists the most common types of tephra:

Ash—Ash is material smaller than approximately a tenth of an inch (2 millimeters) that is emitted from an erupting volcano; it can also contain lapilli (also called cinders or "little stones"), which is between 1 and 25 inches (2 and 64 centimeters). In a large eruption, ash can accumulate to a great thickness and spread out for thousands of miles (usually in the direction of the prevailing winds).

Block—Blocks are solid rock emitted from an erupting volcano. They can be anywhere from the size of a baseball to the size of a boulder as large as a house.

A pair of visitors to the Yellowstone National Park photograph the Old Faithful geyser as it rockets steam and water 100-feet skyward. *AP/Wide World Photos.*

Bombs—Bombs are volcanic rocks that are still molten inside; they are shaped by their passage through the air. (They form the brilliant arcs seen in time-lapse photography of volcanic eruptions.) Typically ranging from baseball to basketball size, they can be as large as a house. Bombs (and blocks) can be ejected from a volcano with initial velocities greater than 1,000 miles (1,609 kilometers) per hour, and can travel more than 3 miles (5 kilometers), with some exploding and gushing molten rock when they eventually strike the ground. There are also certain types of bombs, including spindle bombs (very fluid magma chunks that are streamlined as they fly through the air) and breadcrust, which is formed from viscous magma, creating rounded blobs that often have fractured surfaces.

What are some **major volcanic features**?

There are many features associated with a volcano. The major ones include: the magma chamber (the underground source of magma), volcanic vent (any opening that leads magma to the surface; it can be large or small), central vent (the main, central, long tube conduit that brings the molten rock from the magma chamber to the surface), caldera (the crater, or largest volcanic vent, located at the top of the volcano), lava flow (the magma that flows from the open crater), fumarole (a side vent off the main volcano), splatter cones, side feeders, fissures, and volcanic or lava dome (a steep-sided mound that forms when viscous lava piles up near the caldera).

What are **geysers**?

Most geysers are hot springs that episodically erupt fountains of scalding water and steam. The majority of the world's geysers are associated with regions of volcanic activity, areas that supply the necessary heat to boil shallow ground water. (Although some "cooler" geysers erupt from cold or warm springs powered by gas pressure instead of boiling water.) In general, volcanic-associated geysers need three ingredients: an internal source of heat to warm the groundwater in a confined space, a conduit (a natural pathway) that is almost water and pressure-tight, and enough water pumped in to keep the geyser erupting.

Geysers are actually quite rare, with only about 1,000 known on the planet. The world's premier geyser fields are in Iceland, New Zealand, and Yellowstone National Park, Wyoming, which is home of the famous "Old Faithful" geyser and over 400 others (close to half the geysers that exist in the world).

What is a **phreatic eruption**?

Phreatic eruptions are steam-driven explosions that occur when groundwater or surface water is heated by magma, lava, hot rocks, or new volcanic deposits. The intense heat of such material can cause water to boil and turn to steam, generating an explosion of steam, water, ash, blocks, and bombs.

What **volcanic landforms** form when rising **magma contacts water**?

Two landforms commonly develop when groundwater or ocean water create steam that explodes on contact with rising magma: maars and tuff rings. A maar is a low-relief, flat-floored, broad volcanic crater formed by shallow explosive eruptions. The explosions are caused by the boiling of groundwater when magma invades the groundwater table; the resulting rim of the crater is not new volcanic material, but largely rock from the excavated crater. Maars often fill with water to form a circular lake. The Eifel craters in Germany are examples of maars.

Tuff rings form as hot material erupts in a surge, contacts water, and explodes, building a ring of debris around the explosion crater. The rim material is a combination of surrounding rock and new volcanic material. One of the most famous tuff rings is Diamond Head, a prominent landmark east of Waikiki Beach, Honolulu. This ring

299

What is the connection between volcanic activity and geothermal energy?

Geothermal energy is created as heat from active volcanoes, or geologically young inactive volcanoes that are still giving off heat, warms underground natural fluids. Steam from high-temperature geothermal fluids can be used to generate electrical power or drive turbines; lower temperature fluids can be used to heat greenhouses and for industrial uses, and hot or warm springs are often used at resort spas.

On the average, geothermal energy releases about 1/16th watt from beneath each square yard of the Earth's surface (an area the size of a football field could, for example, run a 60 watt light bulb). Therefore, to make geothermal energy worthwhile, only volcanic regions that are "energy productive" have been developed. Most exist in areas where geysers and hot springs are abundant, including Iceland (where geothermal heat warms more than 70 percent of the homes), Yellowstone, and New Zealand.

formed some 3 million years after eruptions generated by lava interacting with seawater created the island of Oahu.

What is a **pluton**?

Volcanoes are the exposure of magma at the Earth's surface, but sometimes magma remains underground. These under-the-surface movements of magma are called intrusions: molten rock that flows into the countless underground cracks in the Earth's crust. If this magma collects and solidifies, it forms features called plutons. Like jelly in a mold, each pluton takes the shape of its surroundings. The following lists some of the more common types of plutons (categorization is based on shape):

Batholith—A large mass of magma—some extend over a hundred square miles—that collects deep below the Earth's surface. Cooled, solid exposures of batholiths include those found in the Rocky and Sierra Nevada Mountains. Batholiths are called stocks if they cover less than 30 square miles (75 square kilometers).

Laccolith—A large chunk of magma, usually an offshoot from a magma chamber, that fails to reach the surface.

Lopolith—A magma intrusion shaped like a spoon, with both roof and floor sagging downward because the underlying rock sags.

Phacolith—A magma intrusion that forms on top of an anticline (crest-shaped fold in the rock) or at the base of a syncline (trough-shaped fold).

Dike—A dike forms when magma intrudes vertically through rock layers and usually does not reach the surface.

Sill—A sill forms when magma intrudes horizontally through rock layers and usually does not reach the surface.

What **major gases** do **volcanoes emit**?

Volcanic gases contained within the magma (molten rock) are released as they reach the Earth's surface, escaping at the major volcanic opening or from fissures and vents along the side of the volcano. The most prevalent gases are carbon dioxide (CO_2) and hydrogen sulfide (H_2S). Carbon dioxide is a dangerous gas; it is invisible and odorless, and can kill within minutes.

One example in which volcanic gases proved dangerous involved the Dieng Volcano Complex (or Dieng Plateau) in Java, Indonesia. It

Filipino farmers plow rice fields in San Fernando, Philipines, on July 8, 1991, as nearby Mt. Pinatubo erupts with smoke and volcanic ash. *AP/Wide World Photos.*

consists of two volcanoes and over 20 craters and cones, and is noted for its poisonous gas emissions at some craters. In 1979, at least 149 people were killed by poisonous gases as they fled eruptions at two of the craters: the Sinila and Sigludung.

What are **lahars** and why are they so **destructive**?

Lahars are volcanic debris flows that drop poorly sorted material—a mix of sand, silt, and clay—along valleys near explosive volcanoes. Lahars (often called mudflows) often occur when heavy rains wash down loose materials from earlier eruptions or when the material is shaken loose during such events as an earthquake.

For example, the 1991 Pinatubo eruption in the Philippines was the second largest of the 20th century (after Katmai in 1912). The eruption deposited a huge amount of loose material on the steep and gullied slopes of the volcano. Add to this the high precipitation of the country and it is easy to see why hundreds of lahars have been generated since the initial explosion. Since 1991, more people have been killed or injured by lahars near Pinatubo than died or were injured from the original eruption.

What is a **pyroclastic flow**?

A pyroclastic flow (or nuée ardentes, which is French for "glowing cloud") is a ground-hugging, turbulent avalanche of hot ash, pumice, rock fragments, crystals, glass shards, and volcanic gas. The flows can rush down the steep slopes of a volcano as fast as 50 to 100 miles (80 to 161 kilometers) per hour, burning everything in their path. Temperatures of these flows can reach greater than 932°F (500°C). A deposit of this mixture (often the materials are welded together because of heat and pressure) is also often referred to as a pyroclastic flow. An even more energetic and dilute mixture of searing volcanic gases and rock fragments is called a pyroclastic surge, which can easily ride up and over ridges.

There have been numerous pyroclastic flows from volcanoes throughout human history. On May 8, 1902, such a cloud rushed down the volcanic mountain of Mt. Peleé in the West Indies and into the coastal city of St. Pierre. In only a few minutes, everything was destroyed; out of 28,000 inhabitants, only two survived. Mt. St. Helens also experienced a pyroclastic flow in the 1980 eruption, but without the same tragic results.

What is a **fumarole**?

Volcanic gases escape from fumaroles, or vents, around volcanically active areas. They can occur along tiny cracks or long fissures in a volcano, in groups called clusters or fields, and on the surfaces of lava and pyroclastic flows. Fumaroles have been know to last for centuries. They can also disappear in a few weeks or months if their source cools quickly. For example, Yellowstone National Park and the Kilauea volcanoes have many fumaroles and associated deposits; some that have been there for years, while others have just recently appeared.

UNDERWATER VOLCANOES

Are **volcanoes** important features of the **ocean floor**?

Yes, volcanoes are important features of the ocean floor because they create new land that rises from the depths of the sea. In particular, some form along spreading boundaries of the lithospheric plates (called seafloor spreading), expanding and adding to the plate's landmass. This, in turn, keeps the plates in motion around the Earth, a process known as plate tectonics. (For more information about plate tectonics, see "The Earth's Layers.")

Whether they remain in the deep ocean or emerge above the waves, marine volcanoes are also the foundation of unique ecosystems. Active volcanoes are continually adding energy, mostly in the form of heat, to the environment surrounding them. This energy is essential for many organisms, including life that only exists around volcanically active areas.

How much **volcanic activity** occurs **underwater**?

There is a huge amount of volcanic activity taking place underwater—we just can't see it. Some geologists estimate that approximately 80 percent of all Earth's volcanic activity occurs on the ocean floor.

What is **pillow lava**?

Pillow lava forms underwater and is a variation of pahoehoe lava (for more about types of lava, see above). It can be seen on dry land only where the sea floor has been uplifted. Pillow lava is unique to the underwater realm and is the result of colder water temperatures and greater pressures in the deep ocean. As magma flows out of underwater vents, its surface is cooled by the water, forming a flexible crust in the shape of a pillow. Although the crust is solid, the interior remains molten. As more and more magma enters the pillow, it expands until a break occurs somewhere in the crust. The magma then flows out of this opening, another pillow is formed, and the process is repeated until the eruption ends. It is not uncommon for large underwater areas to be covered with a thick deposit of pillow lava during an eruption.

What are **seamounts**?

One of the most spectacular underwater volcanic features are huge, localized volcanoes called seamounts. These isolated underwater mountains rise from 3,000 to 10,000 feet (914 to 3,048 meters) above the ocean floor, but typically are not high enough to poke above the water's surface.

Seamounts are present in all the world's oceans, with the Pacific having the highest concentration: more than 2,000 have been identified, and some are still active. The Gulf of Alaska also has numerous seamounts. The Axial Seamount, an active volcano off the north coast of Oregon, currently rises some 4,500 feet (1,372 meters) from the ocean floor, but its peak is still about 4,000 feet (1,219 meters) below the surface.

Has anyone ever witnessed the **birth of a volcanic island**?

Yes, scientists have witnessed the birth of a volcanic island: Surtsey on the Mid-Atlantic Ridge. Surtsey first appeared on November 16, 1963, off Iceland's southern

Occasionally, a volcano becomes large enough to rise above the ocean surface, forming islands and even chains of islands. One expected to form an island is the Loihi Seamount, currently growing approximately 20 miles (32 kilometers) off the coast of the Big Island of Hawaii. At present, this volcano is approximately 17,000 feet (5,182 meters) above the ocean floor, and about 3,000 feet (914 meters) below the ocean's surface. At the current rate of growth, Loihi will emerge above the waves in several thousand years to become the newest Hawaiian island.

coast. Its emergence above the waves was accompanied by a spectacular display of steam and erupting magma. Over the next four years, it grew to a height of 492 feet (150 meters) above sea level, covering an area of 2 square miles (5 square kilometers). Surtsey is now a permanent island colonized by plants and surrounded by sea life.

Less fortunate were two smaller islands that formed near Surtsey. The Syrtlingur ("Little Surtsey") formed in May 1965, while the Jolnir ("Christmas Island") emerged in December 1965. They both soon disappeared, however, as the eruptions ceased and waves soon eroded them away. Jolnir lasted only 8 months.

What are some **volcanic mid-ocean islands**?

There are many mid-ocean islands of volcanic origin. One of the most famous is Iceland, located along the Mid-Atlantic Ridge just off the southeast coast of Greenland. Eruptions along the Mid-Atlantic Ridge have given rise to approximately 200 active volcanoes on Iceland, all of which helped form this large island.

Also along the Mid-Atlantic Ridge are the Azores Islands, just off the coast of Portugal. Another collection of mid-ocean islands are the islands of the Tristan da Cunha group in the South Atlantic.

What is considered by some to be a **"70,000 kilometer-long" volcano**?

The connected major oceanic spreading centers—including the Mid-Atlantic Ridge, Pacific-Antarctic Ridge, Chile Ridge, Indian Ridges, and Juan de Fuca Ridge—are sometimes considered to be a 43,496 mile (70,000 kilometer) volcano. Lithospheric plates are pulling apart along this ridge system, allowing magma to rise to the surface and create volcanoes.

What is a **hot spot**?

A hot spot is a stationary, localized flow of magma pushing through areas of weakness in an overlying lithospheric plate. Unlike most volcanoes along plate boundaries, this

What is unique about Mauna Loa?

Mauna Loa, a volcano on the Big Island of Hawaii that originated as an underwater volcano, is the most massive volcanic mountain on Earth. It is also the most massive mountain of any type on our planet. Mauna Loa occupies almost 10,000 cubic miles (41,682 cubic kilometers), and rises a total of approximately 30,000 feet (9,144 meters) from the ocean floor. Above sea level, it stands 13,679 feet (4,169 meters) tall.

phenomenon occurs in the middle of tectonic plates, creating volcanoes far away from where they would "normally" be found. The idea of hot spots was first proposed by Canadian geophysicist J. Tuzo Wilson (1908–1993) to explain the presence of volcanoes in mid-plate. Today many geologists are in agreement with this theory, although the exact mechanism is still debated.

Where are some **hot spots located**?

Hot spots are located all around the world, and many of them are quite famous. For example, on land, Yellowstone National Park in Wyoming sits on top of a hot spot. Many hot spots are also located in mid-ocean, where they create underwater volcanoes that often grow high enough to form islands above sea level. For example, the Galapagos Islands, Society Islands, and Hawaiian Islands are all located at hot spots.

What is unique about the **Hawaiian Islands**?

Geologists believe that the Hawaiian Islands are the result of a long-lived hot spot located deep under the Pacific Ocean. Over a period of approximately 75 million years, the plate has moved slowly over the stationary hot spot, resulting in a chain (or "track") of almost 200 volcanoes extending in a northwest direction. Most of the volcanoes are now located beneath the waves, but the ones whose peaks still rise above sea level constitute the present day Hawaiian Islands. The Hawaiian hot spot is still very active: The most recent creation was the Big Island of Hawaii; currently, another volcanic island is forming just off Hawaii: a seamount called Loihi.

What lies **below** the **Hawaiian Island hot spot**?

Geologists have studied the Hawaiian Island hot spot using a technique called seismic tomography. They used seismic waves generated by earthquakes around Fiji and Tonga—waves that traveled through the deep mantle beneath the hot spot, eventually reaching recording stations in California and Oregon.

From this data, geologists discovered that the hot spot responsible for the Hawaiian Islands originates at the boundary between the mantle and the metallic core of our planet. The magma plume is located approximately 1,800 miles (2,900 kilometers) beneath the crust, where rocks at the base of the mantle are heated by the molten outer layer of the core. The seismic data also showed that the magma plume first flows horizontally toward the base of the hot spot, then rises vertically toward the surface.

What **other island chains** are associated with **hot spots**?

Another example of a volcanic track created by a stationary hot spot and moving lithospheric plate is located in the Indian Ocean. What is now India was located above this hot spot approximately 67 million years ago. At this time, large amounts of basaltic lava erupted, producing an extensive volcanic field called the Deccan Traps. The Indian plate moved slowly to the northeast, leading to the creation of the Maldives around 57 million years ago, the Chagos Ridge approximately 48 million years ago, and the Mascarene Plateau some 40 million years ago. Eighteen to 28 million years ago the Mauritius Islands formed, and most recently, in the last 5 million years, Reunion Island, which is made up of the Pito des Neiges and Piton de al Fournaise volcanoes, was created.

What are **lava pillars**?

Lava pillars are underwater columns of black volcanic rock (basalt) found deep on the ocean floor. These cylindrical columns can measure up to 50 feet (15 meters) tall; they range in size from thick and stout to thin and spindly, with some even forming archways. The outside of these pillars is coated with a paper-thin layer of black volcanic glass, while the inside is gray and lined with fine cracks. Lava pillars were first seen by humans using the lights of a deep-sea submersible.

How lava pillars form is still debated. One theory is that during an underwater eruption, water trapped below the growing layer of magma squirts out through gaps in the lava flow. As more magma flows out, the lava thickens and grows upward around the jets, the cold water swiftly "freezing" the lava into pipe-like cylinders.

What are **black smokers**?

Black smokers are actually deep-ocean hydrothermal (hot water) vents, named after the dark, soot-like material ejected from "chimney" formations on the ocean floor. The material is actually superheated water (around 662°F [350°C]) with very high concentrations of dissolved minerals—mostly sulfur-bearing minerals or sulfides from lava on a mid-ocean ridge volcano. As the hot water meets the cold ocean waters, the minerals precipitate out, settling out around the surrounding rock. Over time, the hollowed-out chimneys grow taller as more minerals precipitate out.

Black smokers tend to occur in volcanic vent fields that are typically tens of yards across, with fields ranging from pool-table size (43 square feet [4 square meters]) to tennis court size (8,288 square feet [770 square meters]). For example, vent fields are found on the Juan de Fuca Ridge, a mid-ocean ridge in the Pacific Ocean. Many vents have been discovered since the first site was found in 1977 near the Galapagos Islands (in the small research submersible *Alvin*), and there are probably many more. But scientists have only explored a small portion of the Earth's mid-ocean ridges.

MEASURING VOLCANOES

How are **volcanoes monitored**?

There are numerous ways scientists monitor volcanoes. For example, the shaking of the ground from earthquakes (or seismicity) around the volcano is often an early warning sign of volcanic activity. (Earthquake swarms often immediately precede most volcanic eruptions.) Scientists also monitor other types of ground movements with sophisticated instruments, measuring the changes in the volcano's surface shape caused by the pressure from moving underground magma. (Upward and outward motion above a magma storage area often occurs before an eruption.) Still other movements are monitored with simple techniques, such as measuring volcanic cracking or noting changes in water levels around a crater lake. Measuring changes in electrical conductivity, magnetic field strength, and the force of gravity also helps trace magma movement. These measurements may indicate movement even when no earthquakes or measurable ground deformation occurs.

Another way to monitor volcanoes is to examine compositional changes in emitted gases. Such differences may indicate moving magma, as gases change pathways to escape, or changes in the magma supply or type. Finally, changes in the local hydrology

Can volcanic eruptions affect the global climate?

Most volcanic eruptions do not affect the global climate, although larger ones can cause disruptions—albeit for a relatively short period of time. Large eruptions tend to eject gases and dust high into the stratosphere. From there, prevailing winds carry the particles around the world—sometimes with interesting results.

For example, in 1815, Mt. Tambora on the island of Sumbawa (near Java) erupted, putting out a record amount of ash that briefly changed the world's climate. Huge amounts of volcanic dust rose high into the atmosphere, reaching around the globe. That year (and for some of the following year), volcanic particles screened out some sunlight, causing the global temperatures to fall. In Europe and other parts of the Northern Hemisphere, winter never seemed to end, with frosts occurring throughout the summer.

(water regime) are often monitored, including groundwater temperature and levels, changes in streamflow, sediment content of rivers, lake levels, and even snow and ice levels—all of which can be indicators of possible eruptions in volcanically active areas.

What is a **volcano observatory**?

There are numerous volcano observatories around the world's volcanically active regions. As the name implies, scientists at these observatories keep a keen eye on the local volcanic regime. Most observatories provide continuous and periodic monitoring of the seismicity, geophysical changes, ground movements, gas chemistry, and hydrologic conditions and activity between and during eruptions.

If at all possible, volcano observatories provide a detailed record of all eruptions in progress, especially the type of eruption, what led up to the eruption, and any conditions that might pose a hazard during or even after the eruption. Finally, each observatory is a center for research, exchanging data and observations with other scientists around the world. Such research leads to computer models of volcanic eruption scenarios, showing patterns that might one day be used to help predict volcanic eruptions more accurately.

VOLCANOES AROUND THE WORLD

What is the **Ring of Fire**?

The Ring of Fire, also known as the Circle of Fire, is the name given to the periphery of the Pacific Ocean that contains volcano and earthquake zones—a belt that follows

Snow-covered Mt. Fuji, Japan's highest peak at 12,385 feet and a potentially deadly volcano, is studied using underground blasts that trigger mini earthquakes, which help scientists map the magma bubbling beneath and help gauge the likelihood of an eruption. *AP/Wide World Photos.*

active subducting lithospheric plates. The Ring of Fire, moving counterclockwise, extends from the tip of South America north to Alaska, then west to Asia, south through Japan, the Philippines, Indonesia, then to New Zealand.

This swath is home to more than half of the world's active volcanoes, including 77 subduction-generated volcanoes in Japan, 75 in Chile, 69 in the western United States, 68 in Alaska and the Aleutian Islands, and 65 in the Kamchatka Peninsula. Examples of famous volcanoes along this ring are Mt. St. Helens in Washington, Mt. Katmai and Augustine in Alaska, Galunggung in Indonesia, and Mt. Fuji in Japan.

What are some **famous volcanoes** around the world?

There are many volcanoes around the world. The following lists some of the largest and most famous:

Name; Location	Elevation Above Sea Level (feet/meters)	Comments
Cotopaxi, Ecuador	19,347/5,897	One of the tallest active volcanoes
Etna; Sicily, Italy	10,902/3,323	More than 200 recorded eruptions
Kilauea; Hawaii	4,078/1,243	Currently active

Name; Location	Elevation Above Sea Level (feet/meters)	Comments
Krakatau; Indonesia	2,667/813	Eruption of 1883 was heard more than 2,900 miles (4,700 kilometers) away
Mauna Loa; Hawaii	13,677/4,169	World's largest volcano
Mt. Fuji; Japan	12,388/3,776	Nearly symmetrical cone
Mt. Katmai; Alaska	6,715/2,047	1912 ashflow formed the Valley of Ten Thousand Smokes
Mt. Pelee; Martinique	4,583/1,397	1902 eruption destroyed the West Indies city of St. Pierre in minutes
Mt. St. Helens; Washington	8,366/2,550	Erupted in 1980
Paricutin; Mexico	8,990/2,740	Grew from small crack in field starting in 1943
Stromboli; Italy	3,038/926	Continuous eruptions for at least 2,000 years
Surtsey; Iceland	586/173	First emerged from ocean in 1963
Tamora; Indonesia	9,354/2,851	1815 eruption released the most volcanic material to date
Vesuvius; Italy	4,203/1,281	79 C.E. eruption buried many cities, including Pompeii

What happened when **Mt. St. Helens erupted**?

On the morning of May 18, 1980, a magnitude 5.1 earthquake struck beneath the Mt. St. Helens volcano in Washington state. Later that day, the volcano exploded, initiating a massive avalanche that tore away the northern slope of the mountain and created the largest landslide in recorded history. The conical volcano went from about 9,678 feet (2,950 meters) to 8,366 feet (2,550 meters) in height, releasing a giant plume of ash and gas high into the atmosphere. A lethal pyroclastic flow of hot steam, gas, and rock debris raced down the slope of the mountain, traveling as fast as 684 miles (1,100 kilometers) per hour. In a short time, the ash blanketed central Washington, the prevailing winds carrying an estimated 540 million tons of ash across 22,007 square miles (57,000 square kilometers) of the western United States. The blast devastated 10 million trees in surrounding forests and killed around 60 people. It was one of the few major eruptions in the continental United States during modern times.

The eruption did not take scientists by complete surprise. It was preceded by two months of intense activity, including more than 10,000 earthquakes, hundreds of small phreatic (steam-blast) explosions, and the outward growth of the volcano's entire north flank by more than 262 feet (80 meters).

Mt. St. Helen's erupts on July 22, 1980, in Washington State. Although the resulting quake measured a modest 5.1 on the Richter scale, the eruption also created the largest landslide in recorded history, which reduced the mountain from 9,678 feet (2,950 meters) down to 8,366 feet (2,550 meters) in height. *AP/Wide World Photos.*

What happened **after the eruption** of **Mt. St. Helens**?

The Mt. St. Helens eruption put a great deal of material into the atmosphere. This is typical for most explosive eruptions, but in this case, it happened very close to populated areas. Pyroclastic flows ran down the mountain; mudflows followed, as the melting glacial water mixed with loose volcanic material. The largest mudflow even reached and blocked the shipping channel of the Columbia River about 70 miles (113 kilometers) from the volcano.

A fine ash dust was propelled into the upper atmosphere (stratosphere) at heights reaching up to 15 miles (22 kilometers), spreading to the east by the prevailing westerly winds and eventually reaching all over the world. People in towns nearby (and some as far away as western Montana) were affected by the rain of ash. Car radiators were clogged, upper respiratory problems worsened, air and ground travel were disrupted, and a thick coating of ash particles covered everything outside. The economic loss was also great, not only for the area surrounding the volcano, but also for those states affected by the ash fallout.

Are there many **active volcanoes** in the **United States**?

According to the United States Geological Survey, the United States ranks third (behind Indonesia and Japan) in the number of historically active volcanoes (at least those for which we have written accounts). In addition, of the more than 1,500 volca-

Will Mt. St. Helens erupt again like it did in 1980?

Nearly two months before Mt. St. Helens erupted, there were over 10,000 earthquakes, hundreds of small phreatic (steam-blast) explosions, and the volcano's entire north flank expanded outward by more than 262 feet (80 meters). A magnitude 5.1 earthquake occurred beneath the volcano at 8:32 a.m. on May 18, the start of the devastating eruption.

In the future, geologists will keep a watchful eye on such activity around the volcano. Currently, scientists studying Mt. St. Helens note that the mountain remains active and is therefore potentially dangerous. And this wasn't the only "recent" eruption: Mt. St. Helens has produced four major eruptions in the last 515 years, two of which were within two years of each other. The one that occurred in 1480 was about five times larger than the 1980 eruption. Like most of the volcanic mountains of the Cascade Range, scientists know there is potential for eruption, but they can't say when or even how large it could be.

noes that have erupted worldwide in the past 10,000 years, about 10 percent are located in the United States. Most of these volcanoes are found in the Aleutian Islands, the Alaska Peninsula, the Hawaiian Islands, and the Cascade Range of the Pacific Northwest; the remainder are widely distributed in the western part of the country. Although the Cascade Range volcanoes erupt less frequently than Hawaiian or Alaskan volcanoes, they are considered more dangerous. This is because they are the more explosive types of volcanoes and are near the highly populated areas of Washington, Oregon, and California.

Is **Alaska volcanically active**?

The Alaska Peninsula and Aleutian Islands contain about 80 major volcanic centers—all of which have one or more volcanoes. In all, Alaska has had one or two eruptions per year since 1900. There have been at least 20 catastrophic eruptions in the past 10,000 years, including the most recent at Novarupta in the Katmai National Monument in 1912. And even though most areas of the state are sparsely populated, there are many potential hazards from volcanic activity. Scientists are particularly concerned about volcanoes that may affect the Cook Inlet region, an area that contains 60 percent of Alaska's population.

Why is the **Nevado del Ruiz** volcano famous?

Nevado del Ruiz, a huge volcano in Colombia, brought attention to the dangers of volcanoes in our modern world. In November 1985, the volcano erupted, killing more than 20,000 people. But lava and ash were not the culprits: About 5 to 10 percent of

Why is Krakatau famous?

In 1883, Krakatau, a volcano located in the strait between Java and Sumatra, exploded. It released tons of material—the equivalent of about what is released by mid-ocean ridges in one year—into the atmosphere. The monster explosion killed nearly 40,000 people, and the sound produced by the explosion was heard as far away as 3,000 miles (4,800 kilometers). (For more information about Krakatau, see "Prominent Geological Features in the Southern Hemisphere.")

the ice that capped the volcano's summit melted, mixing with volcanic debris. This created lahars (mudflows) that raced down the mountain's steep slopes and into the town below.

What was one of the **greatest eruptions** of the **20th century**?

One of the greatest—and most surprising—eruptions of the 20th century occurred on the island of Luzon, Philippines. On April 2, 1991, Mt. Pinatubo in the western Samballes range erupted for the first time in nearly 500 years. Volcanologists had believed the volcano was dormant, but it turned out to be very active!

The first eruption was just the beginning. On June 14, 1991, Pinatubo tore itself apart in the greatest volcanic eruption of the century, destroying most of Luzon's west coast. Ash fell like snow across the region; it also rose high into the stratosphere, carried westward by the prevailing winds to mainland southeast Asia. Because of the first eruption, volcanologists watched the volcano closely. Not long after, almost a quarter of a million people were evacuated so that less than 350 were killed in the second and most destructive explosion.

Why are scientists closely watching Sicily's **Mt. Etna**?

The historical record of Mt. Etna's volcanism (the largest and most active volcano in Europe) is one of the longest in the world, dating back to 1500 B.C.E. Two styles of activity are typical of Mt. Etna: explosive eruptions (sometimes with minor lava flows) from the summit craters, and flank eruptions from fissures. Since October 2002, when Etna erupted in one of the most explosive flank eruptions of the past 150 years, scientists have been taking an even closer look at the volcano.

The most recent eruptions continued for three months and two days, ending on January 28, 2003. The tourist complex and skiing areas of Piano Provenzana were almost completely devastated by the lava flows. Heavy tephra falls nearly paralyzed public life in Catania and nearby towns. And for more than two weeks the International Airport of Catania, Fontanarossa, was closed due to ash on the runways.

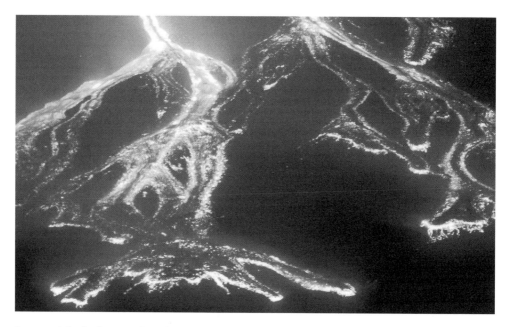

Streams of glowing lava ooze down the slopes of Mt. Etna in Sicily, Italy, in October 1999. *AP/Wide World Photos.*

Much of Etna's eruptions over the past 100 years have been relatively benign and emitted mostly lava, convincing the local population that Etna was not much of a threat. But scientists now realize that Etna's 20th-century, nonexplosive eruptions were unusual.

French scientists recently added another reason to keep watching the volcano: They reported that Etna appears to be undergoing a gradual shift from a "hot spot" volcano (in which magma wells up from within the Earth) to an "island arc" volcano (in which magma is produced from the collision of tectonic plates). Events more indicative of tectonic activity suggest this might be true: Hundreds of small earthquakes occurred in eastern Sicily just before the most recent eruption. If Etna is truly changing, it would be even more dangerous for the surrounding population, as island-arc volcanoes have more explosive, rather than oozing, eruptive events.

What are some **major and notable volcanic eruptions**?

There have been numerous major volcanic eruptions over time; many caused a great deal of death, others did not. The following lists some of the major eruptions since the 1500s, based on information from the United States Geological Survey:

Year	Eruption	Casualties	Major Causes of Death
1586	Kelut, Indonesia	10,000	Unknown
1683	Mt. Etna, Sicily	60,000	Ash and dust

Year	Eruption	Casualties	Major Causes of Death
1669	Mt. Etna, Sicily	20,000	Ash and dust
1784	Lakagigar, Iceland (Mt. Skaptar)	9,800	Starvation; poisonous gases kill livestock and crops
1792	Unzen, Japan	15,000	Volcano collapse; mudslide and resulting tsunami
1815	Tambora, Indonesia	92,000	Starvation
1883	Krakatau, Indonesia	36,000	Tsunami
1902	Mt. Pelee, Martinique	30,000	Pyroclastic ashflows
1980	Mt. St. Helens, Washington	61	Asphyxiation from ash
1985	Nevado del Ruiz, Columbia	25,000	Mudflows
1991	Mt. Pinatubo, Philippines	350	Roof collapse; disease, mudslides from excessive rains after eruption

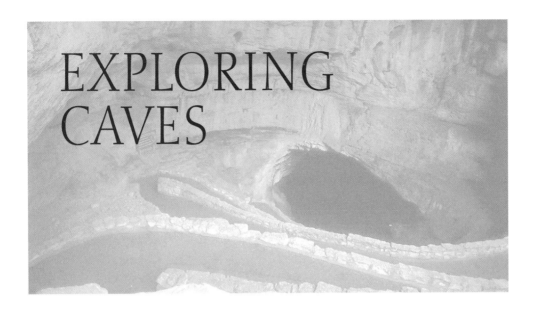

EXPLORING CAVES

ROCKING UNDERGROUND

What is a **cave**?

Simply put, a cave is a hole in the ground—but not just any hole. More specifically, it is a naturally formed, underground, air-filled void with an open area or chamber or a series of chambers. These voids can contain long, complex labyrinths of tunnels with one or more openings or just one main opening to the outside (such as an overhang). A cave may contain specific forms of life and often is large enough to accommodate exploration by humans.

A *cave system* is a network of caves or a complex cave with many "tunnels" (they are not true tunnels, but rather openings in the rocks caused either by physical or chemical weathering). A cave system can also be surface-water and/or groundwater related, although they may not be physically connected.

The term "cave" is also used to describe other geological features that are not as extensive as underground chambered caves. For example, a *sea cave* (or marine cave) is commonly a cavity at the base of a sea cliff that has been worn down by wave action in easily weathered rock. They are usually smaller, most likely at sea level, and are affected by the tides. A *sea chasm* is a deep, narrow sea cave. In addition, coastal caves may be underwater, such as the Sistema Ox Bel Ha in Mexico.

What is a **spelunker**?

Although many caving names are used interchangeably with each other, a spelunker most often denotes a person who enjoys exploring caves for a hobby or recreation; another name is caver. The name for a scientist who studies caves is speleologist, and if they are geologists, they are often called geologist-cavers.

What makes a **cave's interior environment** so unique?

There are two general features that make a cave's interior environment unique (not including overhangs): First, sunlight does not reach the interior of larger caves; second, the layers of rock surrounding the cave insulate the interior from surface weather, usually making the cave's climate relatively stable.

What are the **four zones** found in most caves?

The majority of caves contain four zones based on the amount of sunlight and cave climate variations (mostly temperature and humidity). The most common divisions are as follows:

Entrance—The cave entrance is usually a moist, shaded area just at the beginning of an opening. It is also the place where many animals—from box turtles and spiders to snakes and raccoons—take refuge.

A student collects a sample from a white stalactite while others observe inside Spider Cave at Carlsbad Caverns, New Mexico. Research is conducted on the cave, one of the most hostile environments in the world, in hopes of determining whether life could exist on Mars. *AP/Wide World Photos.*

Twilight zone—Although it is the name of a famous television show, the twilight zone of a cave has nothing to do with eerie events. The term defines the area that extends as far as the unaided human eye can see thanks to the outside sunlight that filters in. It is also the most hospitable part of a cave, with cool temperatures and high humidity, both of which fluctuate with the changing weather outside. Most of the animals found here are birds, frogs, snakes, and many types of invertebrates.

Middle zone—The middle zone—sometimes called the variable-temperature zone—is the part of the cave in which the temperature and humidity can be relatively stable, although in some caves air temperatures vary with the seasons in this zone. This is also the place of utter darkness and is often referred to as Stygian darkness (or blackness), a term associated with the mythical underworld. Animals here include bats, cave invertebrates (such as cave crickets), and various types of crustaceans.

Dark zone—The dark zone, or constant-temperature zone, lies deep inside most caves. Here, the temperature remains a nearly constant 55°F (12.8°C).

Do caves have weather?

Yes, some caves actually have their own weather. In the majority of caves, the air temperatures are cool, with little evaporation and high humidity. Some caves, such as Wind Cave in South Dakota, actually "breathe" as air currents inside the cave switch direction in response to changes in the surface's barometric pressure.

Sometimes "cave weather" is a benefit to humans. For example, in 1901 at Virginia's Luray Caverns, Colonel Theodore Clay Northcott built a sanitarium (Limair) directly above the caves, creating a natural air conditioning system. Installing a five-foot shaft into the cave—and a few fans—the cavern's air cooled the entire structure. Even on the hottest Virginia days, the sanitarium remained a comfortable 70°F (21.1°C).

There is total darkness (the term Stygian darkness is also apt here) and the humidity is often close to 100 percent. An overwhelming majority of the animals are blind, colorless, and unique to this zone (existing nowhere else in the world), including certain types of amphipods, pseudo-scorpions, and springtails.

How many **caves** are located in the **United States**?

It is estimated that more than 30,000 caves exist in the United States, and more are being found each year. Most caves are either run by private groups or are under the auspices of the National Park Service. *Show caves* are those usually open to the public; *wild caves* are those that are not open to the general public and/or remain unexplored.

CAVES DESCRIBED

What are **primary and secondary caves**?

The majority of primary caves develop as the host rock is forming. For example, lava tubes and reef caves are created as the rock forms. Although they are not created as the rock is forming, tectonic caves—those formed by the movement of the crust—are also often classified as primary caves. There are several different types of primary caves, but they are the least common to be found.

The most common caves are secondary caves, those formed by a physical or chemical mechanism that transports material away—in other words, most caves that are found around the world. The major mechanism that produces these caves is water dissolution in which minerals such as limestone, dolomite, gypsum, and marble are dissolved by water.

These are also called solutional or karst caves. Examples include salt (such as Sedom Cave in Israel), gypsum (such as the Barbarossa Cave in Germany), limestone (Lechuguilla Cave, "sister" of Carlsbad Caverns, New Mexico), and limestone karst (such as Carlsbad Caverns).

Is there a difference between **limestone and karst caves**?

It is widely believed that karst and limestone caves are the same, but in reality, there are karst caves in other types of rock and non-karst caves in limestone. Therefore, a good definition for both caves is as follows: Limestone caves are those that form in limestone rock; karst caves are those formed by water in water-soluble rock.

One reason for using the terms interchangeably is obvious: Almost all karst caves are made of limestone, and the majority of limestone caves are karst caves. But karst caves can also include gypsum and salt caves; non-karst caves include tectonic and reef caves.

What are the best **conditions** for **karst caves** to form?

Karst caves require several specific conditions in which to form. In particular, there must be sufficient layers of water-soluble rock (which can be thousands of feet thick); adequate annual rainfall of about 45 inches (114.3 centimeters) to percolate into the karst area; cracks and fissures into the bedrock (such as from faulting by uplifts or seismic activity); and a variable climate.

But there are differences between karst regions. For example, karst areas along a coast differ from those that are inland; temperate and sub-Arctic karst zones do not resemble those in the tropics. There are also differences between landforms in earthquake-prone karst zones versus less seismically active karst areas.

How does a **karst region** and its caves **form**?

Although the formation of a karst region varies depending on climate, type of rock, and sundry other variables, the following lists the major progression of a karst limestone cave region:

1. Rainwater picks up carbon dioxide from the air and carbon (in the form of dead plant debris) in the soil. It then carries the carbon mix (carbonic acid) as it percolates through the soil, eventually working its way into cracks in the mostly limestone bedrock below. (Note: Some karst regions form as a result of sulfuric acid welling up from below instead of carbonic acid percolating

Where did the word "karst" originate?

The name "karst" originated from the Kras region of southwestern Republic of Slovenia (part of the former Yugoslavia). This limestone plateau is located in the hinterland of Trieste Bay along the Adriatic sea, with a small part in western Italy, around the city of Trieste—the area in which it was first discovered. Karst is actually the German word; carso is the Italian word. This area is marked by typical karst terrain, including sinkholes and underground caverns.

One of the first karst studies came from Janez Vajkard Valvasor (1641–1693), a Slovene who explored and described many karst caves in the Kras region. In 1687, he submitted a treatise to the Royal Academy in London on the natural mechanism that fills and empties the intermittent Cerknica Lake, theorizing that it was done through a system of underground streams and reservoirs. (During the heavy autumn rains or the spring melt, water filled this more than 6.2-mile- [10-kilometer-] long by 3.1-mile- [5-kilometer-] wide lake that subsequently disappeared in May or June.) One cave in the region, the Vilenica, also has a grand history: Not only did the local people take refuge from the Turks in this cave, but it is probably the oldest karst cave opened for tourists in the world, with the first visits being recorded in 1633.

down from above; several areas in the western United States are a mix of these two processes.)

2. Soon the bedrock becomes saturated with the water, continually dissolving as the water moves sideways along horizontal cracks and fractures in the rock layers. Over time, as the water continues to move through (by gravity and natural hydraulic pressure), the conduits enlarge, mainly by the dissolving and abrasion of rock. Because there is more water flow, underwater springs often form at the cave entrances.

3. In most karst regions, the groundwater level eventually drops, exposing more of the underground conduits to air. These voids can eventually become passageways scouted by humans when a cave system is discovered. Those areas of karst not exposed to air—and that never develop a natural opening—are often used by humans for drinkable water supplies.

4. During this process, there are several changes in the cave's chemical equilibrium, resulting in cave deposits made of dissolved minerals called speleothems. As time goes on, the cave can intermittently refill with water, continue to dry out, or cycle through both as water levels change. Erosion will continue at

Why are there so many limestone karst caves?

It is estimated that more than 90 percent of all caves around the world are limestone karst caves. The reason for this profuse number is rather simple to understand: Limestone is a very common sedimentary rock that is formed from the remains of all types of (mostly marine) animals. This rock forms (and has formed) almost everywhere on the Earth, including along tropical reefs (limestone corals and sponges), on continental shelf areas (in shallow seas), and in large lakes. Many limestone karst sites originated after rock was exposed to the surface from uplifting processes, such as mountain building; it was then weathered and eventually developed into a karst area.

Where is limestone not so common? On the deep ocean floor (composed mostly of clays), where ocean water absorbs (solutes) the limestone falling from the upper layers of the sea.

various intensities, depending on such factors as local climate and rainfall amounts.

Do **karst regions** provide **drinkable water**?

Yes, many karst regions (especially those made of limestone) contain aquifers that provide large supplies of water to residential, industrial, agricultural, and commercial areas. In fact, more than 25 percent of the world's population either lives on or obtains its water from karst aquifers. In the United States alone, 20 percent of the land surface is karst, with karst aquifers providing 40 percent of the country's potable groundwater.

What **types of limestone** form **limestone karst caves**?

Limestone karst caves can form in the following types of rock: Calcium carbonate (or actual limestone), sometimes referred to as chalk; magnesium calcium carbonate (or dolomite); and metamorphosed limestone (or marble that forms as limestone is exposed to high pressure and extreme heat that is nevertheless not hot enough to melt the rock or change its overall chemistry).

What is a **salt cave**?

Salt (halite, sodium chloride, or NaCl) caves are thought to be the rarest caves on Earth. Similar to limestone caves, these caves are created where salt is formed as it precipitates out of water—usually seawater containing a high amount of salt. Unlike limestone caves, salt caves need to form in a specific geographic location where there are many salts in the water, soil, and rock, and there is an arid climate. One reason they are so rare is because of these specific needs, and because they have a very short life span on the order of several thousand to only a few years.

What is a **gypsum karst cave**?

Like salt caves, gypsum caves are very rare, mainly because gypsum is so soluble in water. This also means that gypsum caves are geologically young (the old ones are easily worn away) and are often very large. The longest gypsum cave in the world is Optimistich-eskaya Cave in the Ukraine, which contains 102.5 miles (165 kilometers) of passageways.

As with salt caves, gypsum caves need special conditions in order to form: Gypsum (calcium sulfate or $CaSO_4$) and anhydrite (alabaster) need a depression to trap and evaporate seawater (like salt, gypsum precipitates out during the evaporation of ocean water); a temporary connection to the sea (so a good deal of gypsum can be deposited); and a location in an arid part of the world (so rapid evaporation can take place).

What **other features** form around **karst areas**?

There are several other features that form around a karst region besides caves. For example, a *sinkhole* forms when the roof of a cave collapses; it can also form when limestone rock underlying the soil is slowly dissolved away by water. Sinkholes are not easily detected, as the erosion occurs mostly underground. Eventually, enough cave material erodes, thinning the roof and causing it to either collapse or slump.

Other features include *natural bridges* and *tunnels* that form when resistant material in a cave collapses. They can also form when a block of bedrock becomes cut off from the main outcrop and is further eroded by wind, ice wedging, and rain. Another feature is a *disappearing (or loosing) stream,* in which a small surface stream runs into holes in the ground and partially or completely ceases flowing on the surface. In actuality, the stream is flowing in and out of a karst landscape. Still another feature is the *spring,* which is any natural discharge of water from rock or overlying soil onto the surface of the land or into a body of surface water. A disappearing stream may reemerge as a spring.

OTHER CAVES

What are **sandstone caves**?

Sandstone caves form along sandstone cliff faces. Most are caused by physical (mechanical) weathering: Water and wind widen cracks and joints, forming long, shallow cavities in the less resistant sandstone. Caves or overhangs are formed, but are not deep and do not have underground tunnels.

Can **caves form** on ancient **coral reefs**?

Yes, caves often form on ancient coral reefs. For example, over time, ancient coral reef sediments that underlie Navassa (an island just off the coast of Haiti in the Caribbean) rose up past sea level. Navassa may have originated as a coral atoll, but about 5 million years ago (during the Miocene Period), the reefs began to rise. This caused the calcium carbonate sediments to change from aragonite to dolomite (a calcium-magnesium carbonate rock). Soon, terraces formed around the island; from there, chemical weathering took place, forming caves in many places.

How do **tectonic caves** form?

Unlike karst caves, a tectonic cave does not form from solution or erosion, and the type of rock is actually not important to its formation. The main reason for the cave's development is a tectonic force—mechanical stresses that move rocks. These caves can be formed by any geological force that causes rocks to move apart, such as earthquakes.

What is the difference between **glacier and ice caves**?

Glacier caves form within ice, unlike ice caves. As meltwater runs from a glacier, it melts ice along its path. If there are cracks in the ice, the water flows out the glacier, often leaving long cave passages within the ice. Most glacier caves are seasonal, including ones in Iceland, New Zealand, Alaska, Greenland, and Norway. In Iceland, glacier caves form as the ice melts around volcanoes or underlying warm springs. This warmth below the glacier often creates caverns inside, and if the conduit is air- and not water-filled, it may produce a glacier cave.

Ice caves are not caves within the ice, but are rock caves that contain ice within their walls. This ice can form from dripping water and low temperatures within anything from a karst to a tectonic cave. Ice caves are very cold (most often below zero). For spelunkers, it is usually not possible to build permanent pathways in the areas with ice, as it moves every year. Overall, the ice cave has to be specially shaped to trap cold air, with the entrance allowing cold air (which is heavier than warm air) to flow down inside. Another criteria is that the average temperature outside is not too much above zero, and there must be a change from warm to cold seasons, which allows the cave to "trap" colder air. Finally (and most importantly), the temperature of the rock in the cave must be above zero, while the cave air remains below zero.

Does **lava** from volcanoes **create caves**?

Yes. The most prevalent types of caves around volcanoes are called lava tube caves. Interestingly enough, many lava tube caves seem to remain rather cold as the tube traps and holds in cold air. Collapsed parts of a lava tube invite water to drip in, creating ice formations on the walls and floor, many of which often last until midsummer. For example, in the area around Mt. Fuji, the Narusawa Ice Cave was originally a lava tube cave that was made larger from erosion by ice formations.

COMMON CAVE FEATURES

What are **speleothems** and how do they form?

Another word for the formations that decorate caves is speleothem, which comes from the Greek words *spelaion* (meaning "cave") and *thema* (meaning "deposit"). These formations are mostly calcite, the mineral that makes up limestone.

Two factors are important in the growth of speleothems are water and temperature. In the case of water, the more rainfall, the faster the growth of the speleothems. In addition, because the outside temperatures affect the decay rate of plants and animals (higher temperatures result in faster decay), it also affects the amount of carbon dioxide in the soil. As the water seeps through the soil, it picks up this gas, making the water more acidic.

For example, in limestone caves, a weak solution of carbonic acid in seeping rainwater dissolves a small amount of the limestone rock as it passes through cracks and pores on its journey down into the cave. As this water drips into the air-filled cave, dis-

Visitors to Onondaga Cave in Leasburg, Mossouri, wind along a path in the midst of countless stalactites and stalagmites, which are found inside caves around the world. Water seeping from above assists in making a variety of other formations inside caves, as well, including pools and streams. *AP/Wide World Photos.*

solved carbon dioxide is given off. Because the calcium-rich water has lost carbon dioxide, it can't hold as much dissolved calcium. This excess calcium is then precipitated (most often as the mineral calcite) on the cave walls and ceilings, often creating a plethora of speleothems such as stalactites and stalagmites.

What are **stalactites** and **stalagmites**?

Calcite-rich waters that concentrate along cracks in a cave are deposited as flowstone or dripstone. This includes such features as stalactites and stalagmites, which are two of the most prevalent and recognized features in caves. A stalactite is a calcite formation attached to the ceiling of a cave that resembles a rock-hard icicle. A stalagmite is made of the same material, but it forms as minerals precipitate on the ground, forming an upside down icicle shape (pointed-side up).

What **other calcite forms** are commonly found in **limestone caves**?

There are numerous calcite forms that can be found in limestone caves under the right conditions. The following lists some of the more common ones:

Popcorn—Small, knobby growths of calcite on cave walls are often called popcorn. This feature forms in two ways: either when water seeps uniformly out of a limestone wall and precipitates calcite growths, or when water drips from the cave walls or ceilings and splashes on the floor or ledges along walls, creating these calcite nubs.

Frostwork—These calcite growths are shaped much like delicate needles. They may be made of calcite or aragonite (related minerals). It is unknown how these form. One theory is that frostwork forms where wind flows in a cave, allowing the water to evaporate more quickly into the needle-like formations.

Helictite bushes—These calcite growths are shaped like a gnarled tree, with twisted and curled "branches." They often grow from the cave floor and are thought to form when water seeps into the cave through pores so small that flow is controlled by capillary action instead of gravity like most other formations. (Such an action allows water to move "uphill" to deposit calcite against the force of gravity.) Another theory states that the bushes might have formed underwater, as the water rising from below mixed with cave waters of a different chemistry.

Dogtooth spar—Spear-shaped crystals of calcite called dogtooth spar frequently line small pockets in the limestone rock. These crystals are a prominent feature of caves in the Black Hills, most notably, Jewel Cave of Jewel Cave National Monument.

Flowstone—These are any mineral deposit that forms on the walls or floor of a cave as a result of water flowing over the surface; flowstone is often called travertine.

CAVING THE EASY WAY

How many **caves and karst areas** are in **United States national parks**?

Caves and karst features occur in 120 National Parks; 81 contain caves and an additional 39 contain karst. Over 3,900 caves are currently known throughout the system, with 11 parks providing some type of regularly guided tour of the caves.

Some examples of popular karst caves include Carlsbad Caverns National Park, New Mexico; Great Basin National Park, Nevada; Jewel Cave National Monument, South Dakota; Mammoth Cave National Park, Kentucky; Russell Cave National Monument, Alabama; Sequoia National Park, California; and Timpanogos Cave National Monument, Utah. Lava tube caves include those found at Craters of the Moon National Monument, Idaho; Hawaii Volcanoes National Park, Hawaii; Lava Beds National Monument, California; and Sunset Crater Volcano National Monument, Arizona. An example of a sea cave can be found at Point Reyes National Seashore, California.

Popular sites with karst features include the Colonial National Historic Park, Virginia; Everglades National Park, Florida; Ozark National Scenic Riverways, Missouri; and Valley Forge National Historic Park, Pennsylvania.

What cave was mentioned in the 1988 Guinness Book of World Records and why?

The limestone cave system known as Luray Caverns was discovered in 1878. The caverns include early Ordovician dolomite, and in 1956 electronics engineer and accomplished organist Leland Sprinkle (1908–1990) used 37 stalactites in the "Cathedral" section of Luray to create a remarkable organ that produced perfect pitch. Played from a keyboard that resembles a normal organ, the musical tones are produced by a series of rubber-tipped plungers that gently strike each stalactite. This remarkable instrument was highlighted in the 1988 edition of the *Guinness Book of World Records* when the Stalacpipe Organ was declared the world's largest natural musical instrument.

Carlsbad Caverns National Park in New Mexico is the state's most popular tourist attraction. The caverns were carved out of limestone by the slow drip of acidic water. *AP/Wide World Photos.*

What is the difference between **Lechuguilla Cave** and **Carlsbad Caverns**?

Lechuguilla Cave was always thought to be just a "sister" cave to Carlsbad Caverns, an insignificant opening located in the park's backcountry. Its entrance was mined in 1914 for bat guano, but only for a year. Almost everyone thought the cave led to mere dead-end passages. It took until the 1950s before someone truly noticed the cave when cavers heard wind roaring up from the depths. There was no obvious way the cave could "breathe," but it was theorized that more passages lay below rubble not far from the cave entrance. It took until 1984 before the cave was truly examined when the National Park Service granted permission for a group of Colorado cavers to explore the cave.

The results startled everyone. Since 1984, explorers have mapped more than 100 miles (160.9 kilometers) of passageways, discovering the true depth of the cave to be 1,567 feet (477.6 meters). This makes the Lechuguilla Cave the fifth longest in the world and third longest in the United States; it is also the deepest limestone cave in the country.

Overall, Lechuguilla Cave has surpassed Carlsbad Cavern in size, depth, and variety of speleothems. The only claim the sibling has left is that Carlsbad's Big Room does not yet have a rival in Lechuguilla Cave.

What place holds some of the **most extensive lava caves**?

Lava Beds Caves at Lava Beds National Monument in California covers a wide area dotted with cinder cones and covered with lava flows. Throughout the monument there

Why is Wind Cave so special?

Wind Cave in Wind Cave National Park, South Dakota, was the seventh national park to be created, and the first cave to be protected within a national park. Wind Cave has interesting characteristics relating to its name. For example, it actually "breathes" air because of an interplay of tunnels, cave openings, and atmospheric pressure differences between the cave and the surface.

Wind cave is also one of the world's longest and most complex caves, with 108.10 miles (173.97 kilometers) of known caverns fitting under just over one square mile (2.6 square kilometers) of land. Estimates put the cave at over 300 million years old—about the time when the dinosaurs were just beginning to evolve—making it one of the oldest known caves in the world.

are over 500 lava tubes, all of which formed at a time when various volcanoes and cinder cones were active there.

Why are **karst regions and caves** easily **polluted**?

Karst regions and caves are easily polluted for a good reason: All the cracks, crevasses, and openings make it easy for water to flow underground into karst areas. If the groundwater is polluted in a certain area, the karst openings act like pipes, carrying the pollution swiftly away from the original site. In addition, karst areas are not good natural filtration sites, so it's easy to see why they can become so vulnerable to pollution. In some rural areas, cave openings essentially become natural sewer lines, and sinkholes become garbage dumps, often causing local drinking water (usually well water) to become polluted. In urban areas built on karst, the land might not support the development of numerous houses and landfills needed to provide a dumping ground for garbage. In addition, there are problems with putting in septic tanks, underground pipelines, and cables in such rock.

What are some **cave records** around the world?

There are several cave records around the world—although the claims of longest and deepest changes as more cave systems are explored. The following lists some of the latest records:

Longest karst cave in the world—The Mammoth Cave (actually Mammoth Cave-Flint Ridge System) in Kentucky is the longest known karst cave in the world, measuring about 1,847,999 feet (563,270 meters), which is more than 348 miles (560 kilometers). It is doubtful that any other cave will surpass this one in length, as this system is so much longer than any other like it on Earth.

Deepest cave in the world—Currently, the deepest cave is the Voronya (Krubera) Cave in Abkhazia, Georgia (near Russia), with a depth of 5,610 feet (1,710 meters). At one time, the Lamprechtsofen Cave in Austria was thought to be the deepest cave, measuring in at 5,354 feet (1,632 meters). But in 2001, new discoveries in the Voronya Cave system showed that Lamprechtsofen was number two.

Biggest single chamber in a cave—The Sarawak Chamber in the Gunung Mulu National Park, Sarawak, Malaysia, has the largest single chamber in the world. This huge chamber (also called a cavern) is about 2,297 feet (700 meters) long, 984 feet (300 meters) wide, and 230 feet (70 meters) high.

Deepest and longest lava tube in the world—The deepest and longest lava tube happens to be the same one: the Kazumura Cave located along the east flank of the Mt. Kilauea volcano, Hawaii. It seems to go on forever, measuring about 214,895 feet (65,500 meters) in length and with a height difference (the distance from the deepest part of the volcano to its highest point) of 3,612 feet (1,101 meters).

GEOLOGY AND THE OCEANS

DESCRIBING THE OCEANS

What is **marine geology**?

The study of ocean geology is usually referred to as marine geology (or geological oceanography). There are also many divisions of marine studies that include geology. For example, both physical and chemical oceanography often study how the geology of the ocean changes due to physical and chemical means; marine biologists often cross over into geology, describing how various large and small organisms affect (and even create) ocean features. Other geoscientists study how earthquakes or volcanoes form and affect the marine environment; coastal geologists determine how coastlines change over time; and still other scientists collect and analyze marine data to uncover new pockets of oil or mineral resources.

What **percent of the Earth** is covered by **oceans**?

In general, the Earth can be divided into ocean and land. Land represents about 29.22 percent of the surface; the oceans represent 70.78 percent. This is equal to about 57.5 million square miles (148.9 million square kilometers) and 139.4 million square miles (360.9 million square kilometers), respectively. The distribution of land and ocean also differs between the hemispheres. In the Northern Hemisphere, the water to land ratio is about 3:2 (or about 61 percent oceans, 39 percent land); in the Southern Hemisphere, it is about 4:1 (or about 81 percent ocean, 19 percent land).

How does **ocean water** affect **marine geology**?

Ocean waters are responsible for creating marine geology. For example, waves, currents, tides, and storms carve parts of continental coastal edges and deposit sediment in others. Waves and currents also carry particles of rock and organic debris throughout the oceans, creating a "snowfall" of sediment that continually coats the ocean floor. In the deep oceans, strong currents flow up and down gullies, ridges, and rises, carrying particles with them. These movements also cut through cracks and fissures in rocks, extracting elements that often change the chemistry of the ocean waters, as well as carving the ocean landscape.

Where do **dissolved particles** in seawater **originate**?

The dissolved particles (ions) in seawater most commonly originate from two processes in the ocean: chemical weathering (mostly sodium, calcium, magnesium, sulfur, potassium, and bromine) and the degassing of the mantle's magma under the oceans (chloride and sulfur). Concentrations of these ions would be higher, but most of these particles are removed by chemical precipitation, the biological actions of plants and animals, and through absorption into clay minerals.

Does **salt vary** in the oceans?

Because sodium and chloride ions are so prevalent, a great deal of salt is formed in the world's oceans. But this salinity is not consistent throughout the oceans. In general, salinity is lower near the equator because precipitation is higher; at the mid-latitude oceans, salinity is higher because evaporation is much greater than precipitation; and there is less salt near the mouths of major rivers because of the influx of fresh water.

What are some **resources** that are obtained from oceans?

Geologically speaking, the ocean floor and seawater contain rich resources. The ocean floor has a long history of being "mined." Shallow continental shelves are often mined for sands and gravels used in buildings and infrastructure materials. Deposits of petroleum are often extracted in offshore areas, particularly along the Gulf of Mexico and California coasts of the United States and in the Persian Gulf. Manganese nodules (comprised of manganese and iron oxides)—formed by the precipitation of manganese oxides and other metallic salts around a nucleus of rock or shell—are commonly harvested from the ocean floor.

Seawater is often processed to extract such compounds and elements as salt, magnesium, and bromine. For example, solar sea salt (in which seawater evaporates leaving behind sea salt) is produced in about 60 countries, including China, Australia, Mexico, and India. But not all elements and compounds are worth extracting. Of the approximately 60 valuable chemical elements known to exist in the oceans, most are so diluted that they can't be economically harvested.

Ocean water itself is commercially useful, especially through the process of desalinization that provides fresh water in several arid places in the world, including Israel and Kuwait. And in the future, if nuclear fusion reactors are ever developed, the ocean's large amounts of deuterium would provide an almost limitless source of energy.

OCEAN TIDES

What **causes tides**?

Tides are the natural, regular rise and fall of the ocean surface caused by the gravitational attraction of the Moon and, to a much lesser degree, the sun. This pull causes the oceans to bulge out slightly on the side facing our satellite; at the same time, a bulge occurs on the opposite side of the Earth due to inertial forces. As the Earth rotates, the bulges remain virtually stationary. This creates the up-and-down movement of water along most ocean shorelines that we call high and low tides. Although there is a lesser pull from the sun (see below), the Moon has a much stronger effect. Thus, tides on Earth follow the Moon's cycle.

Tides simply "follow" the Moon, which rotates around the Earth every 29.5 days. As the Earth spins and the Moon orbits the planet, the overall effect is that the Moon is almost an hour "behind" each following night. This lag in the Moon's position also causes most tides to occur about an hour later each day.

What are **spring and neap tides**?

Spring and neap tides occur depending on the orbital position of the Moon, Earth, and sun. When the Moon and sun align during the full or new moon phase, the gravitational effects of both planetary bodies create spring tides—the highest tides. If the Moon is near first or third quarter (what we see as a waxing or waning half moon), the smallest, or neap tides occur.

Beachcombers stretch out along a narrow, usually water-covered reef off Seattle's Alki Beach Park during an especially low tide. The highest and lowest tides occur in this area, with a 16-foot difference between water levels. *AP/Wide World Photos.*

Why are **tides important** to geology?

Tides are most important to geology because of their ability to transport sediment and carve coastlines. Over a long period, shorelines shift, estuaries and rivers are filled in with sediment, and beaches change from the rise and fall of the tides. This is what is meant by the constant shifting of the shoreline, and why it has been so difficult for humans to live in these areas where change is constant.

What are the different **types of tides**?

Not all tides are the same around the world. *Diurnal tides* occur when there is one high and one low tide per day, such as along coastal regions of the Caribbean Ocean; *semidiurnal tides* occur where there are two high and low tides each day, such as in many regions of southwest Florida; *mixed tides* occur in an area affected by both semidiurnal and diurnal tides, such as in northwest Florida's Apalachicola Bay; and *unequal tides* occur in areas in which two high tides don't reach the same height on any given day.

How are high and low **tides computed**?

There are many factors that go into determining the local high and low tides. In particular, scientists use the orbital characteristics of the Moon and sun; the Coriolis effect (this force causes the tides to move to the right in the Northern Hemisphere and to the left in the Southern Hemisphere); the depth of ocean basins; the shape of the ocean basins; and sundry other parameters. The resulting tidal almanacs are used not only by scientists, but by the military, commercial fishermen, ships, and other nautical occupations.

Do the **sun and other planets** in our solar system **affect tides**?

Yes. In particular, the sun contributes to the Earth's tides to a certain extent. But although the sun is 200 million times more massive than the Moon, it is 400 times farther away, so it is estimated that the Moon has 2.2 times more tidal-raising power than the sun. To put it another way, on the average, solar tides are less than half the effect of lunar tides. The greatest influence on the Earth's tides is when the Moon and sun align during a full or new moon, creating what are

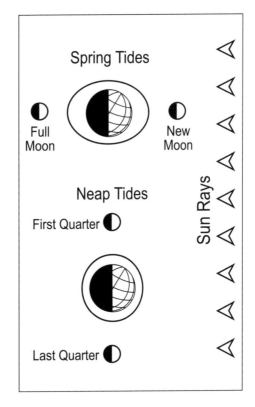

The gravitational pull of the Moon influences the Earth's tides depending on where the Moon is located in its orbit.

known as spring tides (see above). There are also over fifty other components that influence the Earth's tides, but all of these have much less influence on tides that either the Moon or the sun.

As the planets travel around the sun, they do have minute effects on our tides. If you consider the gravitational forces to be 2.2 for the Moon and 1 for the sun, then the "pull" of Venus is about 1.13×10^{-4}, Mars is about 2.3×10^{-6}, Jupiter is about 1.31×10^{-5}, and the farthest planet, Pluto, is about 1×10^{-13}! Thus, contrary to popular belief, a lining up of the planets on one side of the Earth will not cause catastrophic destruction in the form of earthquakes, flash floods, and other disasters. Such lineups of the planets, the Earth, Moon, and sun occur quite frequently, in fact. For example, on February 4, 1962, the sun, Moon, Mercury, Venus, Earth, Mars, and Saturn were in one place in the sky—plus a total eclipse of the sun occurred—all with no ill effects to our world. Also, on February 26, 1953, as a very tight collection of naked-eye planets gathered in the sky, scientists measured the ocean tides and discovered that the average tide rose less than an additional 1/100th of an inch (0.003 centimeters).

Do other planets or satellites have tides?

Yes. It is thought that all planets that have a moon (or many moons) experience some type of tidal influence. This includes all the planets in the solar system except Mercury and Venus. Scientists know that the Moon has a great effect on Earth because of its comparatively large size. The only other planet that might be similar in terms of tidal influence is Pluto, the solar system's smallest planet. It, too, has a relatively large moon in comparison to the parent planet. But because of Pluto's small size and great distance from the Earth, scientists have not yet detected any tidal movements.

For Jupiter, tidal influence seems to act in the opposite direction, mainly because the gas giant is so much larger than its moons. For example, Io, one of Jupiter's largest moons, is being ripped apart by the gravitational pull of this gas giant. These tidal influences cause Io to release a great deal of molten, sulfuric material, and it is now thought to be one of the most volcanically active bodies in our solar system.

Does the **Moon's tidal pull** have **other effects** on our planet?

Many features on Earth are affected by the Moon's gravitational pull. For example, the Earth's land surface is affected, but only to a small degree compared to the oceans. These land tides usually occur twice a day; on the larger continental masses, the land can rise and fall by as much as 6 inches (15 centimeters) when the Moon is directly overhead. Scientists have also discovered that trees are affected by the tides. Many trees bloat, then shrink, with the rhythm of the tides twice a day, expanding and contracting often by several hundredths of an inch (a few millimeters).

What regions have the **lowest and highest tidal range**?

Tidal range is the difference in height between a high and low tide in a certain region; it varies around the world, depending on the seafloor surface, shape of the coastline, and other variables such as regional winds and ocean currents. In general, the highest tidal ranges occur in bays and estuaries where the tides push water into narrow inlets, creating higher and lower tides than along an open coastline. Water levels may rise and fall only a few fractions of an inch (centimeters), or they may vary more than 32 feet (10 meters) per day.

The lowest tidal range occurs in enclosed seas, such as the Mediterranean Sea, a region that is virtually tideless; the highest tidal range occurs in Burntcoat Head, Nova Scotia, at the Bay of Fundy, where a record range of 53.38 feet (16.27 meters) was measured at the Minas Basin. (The average tidal range for the Bay of Fundy is 47.5

A professional surfer successfully negotiates a 60-foot wave at Cortes Bank, located 100 miles off the California coast, in January 2001. This wave was the largest ridden during the entire year. *AP/Wide World Photos*.

feet [14.5 meters].) Other areas with high tidal ranges include Rance Estuary, France (44.3 feet [13.5 meters]); Anchorage, Alaska (29.6 feet [9.03 meters]); Liverpool, England (27.1 feet [8.27 meters]); St. John, New Brunswick, Canada (23.6 feet [7.2 meters]); and Dover, England (18.6 feet [5.67 meters].)

Is there such a thing as a **tidal wave**?

Yes, tidal waves actually do occur. In fact, the conditions at Minas Basin in Nova Scotia create *tidal bores* (also often called tidal waves). As high tide occurs, ocean water surges upstream; where it meets the outflowing river water, a tidal bore forms, resembling a wave moving against the natural flow of the river. Many such bores form along the numerous rivers of the Minas Basin as high tide occurs. Tidal waves about 25 feet (8 meters) in height can also form as water races over 15.5 miles (25 kilometers) per hour up the Fu Ch'un River in northern China.

ALONG THE SHORE

What causes most **ocean waves**?

The majority of ocean waves are generated by wind. Ocean waves are oscillations of the sea surface most often caused by the frictional drag of wind, with the waves traveling in the wind's direction. In a cross section, the water within the wave is moving in roughly a circular path, rising up on the crest, advancing, descending, and then retreating in the trough as the wave passes.

Not all ocean waves form from the movement of the wind, however. *Rossby waves* (or planetary waves) form because of the Earth's shape and rotation. They always travel in an east to west direction around the planet at about 2 inches (10 centimeters) per second or a few miles a day (at mid-latitudes, the waves can take months—or even years—to cross the Pacific Ocean). They are only a few inches deep, but can be thousands of miles across. Also less common are ocean waves caused by earthquakes, landslides, asteroid impacts, density differences (mostly caused by differences in water temperatures), and changes in atmospheric pressure.

How do **waves affect** the geology of **coastlines**?

Waves affect coastal geology by continuously eroding and depositing sediment and other material along a shore. Erosion removes material by the action of waves and abrasion caused by rocks and sand moved by the waves. Depending on the material that makes up a coastline, wave erosion can move either a little or up to tons of material each year. For example, soft chalk cliffs and sandy shores can often be worn away at a rate of about 7 feet (2 meters) annually. Just as important is deposition by ocean waves, as the water carries and drops eroded sediment and other particles elsewhere along the shore.

What are the various **types of shores**?

Depending on the local rock, shorelines can come in a wide variety of types. For example, along the Louisiana coastline, rivers drop sand, silt, and mud, creating a muddy shoreline. Along the Oregon and Maine coastlines, resistant rock breaks into smaller chunks as it weathers; the ocean waves further round the rocks, leaving cobblestones along the beaches. Some shorelines, including those surrounding some small islands off the coast of Florida, are often comprised completely of small shells. Sandy shores result from the action of the waves breaking down rocks into finer particles. Depending on the area, the sands may consist of light-colored quartz and feldspar grains, pinkish sands from coral areas, black sands from volcanic rock, or green sands from the erosion of a volcanic rock that contains the mineral olivine.

What are some **geologic features** along a **coastline**?

There are many geologic features along a coastline, all of which are formed and continually affected by the action of wind, water, weather, and currents. Depositional features include beaches (gently sloping areas usually composed of small sand particles); deltas (triangular deposits of sediment that waves can't remove; they most often are seen at the mouths of rivers); spits (elongated deposits of sand or gravel that project from land into the open water); bay barriers (a spit extending across a bay); tombolos (a spit that connects an offshore island to the mainland); and barrier islands (constantly changing islands formed parallel to the shore by deposition of sand due to beach and longshore drift). Erosional features along a coastline include cliffs, sea caves, sea stacks, and sea arches.

How are **sand and other sediment** deposited on a **beach**?

Sand and sediment are mainly deposited in the swash zone, which is the sloping part of a beach that is alternately covered and uncovered by ocean waves. The swash—waves that occur after a wave collapses onto shore—pushes sand and gravel landward (up the beach); backwash—the water that returns to the ocean—moves the sand and gravel seaward on a beach. Sideward movement created as the waves move back and forth is called beach drift. This motion causes the sand to move in a series of arched paths; thus, it can move material long distances along the shoreline.

Do **waves affect** the geology of the **deep oceans**?

No. In the deeper oceans sediment is not as readily moved by ocean waves as it is in shallower waters. It is known that water deeper than one half an ocean wave's wavelength (the distance from crest to crest) is not moved by the wave (this depth is called the wave base). Therefore, below this wave base, bottom sediment will not move or erode. For example, in the case of ocean waves in the Pacific Ocean, which have been observed to have wavelengths of up to 1,969 feet (600 meters), water deeper than 984 feet (300 meters) will not experience the passage of these waves. But when these same ocean waves occur along the shallower continental shelves (where the seafloor is around 656 feet [200 meters] deep), a great deal of erosion can occur.

What **larger waves** can affect ocean geology?

There are many large waves found in the ocean (and other bodies of water) that can affect marine geology. The following lists some of the more common ones:

Rogue waves—These are huge waves in the open ocean. Reaching some 100 feet (30 meters) high, they seem to appear out of nowhere. Their origins are poorly understood. One theory maintains that they are formed as several individual waves combine to create one huge wave. Rogue waves are thought to rarely cause problems on land, because they usually dissipate before striking a coast. But in open waters, they can become destructive to oceangoing vessels. As to their effects on the ocean geology, no one really knows what these might be.

Tsunamis—Another large wave that occurs not only in the open ocean but also along coastlines is the tsunami. Often erroneously called tidal waves, tsunamis are actually seismic waves thought to be caused by the sudden displacement of water during an ocean floor earthquake. (For more information about tsunamis, see "Examining Earthquakes.") These waves cause few prob-

What was one of the largest storm surges recorded in the United States?

One of the largest storm surges in the 20th century took place on August 17, 1969, when Hurricane Camille (a category 5 storm, the highest number for a hurricane) pushed a 24-foot (7.3-meter) dome of water into Pass Christian, Mississippi. At least 3 feet (1 meter) of storm water reached inland as far as 125 miles (201 kilometers) east and 30 miles (48 kilometers) west of the town; thousands of homes were destroyed or seriously damaged.

lems in the open ocean. But if a tsunami does not dissipate before reaching the shore, it can cause major damage along a coast both to humans and to the geology of the coastline.

Internal (density) waves—Still another type of large wave are internal waves, which occur in layers of seawater having different densities. These interior ocean waves often break along the ocean's edge and might play an important role in upwelling and the geology of ocean shorelines.

Why do **hurricane winds** cause so much **destruction** along a coastline?

Hurricane-force winds can cause a great deal of destruction along a coast for two major reasons. First of all, hurricane winds reaching greater than 74 miles (119 kilometers) per hour can create waves often reaching 35 to 45 feet (10 to 13 meters) in height from crest to trough. These waves can crash into the coast at speeds of up to 25 miles (35 kilometers) per hour, tearing away at coastal structures and churning up sediment along the shoreline.

Second, a hurricane may also generate a *storm surge,* a massive raised dome of water that precedes the storm heading for land. Like a powerful vacuum cleaner, the extremely low pressure of the hurricane interacts with the ocean surface, literally sucking up a huge dome of water at the center of the storm's rotating arms. As the dome reaches the shallower shoreline, and the winds pile up the water even more, a huge influx of water reaches the coast. In general, the gentler the slope of the shoreline, the more chance there is for a storm to occur and cause extensive damage.

What is a **longshore current**?

A longshore current (also called an along-shore current) travels parallel to the shoreline. These currents form as a line of waves hits a beach at an angle, creating the weak current along the shore. They are usually not dangerous, but they are instrumental in erosion and deposition along a shoreline, carrying sand and silt down the coast.

These and other currents are why groins constructed along beachfront property often create more problems than solutions. Groins are usually built perpendicular to the shore to prevent the movement of sand along a shore. But as the sand builds up on one side of the groin from such a current, it also scours out sand on the other side.

What is **littoral drift**?

Littoral drift is essentially the combined efforts of waves (swash and backswash) and currents (longshore currents) to move sediment along a shoreline, shaping the appearance of the coast. A straight or curved shoreline usually means that the prevailing winds cause the sands to move generally in one specific direction. In addition, certain features form from littoral drift, such as sand spits, which are low sand or gravel tongues of land with one end attached to the mainland and the other jetting out into the open water.

BEYOND THE SHORE

What is meant by the **continental margin**?

Scientists consider the interface between the land and sea to be the continental margin. It includes the continental shelf, continental slope, and continental rise, all of which vary in size and extent, depending on their location. Overall, the shapes of the world's continental margins are influenced mainly by rock type, tectonic history, changes in sea level over time, and the forces of water, ice, and wind.

What is the **continental shelf**?

The continental shelf is a gently sloping, submerged plane just beyond the main coastline of all the continents; it is underlain by continental (not oceanic) crust. This underwater extension varies greatly in width, with the average being around 50 miles (78 kilometers) wide. For example, the Siberian continental shelf in the Arctic Ocean is about 930 miles (1,500 kilometers) wide; whereas the continental shelf along South America's west coast is less than a mile (1.6 kilometers) wide. Continental shelves are mostly flat, with the edge of the shelves (called *shelf breaks*) marked by a "lip" where there is a sudden increase in slope averaging about 4 degrees.

Continental shelves are also well known for their commercial resources—from fishing ventures to petroleum, sand, and gravel extraction. This is mainly because continental shelf areas are the most accessible to humans, averaging 430 feet (130 meters) in depth; overall, depths range from 60 to 1,800 feet (20 to 550 meters). One of the deepest continental shelves occurs around Antarctica, in which the ice depresses the shelf down to a depth of about 1,148 feet (350 meters).

Are **continental shelves** part of the **continents**?

Yes, some scientists consider continental shelves to be part of the overall continents. When the continental shelves are added to the continents their combined landmass covers about 47 percent of the Earth's surface. But in reality, when most of us talk about the land surface of the Earth, we mean the just over 29 percent that is not submerged—that is, the land without the continental shelves included.

What are **continental slopes**?

Continental slopes begin where continental shelves break, plunging downward to the great depths of the continental rise. They represent the boundary between the continental and oceanic crust and have average slopes of about 3 to 6 degrees. The origin of the continental slopes is still a mystery. One theory states that the slopes developed as continental landmasses pulled apart at fractures, creating slopes along large and small ocean basins over time. Another theory is that the slopes were formed by the uplift of land at the edge of the continental shelf—perhaps as a result of volcanism.

What are **submarine canyons** and how do they form?

Submarine canyons are just what the term implies: They are winding canyons—some comparable in size to the Colorado River's Grand Canyon—found underwater on the continental slopes of the major continents. These deep submarine canyons (many

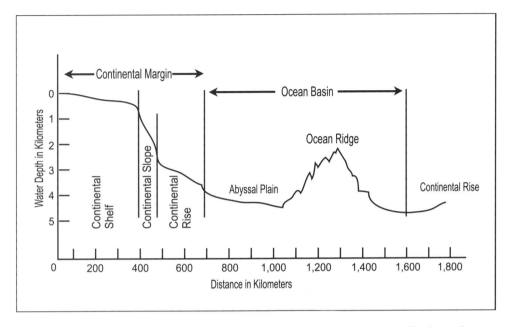

A diagram indicating typical depths and distances for the various features of a continental shelf and ocean bottom.

complete with tributaries) are sometimes found cutting through the continental shelves and slopes, often as extensions of the mouths of larger rivers. For example, the Hudson (North America), Amazon (South America), and Congo (Africa) all have canyons extending through the continental shelf and slope and on to the deep ocean floor. In these cases, the canyons are named after their respective rivers.

Submarine canyons form in much the same way as canyons on land: a mix of water and sediment slides down a slope, carving a V-shaped canyon. It is thought that many of the underwater canyons on the continental shelf formed during the last ice ages. As the oceans receded and exposed the land during that time, runoff carved the canyons. Since that time, sea level has risen, covering the canyons with water. Currently, these canyons continue to be carved and modified by turbidity currents—a type of underwater landslide (see below).

What are **submarine landslides** and **turbidity currents**?

Although it is a fine distinction, there are two major types of landslides in the ocean. Most submarine landslides (sometimes called debris flows) form as unstable sediment builds up along a slope. The mass movement of the material can be triggered by the oversteepening of the slopes, the release of trapped gas below the sediment, storm waves, and earthquakes. (Long ago, some scientists suggested that larger marine animals hitting the side of a slope caused the slides, but because most large creatures live in the open oceans, this is probably not true.) These underwater slides appear to be

343

What happened in Papua New Guinea in July 1998?

On a July 1998 morning, most villagers along the northwest coast of Papau New Guinea were sound asleep. Suddenly, an earthquake measuring 7.0 on the Richter scale shook the land. Almost immediately—and unknown to the villagers—the quake disrupted the nearby unstable volcanic sediment along the coast, creating a huge underwater landslide. This chunk of land displaced water, generating a 50-foot (15-meter) tsunami (seismic ocean wave) close to shore. Only 10 minutes later, the wave struck the coast, killing approximately 3,000 people.

Scientists now know this was not the only time a submarine landslide has affected a coastal region. Digging back into historical records and geological evidence, they have discovered several other catastrophic tsunamis precipitated by submarine slides, such as one near Santa Barbara, California, in 1812, and those affecting Puerto Rico in the late 1800s and early 1900s. A new list was recently developed, highlighting North American sites with the potential for slide-generated tsunami. These sites include areas in Oregon, Washington, and northern California.

most active near deltas, where loose sediment piles up at the mouth of a river. For example, such slides in the Gulf of Mexico (an area in which the Mississippi River has dropped huge quantities of material) often pose a risk to offshore oil rigs when the sediment gives away.

Turbidity currents (or density currents) are also responsible for moving sediment along the coastline. These tongue-shaped slurries of sediment and water form much the same way as slumping sediment, but they are much faster and shorter-lived. They are also thought to be triggered by much the same mechanisms as regular underwater slides. Turbidity currents are usually associated with submarine canyons and are thought to be responsible for carving the canyons deeper since the end of the last Ice Age.

What is the **continental rise**?

The gently sloping, smooth-surfaced continental rise occurs where the continental slopes end. These areas often have an inclination of only 0.5 degrees or less and consist of thick sedimentary deposits. It is thought that these deposits formed after thousands of years of sediment slumping and turbidity currents from off the continental shelves and slopes.

Features such as deep-sea fans or cones are also often found on continental rises, especially at the mouths of the larger rivers, such as the Indus, Ganges, Amazon, and Mississippi. Similar to debris fans and channels on land carved by water, these deep ocean fans also contain such features as meander patterns, oxbow cutoffs, and levees. The continental rises also contain larger features, such as drifts and ripples. For example, the deep Western Boundary Undercurrent (the largest deep ocean current) carries suspended sediments and often dropping the material in the form of drifts (long ridges), furrows (cut by the current), and ripples (huge undulations of sediment).

THE DEEP OCEANS

An image of where a church mission once stood in the village of Sissanoi in Papua New Guinea. In July 1998, a 23-foot wall of water crashed into the northwest coast of Papua New Guinea, destroying three villages and killing thousands. *AP/Wide World Photos.*

What are the **major** surface and sub-surface **currents** in the oceans?

The major warm ocean surface currents are the North Equatorial, Kuro Siwo, Gulf Stream, and South Equatorial Currents, plus the Equatorial Countercurrent and North Atlantic Drift. The major cold ocean surface currents are the Oyashio, California, Labrador, Peru (Humboldt), Benguela, Canaries, and Antarctic Circumpolar Currents. The major subsurface currents—all of them cold—are the Cromwell Current, Weddell Sea Bottom Water, Deep Western Boundary Current (the largest deep-ocean current), and the North Atlantic Deep Water.

What causes **deep ocean currents** to **flow** around the world?

The deep ocean currents are driven by thermohaline circulation, which is movement caused by differences in the temperature and salinity content of the water. Because cold, salt-laden water is heavier than warm water, it sinks to the bottom of oceans. To replace it, warmer water fills in, and as it subsequently cools, the rotation is repeated. This constant movement of water has often been referred to as a giant global conveyor belt or pump that slowly circulates water all over the world's oceans.

For example, the warm Gulf Stream current is heated by the sun, "starting" in the Caribbean. It then flows north along the east coast of North America (mostly along the United States coastline) until it reaches sub-polar waters in the North Atlantic. Between Greenland and Norway, the cold Arctic winds cool the salt-laden water almost to the freezing point. Huge amounts of the now-cold, heavy salt water sink at this point to depths of around 3 to 4 miles (5 to 6.5 kilometers) and begin the next phase of the journey, traveling southwards through the Western Atlantic Basin to the Antarctic Circumpolar Current, and then into the Indian and Pacific Oceans. This trip takes

345

many years. Off the coasts of Peru and California, for instance, upwellings often consist of ocean waters that sank to the depths centuries before.

What are **deep sea channels**?

Deep sea channels (or mid-ocean canyons) occur on the deep ocean floor beyond the continental rise. They are usually U-shaped, shallow troughs with few tributaries, and are either parallel or at an angle to the continental margins. For example, the Northwest Atlantic Mid-Ocean Canyon is a deep sea channel that runs under the Labrador Sea just off Newfoundland.

What is the composition of **sediment** on the **ocean floor**?

Ocean sediment is composed of organic and inorganic matter—from grains of sand and mud to pieces of shells. The size, density, and shape of each grain determine how it moves within the ocean: Larger and more dense pieces fall to the bottom of the ocean more quickly than rounded, mudlike particles that are easily swept away by currents. Still other material is carried by winds (such as a hurricane), ice (such as via melting icebergs), and water (such as particles in rivers).

In shallow seas, sediment accumulates rapidly—about 2 to 12 inches (5 to 30 centimeters) per 1,000 years; in the deep sea, the rate slows down, with sediments accumulating an average of 0.004 to 1 inch (1 to 25 millimeters) per 1,000 years. Most material that doesn't stay suspended forever—or that doesn't dissolve at the surface or elsewhere in the water column—eventually falls or settles to the ocean bottom. With time (millions of years), the particles compact, crystallize, and cement together to form hard rock.

What types of **organic and inorganic matter** help form ocean **sediments**?

Much of the ocean sediment forms from materials with organic origins. This includes shells, skeletons, and teeth—mostly the hard parts that eventually settle on the ocean floor—from dead marine organisms such as fish, corals, and crustaceans, as well as algae. In the deep oceans, organic sediment contains mostly small shells made of silica or calcium carbonate, creating what is called a deep sea ooze. These oozes are named after the dominant organisms that comprise them, such as foraminifera ooze—a calcium carbonate ooze made up of the tiny shells of foraminifera, one of the most prevalent organisms in the ocean.

As expected, most inorganic material that forms ocean sediments comes from the land, and most of it originates near the shore. In general, only a small amount of land-derived clay particles are found in the deepest oceans. Of these clays, most are thought to originate from the physical, chemical, or mechanical weathering of parent rock on land.

Besides organism and land-derived ocean sediments, scientists believe some ocean sediments form from volcanic ash and geological particles created by chemical and biological reactions in the oceans. Most of these sediments are found close to the continental margins; in the deep oceans, there is less sediment accumulation. Even more interesting, deep sea "sediments" are small, glassy particles thought to be micrometeorites from space. It is estimated that these tiny particles accumulate very slowly in the oceans at a rate of only a fraction of an inch (2 millimeters) per 100 million years.

What is the **deep ocean floor or basin**?

The deep ocean floor (basin) extends from the oceanward edge of the continental rise (or a trench if one is present) to the base of the mid-ocean mountains. This basin contains many relief features, including the abyssal plains and hills.

What are the **abyssal plains and hills**?

The abyssal plains of the deep ocean basins are the most vast, expansive, and flat areas found anywhere on Earth. They are usually located at the base of the continental rise and are underlain by the oceanic crust. The abyssal plains cover about 30 percent of the Atlantic Ocean and nearly 75 percent of the Pacific Ocean floors. Around the world's oceans, most lie at depths averaging between 13,123 to 19,685 feet (4,000 to 6,000 meters). Many scientists believe the plains formed as turbidity currents carried fine sediment to the deep ocean floor. Such deposition over millions of years no doubt helped to smooth out most of the curves and irregularities.

The abyssal plains are often broken by low, oval-shaped mounds called abyssal hills, averaging about 330 to 660 feet (100 to 200 meters) in height and 33 feet (10 meters) in diameter. These low-relief features are actually volcanic (basaltic) hills. The younger abyssal hills are nearer the mid-ocean ridges and less covered with sediment than the abyssal plains; the older hills are usually covered with more sediment, and are usually found farther away from the mid-ocean ridges within the plains and basins.

What are **mid-ocean ridges**?

Mid-ocean ridges are literally volcanic mountains on the ocean floor that form from the movement of continental plates. In a process called plate tectonics (for more information about plate tectonics, see "Through the Earth's Layers"), the lithospheric plates pull apart along the ridge system, causing magma to rise to the surface. This movement also "pushes" the mountains apart, creating what is called a spreading center that can move fractions of an inch to several inches per year. The major oceanic spreading centers include the Mid-Atlantic Ridge, Pacific-Antarctic Ridge, Chile Ridge, Indian Ridges, and Juan de Fuca Ridge. (For more information on volcanoes, see "Volcanic Eruptions".)

How does an **ocean trench originate**?

As the ocean floor spreads at seafloor spreading centers, older parts of a plate may subduct underneath another plate. This action causes an upwelling of magma, often forming volcanoes, while the subducted part of the plate creates deep, V-shaped ocean trenches. Many trenches are thousands of miles long, generally hundreds of miles wide, and extend 2 to 2.5 miles (3 to 4 kilometers) deeper than the surrounding ocean floor. (For more information on trenches, see "Through the Earth's Layers"). For example, when the Pacific Plate encounters the Philippine Plate, it subducts because it is older and more dense. This creates an ocean trench—the deepest one known in the world—called the Mariana Trench.

What are the more **well-known ocean trenches**?

Because of the active lithospheric plate movement in the Pacific Ocean, there are many ocean trenches, some of which follow the "Ring of Fire," a belt of seismic activity and volcanoes rimming the region. The following lists some of the major trenches in the Pacific Ocean:

Trench	Depth (in feet)	(in meters)
Mariana	36,201	11,033
Tonga	35,505	10,822
Kermadec	32,963	10,047
Japan	34,626	10,554
Kurile	34,587	10,542
Mindanao	34,439	10,497
Bougainville Deep	29,987	9,140
Peru-Chile	26,460	8,065
Aleutian	25,663	7,822

The Puerto Rico Trench (30,184 feet [9,200 meters]) near the small island of Puerto Rico and the South Sandwich Trench (27,663 feet [8,428 meters]) near Antarctica are found in the Atlantic Ocean.

Some of the record trenches around the world are the Tonga-Kermadec Trench in the Western Pacific between New Zealand and Samoa, which is the largest (it could contain six Grand Canyons), narrowest, and straightest; the Mariana, the deepest-known trench on the planet; the Kurile Trench is the widest; the shortest is the Japan Trench measuring 150 miles (241 kilometers) long; and the Peru-Chile Trench off the coast of South America is the longest at 1,100 miles (1,770 kilometers).

What are **seamounts** and how do they form?

Similar to stand-alone volcanoes that form on land, seamounts are singular volcanic islands rising from the ocean floor. Many occur as isolated mountains, but they can also be found in chains or clusters. In most cases, seamounts are about 3,000 to 10,000 feet (914 to 3,048 meters) in height above the ocean floor, but they remain submerged below the water. They are found in all the oceans of the world, but the Pacific Ocean contains the most, with many of its 2,000 seamounts being active.

Seamounts are thought to be remnants of now-extinct underwater volcanoes. This is because many exhibit a typical volcanic cone shape, including a small crater (depression) at the summit called a caldera. A guyot (or tablemount) is also a seamount, but one with a flat top and usually over 30 million years old. Their formations are still debated. One theory is that a constant flow of magma accumulated, filling the top caldera of the seamount and creating a level top.

GLOBAL WARMING, OCEANS, AND GEOLOGY

What is **global warming**?

It is known that global warming is necessary to a certain point: Without the ability of certain gases in the Earth's atmosphere to help retain radiation from the sun, our planet would be a cold ice ball in space. These gases act like the glass of a greenhouse (thus the name greenhouse gases), trapping much of the energy emitted from or bounced off the ground and keeping our world warm. This, in turn, allows organisms—plants, animals, and otherwise—to live.

More recently, global warming has been used to describe the unnatural increase in the average surface temperatures around the world. Many scientists (and others) believe humans have pumped excess amounts of greenhouse gases, such as carbon dioxide, methane, and nitrogen oxides, into the atmosphere, causing temperatures to

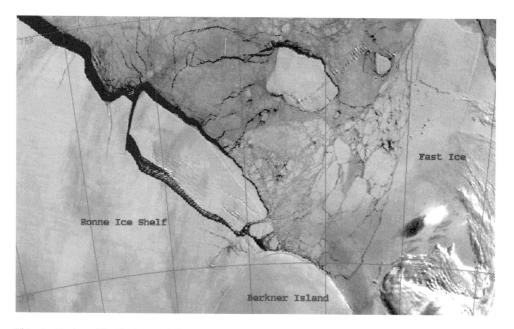

This giant iceberg, 92 miles long and about 30 miles wide, is larger than the state of Delaware. It broke off an Antarctic ice shelf. Some scientists believe that the breaking off of icebergs might be an indicator of global warming. *AP/Wide World Photos.*

rise. Something has already raised the surface temperatures about 0.5°C (1°F) in the past 100 years, and scientists believe it to be human-induced.

Is there a connection between **global warming and geology**?

Yes, there is an indirect connection between global warming and geology because changes in one part of the Earth's complex systems affect other parts. In particular, a change in the global atmosphere (and biosphere) can affect the rock cycle: A rise in sea level can change the expanse of glacial ice, change positions of deserts, and cause ocean waters to inundate coastlines—all of which would change rates and types of weathering taking place on our planet.

Another important connection involves dissolved carbon dioxide in the oceans and surface water. Undoubtedly, the gas is taken up by organisms, but it also precipitates out of the ocean and surface waters to form certain sedimentary rocks. Carbon dioxide then returns to the system from a multitude of places, including the dissolution of carbonate minerals in rocks and shells, weathering of carbonate minerals, volcanic eruptions or hot springs, reactions with the atmosphere, respiration of organisms, and through streams and groundwater. Most scientists agree that a major change in the amount of carbon dioxide in or out of the environment can affect us all. Not only would humans and other living organisms be affected, but also the natural cycles connected to the world's geology.

How high would the oceans rise if all the ice sheets melted?

It is estimated that if all the ice sheets melted, including the polar ice caps, ice on Greenland and Iceland, and all the glaciers on the planet, the average worldwide sea level would rise by about 250 feet (76 meters). Such a rise would decimate almost all coastal cities, shrink the living space on all the continents, cause massive climate fluctuations, and change our lives forever.

Still another geologic–global warming connection may be found in the weather and climate. If global warming continues, more powerful and intense weather systems may develop. Such events as superhurricanes would cause a great deal of erosion along coastlines, not to mention deluging countless rivers and creeks inland. In terms of climate change, variations in vegetation patterns could contribute to more erosion in various areas; glacial, polar, and sea ice would change drastically, altering the amount of radiation reflected back into space—thus enhancing the warming effect; and changes in the hydrologic cycle would alter stream flow and groundwater levels.

Do we have any **evidence of global warming** in the **past**?

Yes, there is a great deal of evidence for global warming events in the past. The following lists two of the more well-known ones:

Mid-Cretaceous Period—During this period (between about 120 and 90 million years ago), new ocean crust was produced at about twice the normal rate. Large volcanic plateaus were forming in the ocean basins, ocean temperatures were very high, and there was a peak in worldwide petroleum formation. Just as startling was the sea level, which was about 330 to 660 feet (100 to 200 meters) higher than at present. The reasons for the high temperatures were probably numerous, including the release of carbon dioxide (a greenhouse gas) by volcanic eruptions, creating a "supergreenhouse" effect. This led to temperatures about 20 to 22°F (10 to 12°C) above our current average global temperatures. Interestingly enough, it is thought that the large volume of basalts that erupted on the ocean floor displaced a great deal of ocean water, causing sea levels to rise. And with the rise in sea level and temperatures, organisms flourished, eventually providing material necessary for petroleum formation.

Eocene Period—During this time (between about 55 to 38 million years ago), temperatures also increased, with tropical vegetation reaching about 45 to 55 degrees north and south of the equator, or about 15 degrees higher than today. Based on rock samples, it appears that the Earth had between 2 and 6 times the amount of carbon dioxide we have today. Scientists believe this

351

global warming was caused by continental collisions, events that released large amounts of this greenhouse gas into the atmosphere. (This also shows how the rock cycle and tectonic processes can affect atmospheric conditions.)

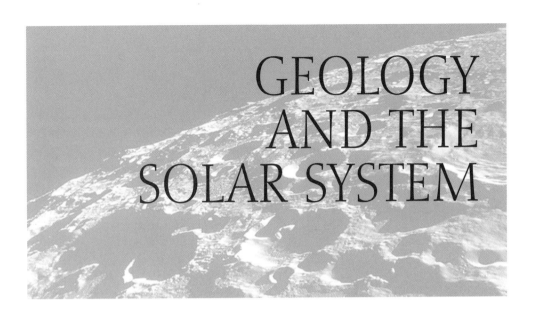

GEOLOGY AND THE SOLAR SYSTEM

MAKEUP OF THE SOLAR SYSTEM

What are the **nine known planets** that make up our solar system?

The nine known planets, in order from the sun outward, are Mercury, Venus, Earth, Mars, Jupiter, Saturn, Uranus, Neptune, and Pluto. The four inner planets are also called the terrestrial planets because they are mostly rock like the Earth ("terra" is the Latin name for earth). The outer planets are also called the gas giant planets because they are mostly composed of gas, with the exception of Pluto, which is now considered by many scientists to be an icy satellite and not a gaseous or terrestrial planet. There are also some newly discovered planetary bodies located beyond the orbit of Neptune called Edgewood-Kuiper Belt Objects.

What is an **astronomical unit (AU)**?

An Astronomical Unit (AU) is equal to about 93,000,000 miles (150,000,000 kilometers), or slightly less than the average distance between the sun and the Earth.

What **planet is closest** to the Earth?

Venus is the closet planet to Earth, with an average orbit from the sun of approximately 0.7 Astronomical Units (AU). Venus can come as close to Earth as 0.267 AU (in comparison, one of Mars' closest approaches was 0.373 AU on August 28, 2003).

What **planet is farthest** from the Earth?

Pluto is the farthest planet from Earth. At an average distance of 2.7 billion miles (4,345,228,800 kilometers, or about 29 AU) from our planet, Pluto is a dim speck of light in even the largest of telescopes. (The planet is 3.67 billion miles [5.9 billion kilometers]

The northern hemisphere of Venus is displayed in this global view of the planet's surface. The closest planet to Earth, Venus is visible to the naked eye on a clear night, and appears as the first and brightest star against a dark sky. *AP/Wide World Photos.*

from the sun.) It takes almost 247.7 Earth years for Pluto to make one swing around the sun; it does so in a long, looping orbit that takes it above and below the plane (ecliptic) of the other planets.

Pluto is often called the ninth planet from the sun, but sometimes this changes. Between 1979 and 1999, Pluto's eccentric, ellipse-shaped orbit brought it inside the orbit of Neptune, making Pluto the eighth planet for two decades. (There is no danger of collision between the two planets, as Pluto's orbit is inclined more than 17 degrees to the ecliptic and never actually crosses Neptune's path.) In the years 2231 to 2245, the same thing will happen again, giving Neptune the farthest planet status for a while.

What are the **Edgewood-Kuiper Belt Objects**?

The Edgewood-Kuiper Belt Objects (KBOs; also called Trans-Neptunian Objects) are thought to be a large, disk-shaped swarm of icy bodies located just beyond the orbits of Neptune and Pluto. They are thought to orbit the sun at distances of 50 to 100 Astronomical Units, or just over 3 billion miles (5 billion kilometers) beyond the orbit of Neptune.

Over the past decade, more than 500 icy KBOs have been found. With a few exceptions, all have been significantly smaller than Pluto. The record holder is the KBO named Quaoar (see below for details). Other larger KBOs include Varuna and an object called 2002 AW197, each approximately 540 miles (900 kilometers) across. Though it is less certain than direct measurements, scientists usually determine a KBO's diameter by measuring the object's temperature; they then calculate the size based on assumptions about the KBO's reflectivity.

Are KBOs truly planets? So far, most scientists agree that all the KBOs discovered to date are too small to be called planets. But if you ask a dozen astronomers how big a space body has to be to merit being called a planet, you will get a dozen different answers. Planetary status depends on more than just size, but for now, our solar system officially has nine planets.

What is the **largest Edgewood-Kuiper Belt Object** found to date?

In June 2002, astronomers discovered a large object in a nearly perfect circular orbit about a billion kilometers beyond Pluto. Swinging around the sun every 288 years, the space body named Quaoar (after a Native American god) is estimated to be about 808 miles

Is the planet Pluto truly a planet or is it an Edgewood-Kuiper Belt Object (KBO)?

In recent years, scientists have debated whether the planet Pluto—the smallest and farthest planet in the solar system—is actually a planet or an Edgewood-Kuiper Belt Object (KBO). The reasons are many: First, Pluto is very small—six times smaller than the Earth and even smaller than seven of the solar system's moons (our Moon; Jupiter's Europa, Ganymede, Io, and Callisto; Saturn's Titan; and Neptune's Triton). Pluto's moon Charon is larger in proportion than any other planet-satellite mix in the solar system. Pluto's orbit is highly elliptical and tilted with respect to paths of all the other planets. And finally, Pluto appears to be a mix of rock and ice; all the other rocky planets are located in the inner solar system.

Currently, many scientists have solved this dilemma by saying that Pluto is both a planet and a member of the Edgewood-Kuiper Belt. The debate about Pluto will probably go on for decades, especially if astronomers discover any KBOs larger than Pluto.

(1,300 kilometers) wide, which is about 250 miles (400 kilometers) wider than Ceres, the largest main-belt asteroid, and more than half the diameter of Pluto. To date, it is the largest object to be discovered in the solar system since Pluto was found in 1930. Scientists speculate that Quaoar is made up of ice and rock, which makes it similar to most comets.

IMPACTS EVERYWHERE

Is the **Earth** the only planet with **impact craters**?

No, Earth is not the only planet with impact craters. These huge holes are found on almost all the planets and satellites in the solar system, with some planets having more craters than others. Impact craters form when huge space rocks (such as asteroids or comets) strike a planet or satellite, excavating dirt, soil, and rock at the collision site.

Impacts are readily seen on our Moon; they also are observed on some of the smallest members of the solar system, the asteroids. The South Pole-Aitken Basin on the farside of the Moon is the largest (1,616 miles [2,600 kilometers] in diameter) and deepest (over 7.5 miles [12 kilometers]) impact basin known in the solar system.

What is an **asteroid**?

An asteroid is a large space body that ranges from about boulder-size to 584 miles (950 kilometers) in diameter. (The latter is the size of the largest asteroid, Ceres, or about

This photo, taken during the *Apollo 11* mission, shows the heavily cratered far side of the moon. International Astronomical Union crater no. 308, at the bottom of the photo, is about 58 miles in diameter. *AP/Wide World Photos.*

one-quarter the diameter of our Moon.) In general, asteroids are classified as carbonaceous, stony, or metal (mostly iron). The majority of the asteroids are found between the orbits of Mars and Jupiter in an area called the Asteroid Belt. Occasionally, they will stray from the belt, either because of the gravitational attraction of Jupiter or by collision, with some eventually working their way into the inner solar system.

What are **comets**?

Comets—once called dirty snowballs in space—are a mix of dust, gases, rock, and ice orbiting the sun. After several satellite missions to and close encounters by comets, scientists now realize these objects resemble mudballs, as they carry more dust than ice. The average comet also carries carbon dioxide, frozen water, methane, ammonia, and materials such as silicates and organic compounds.

Comets originally formed in the cold, outer planetary system. Short period comets (those with orbits ranging from a few to 200 years) are thought to have originated in the Edgewood-Kuiper Belt, a disk of objects beyond the orbit of Neptune. Long-period comets (those with orbits that travel through the solar system once every few thousand of years or never return at all) are thought to originate in an area called the Oort Cloud. This theoretical collection of comets is thought to surround the solar system about 100,000 Astronomical Units from the sun.

What are **near-Earth objects**?

Near-Earth objects (NEO) are just what the term implies: Space objects—comets or asteroids—that come close to the Earth's orbit. NEOs have been nudged by the gravitational attraction of nearby planets (especially Jupiter), allowing the bodies to enter

Have any asteroids been found inside the Earth's orbit?

Yes. In early 2003, astronomers found the first object—other than the planets Mercury and Venus—whose orbit about the sun is completely inside the Earth's orbit. The object, labeled 2003 CP20, was detected by a sky survey and is thought to be only a few miles in diameter.

the inner solar system and eventually the Earth's neighborhood. These comets and asteroids are relatively unchanged debris remnants from the solar system's formation some 4.55 billion years ago.

Near-Earth Comets (NECs) are further restricted to include only short-period comets. The vast majority of NEOs are asteroids—referred to as Near-Earth Asteroids (NEAs)—that come close to the Earth or cross the Earth's orbit. NEAs are subdivided into groups—Aten, Apollo, Amor, and several other groups not yet agreed upon—according to the asteroid's orbit, especially its closest and farthest distance to Earth's orbit.

How many **Near-Earth Objects** have been **discovered** to date?

As of February 2003, about 2,269 Near-Earth Objects (NEOs) have been discovered; 642 of these NEOs are asteroids with diameters of just over a half mile (1 kilometer) or larger. Scientists believe there may be close to 2,000 such asteroids orbiting nearby.

Comet Ikeya-Zhang heads towards the tree line over Palm Beach Gardens, Florida, on March 19, 2002. It is believed that this comet's last trip through our solar system was in 1661. *AP/Wide World Photos.*

What is a **Potentially Hazardous Asteroid**?

Potentially Hazardous Asteroids (PHAs) are asteroids with the potential to make threateningly close approaches to the Earth. One guideline includes all asteroids with an Earth "minimum orbit intersection distance" of 0.05 Astronomical Units or less, which is about 4,650,000 miles (7,480,000 kilometers) from the Earth. Another criterion is that the asteroid is larger than about 500 feet (150 meters) in diameter. As of this writing, 495 NEOs have been classified as Potentially Hazardous Asteroids.

This potential to come close to the Earth does not mean the PHAs will impact our planet; it only means there is a possibility for such a threat. Many groups are monitoring the PHAs, updating the orbits and seeking other close space objects. This is done in order to better predict the possibilities of NEOs striking the Earth.

How have **asteroids and comets affected Earth** in the past?

Asteroids and comets have definitely affected the Earth in the past. Asteroids were not readily seen by humans until the advent of telescopes, but people have observed comets for thousands of years. To ancient civilizations, seeing a comet swing by the sun was often regarded as an ominous sign of approaching doom and destruction. Sometimes, this has indeed been the case.

Scientists now know that huge asteroids and comets have collided with our planet, often creating gaping holes called impact craters. Large impacts have been associated with the extinction of a large number of species over time. This includes the "Permian extinctions," a point in geologic time about 250 million years ago, in which about 90 percent of all species on land and in the oceans became extinct. Some scientists believe another extinction caused by an impacting space object came at the end of the Cretaceous Period, which helped contribute to the demise of the dinosaurs.

Where are some of the **major impact craters** located on the **Earth**?

The following lists the largest known terrestrial impact craters on Earth:

Place	Diameter	
	miles	kilometers
Vredefort, South Africa	186	300
Sudbury, Ontario, Canada	155	250
Chicxulub, Yucatan, Mexico	105	170*
Manicouagan, Quebec, Canada	62	100
Popigai, Russia	62	100
Acraman, South Australia	56	90
Chesapeake Bay, Virginia, USA	53	85
Puchezh-Katunki, Russia	50	80
Morokweng, South Africa	44	70
Kara, Russia	40	65
Beaverhead, Montana, USA	37	60

* Thought to be associated with the extinction of the dinosaurs.

What could we do to **prevent a large asteroid or comet** from **colliding with Earth**?

This question is highly debated. Most scientists agree that the "movie solution"—blowing up a large approaching Near-Earth Object (NEO)—is not the answer. Exploding the space object would create even more problems, because many smaller pieces would rain down on the Earth. Plus, many astronomers believe that blowing up an NEO might not work, especially when it comes to asteroids. Recent flybys of asteroids by spacecraft (and telescopic studies of asteroids flying by the Earth) show that these bodies may be "piles of rocks" in space, making them difficult to break apart.

The trick of eliminating an advancing NEO is not to blow up the object, but to gently nudge it away from Earth. Scientists are counting on plenty of forewarning if an NEO is on a collision course with the Earth. If we do have several years of warning, existing technology might be able to deflect the object away from Earth. This would have to be an international effort, with countries sending numerous spacecraft to the offending object. Some scientists suggest that nuclear fusion weapons set off above the surface would change the NEO's speed without fracturing it. This would cause a modest change in the NEO's velocity—only a few fractions of an inch per second—but it would be enough to change its course to miss Earth. Another suggestion involves installing solar sails placed on the object. The solar wind—high energy particles from the sun—would then push the object, redirecting it from striking the Earth.

Are any **Near-Earth Objects** currently **threatening Earth**?

At this writing, there are no known Near-Earth Objects (NEOs) in danger of striking the Earth. Many scientists note that the threat to any one person from disease, auto accidents, or other natural disasters are much higher than threats from NEOs.

The most difficult part in determining an NEO impact threat is trying to determine accurate orbits for the objects. Until scientists can pinpoint the location and motion of an object, it is difficult to determine the threat. That is why the media will often report the discovery of a possible threatening NEO. For example, the over half-mile- (1-kilometer-) wide asteroid 1997 XF11 received a great deal of media attention in March 1998, when orbital data indicated for a time that the asteroid would come remarkably close to the Earth on October 26, 2028. But more data and orbital measurements showed the probability of impact was essentially zero. In 2002, several researchers proposed that another asteroid, 1950 DA, measuring more than a half mile (about 1.1 kilometers) in diameter, had the potential to collide with Earth on March 16, 2880, which is too far ahead to worry about, if it happens at all. Still another asteroid, 2002 NT7, was thought to have potential for a collision on February 1, 2019; more data again showed there was no chance of collision with the Earth.

What would happen if a large **comet or asteroid struck the Earth**?

Around 70 percent of the Earth is covered with oceans, giving us about a 30 percent chance that an asteroid or comet would hit land. If a small space body struck the oceans, there might not be too much of an effect (it would be like dropping a molten rock in a pond). Steam would result and local wildlife would be killed, but not much other damage would occur.

But if a small asteroid or comet of just under a half mile in diameter were to strike land, the results would be disastrous to people, animals, vegetation, and structures in

The mile-wide meteor crater near Winslow, Arizona. The crater was made 500 centuries ago, when a 10,000,000-ton meteor impact dislodged 300,000,000 tons of rock. The 600-foot-deep crater is three miles in circumference. *AP/Wide World Photos.*

the local environment. For example, the asteroid that carved out the three-quarter-mile-wide Meteor Crater in Arizona just 49,000 years ago most likely destroyed all life for miles in every direction.

It gets worse. Scientists believe that if an object with a diameter larger than about six-tenths of a mile (1 kilometer) hit Earth it could trigger huge ocean waves and kill billions of people. Scientists estimate that there are between 700 to more than 1,000 near-Earth asteroids—those with the greatest potential to hit the Earth—about this size, and it appears that they hit our planet about every 500,000 to 10 million years.

One result of such an impact would be that the impact debris would spread throughout the Earth's atmosphere, causing changes to the climate. Plant life would suffer from acid rain, partial blocking of sunlight, and firestorms resulting from heated impact debris raining back down on the Earth's surface. An asteroid with a diameter of 6 miles (10 kilometers) or more could even eliminate all life on Earth. But don't increase your life insurance policy yet—this type of impact seems to occur only every 100 million years or so.

Is anyone keeping track of the asteroids and comets that come close to the Earth?

Yes, there are numerous programs involved in watching the skies for larger asteroids that come close to the Earth. Scientists are watching out for Potentially Hazardous Asteroids (PHAs). Although no pending collision with an asteroid is known or has yet

Have we ever seen an impact on another planet?

Yes. From July 16 through July 22, 1994, at least 21 discernible fragments of a broken comet collided with the planet Jupiter, marking the first time humans have ever witnessed a collision of two solar system bodies. The comet, designated Comet P/Shoemaker-Levy 9 (after the team of discoverers), made a spectacular display as its chunks—some just over a mile (2 kilometers) in diameter—slammed into the upper atmosphere of the gas giant. Dark-colored "rings" developed at the points of impact, many of which lasted for weeks. Not only did numerous spacecraft (such as the Galileo, Hubble Space Telescope, and Ulysses) record images or data of the event, but hundreds of amateur and professional astronomers were able to witness and study the results of the impacts.

been predicted, the objects on this list might impact Earth at some point in the distant future. Scientists have also established the Torino Hazard Scale, a scale from 1 to 10 that gives scientists and the public alike a way of knowing the potential destructive capabilities of a known asteroid. An asteroid classified at 1 would do little (if any) damage; a classification of 10 means the object could destroy all life on Earth.

Have we ever seen an **impact on the Moon**?

Yes. In 1956, amateur astronomer Leon H. Stuart observed and photographed a flash on the Moon. So far, this is the only verified record of a crash of an asteroid-sized body on the lunar surface. In 1994, the Clementine spacecraft took images of the Moon, including the area where Stuart reported the impact site. By the early 2000s, scientists believed they had located the nearly 1-mile (1.5-kilometer) impact crater left by the event. They noted a fresh-looking ejecta blanket at the location of the flash, a site seemingly fresher than other surrounding young craters. Estimates put the colliding space rock at about 130 feet (40 meters). A collision this size may occur on the Moon every 10 to 50 years.

HOW EARTH IS SIMILAR
TO OTHER PLANETS

Is the Earth the **only planet with a moon**?

No. Almost every planet in our solar system has at least one moon. The only exceptions are Mercury and Venus. Planets in the solar system have a total of 124 (probably more) orbiting moons (satellites). Earth has one moon, as does Pluto; Mars has two; Jupiter has 58 (so far the most in the solar system, and with improved telescopes, scientists continue to find more); Saturn has 30; Uranus has 21; and Neptune has 11 moons. The largest satellites are Jupiter's moon Ganymede (larger than the planets Mercury and Pluto); Saturn's moon Titan (larger than the planets Mercury and Pluto, but smaller than Ganymede); and Jupiter's moon Callisto (about the same size as Mercury). The smallest satellites (only a few miles across) are found in Saturn's ring system.

Do all **planets "wobble"** as they rotate?

Yes, all planets slightly wobble like a top as they rotate. Scientists believe such wobbling, including the wobble of the Earth, is caused by how material is distributed inside the planet. For example, the composition of a planet's core and its size could cause the planet to wobble. In the case of Mars, scientists believe the planet wobbles because of an iron central core that occupies about half of the planet.

Do any **planets** besides the Earth have **polar ice caps**?

The Earth, Mars, and perhaps Pluto seem to be the only planets with polar ice caps. The Earth's are located at the North and South Poles. The southern ice cap covers the continent of Antarctica and has about 90 percent of all the freshwater in the world—of course, it is in the form of ice. (For more information on ice caps, see "Ice Environments.")

Mars also has permanent ice caps at both the north and south poles. They are composed mostly of solid carbon dioxide (also called "dry ice"). Both polar caps have alternating layers of ice and dark dust, though no one knows the reasons behind the layering (one theory is that the layers represent long-term climate changes on Mars.) In the northern summer, the carbon dioxide of the cap sublimes, leaving a layer of water ice. Scientists do not know if it is the same in the southern ice cap, since its carbon dioxide layer never truly disappears.

Pluto might also have ice caps. But so far, images taken by the Hubble Space Telescope only reveal a possible ice cap on the north pole of the tiny planet, not the south pole. Currently, no one really knows the composition of the ice cap(s), but many scientists do not believe it is water.

Is Earth the **only planet** with an **atmosphere**?

No, most of the other planets have some type of atmosphere, but so far, Earth is the only planet that has an atmosphere able to sustain carbonate life forms. The following details the other planetary atmospheres:

Mercury—Mercury is thought to be the only planet in the solar system without an atmosphere. Scientists believe any atmosphere on the planet was burned off after the formation of the solar system.

Venus—Venus has a thick atmosphere made up of carbon dioxide (97 percent), nitrogen (just under 3 percent), and some trace chemicals. The thick

362

Why did scientists once think Jupiter had potential to be the solar system's second star?

At one time, scientists believed Jupiter had the potential to become a star. But data from various spacecraft have shown that this would be impossible. It is true that Jupiter radiates more energy into space than it receives from the sun, with the interior of the gas giant close to 20,000°K (35,540°F and 19,727°C). This is probably due to the gravitational compression of the planet. But Jupiter does not produce energy by nuclear fusion as the sun does; it is much too small, and thus, its interior too cool to ignite nuclear reactions in the core. Plus, scientists know Jupiter would have to be at least 80 times more massive to become a star.

layer of the greenhouse gas carbon dioxide causes the planet to keep most of the heat it receives from the sun, creating surface temperatures close to 890°F (477°C).

Mars—Apparently, the Martian atmosphere once had the chance to be similar to the Earth's, but differing conditions—especially the size and orbital placement of the red planet—caused the atmosphere to develop differently. Today's very thin Martian atmosphere is made up of carbon dioxide (95.3 percent), nitrogen (2.7 percent), argon (1.6 percent), traces of oxygen (0.15 percent), and water (0.03 percent).

Jupiter—Jupiter's atmosphere is composed of hydrogen (about 90 percent) and helium (about 10 percent), with traces of methane, water, ammonia, and dust. The planet's composition is thought to be close to that of the primordial solar nebula from which the entire solar system was formed.

Saturn—Like Jupiter, Saturn's atmosphere is mostly hydrogen (about 75 percent) and helium (about 25 percent), with traces of water, methane, ammonia, and "rock" at its core. Again, like Jupiter, Saturn's atmosphere is thought to be similar to the composition of the primordial solar nebula.

Uranus—Uranus's atmosphere is composed of hydrogen (83 percent), helium (15 percent), and methane (2 percent).

Neptune—Neptune's atmosphere is mostly hydrogen and helium with a small amount of methane. The planet's blue color is mostly due to the absorption of red light by methane, but scientists believe there is also some additional as-yet-unidentified element that gives the clouds their extreme blue tint.

Pluto—Pluto's atmosphere probably consists of primarily nitrogen, with some carbon monoxide and methane. Because the planet is so small, the atmosphere is very tenuous and might only exist as a gas when Pluto is closest in its

A view of the planet Mars taken by the NASA Hubble Space Telescope. *AP/Wide World Photos.*

orbit to the sun. Some of the gases may even escape into space, interacting with Pluto's tiny moon, Charon. But at most times during Pluto's long year, the atmospheric gases might be frozen.

Did **Mars resemble Earth** in the past?

Yes, Mars was much more like Earth in the past. In its early history, Mars used almost all of its carbon dioxide to form carbonate rocks similar to those on the early Earth. But because the Martian surface did not break apart into plates and move around the planet (as on Earth), Mars was unable to recycle any of its carbon dioxide back into the atmosphere. Because of this, it could not sustain a significant greenhouse effect—causing the Martian surface to remain much colder than the Earth would be if it were at the same distance from the sun.

In what other ways does **Mars resemble Earth now**?

Mars is probably the most similar planet to the Earth—although it is only just over half the size of our planet. And though Mars is much smaller than Earth, its surface area is about the same as the land surface area of Earth. In addition, some of the weather events on Mars are similar to those on Earth, including those that affect the surface morphology. For example, Mars has massive dust storms that scour the land and often cover the entire planet. There are also 5-mile-high (8-kilometer) tornadoes (called dust devils) that often rip across the Martian landscape.

Mars also apparently had water scouring its surface in the past. Evidence from the *Viking Orbiter* and other spacecraft on the red planet show huge channels that might have been carved by rivers. Several areas show layered rock, possibly indicating sediment laid down by water.

Is it true that **Mars currently has water** similar to Earth?

In 2000, scientists looking over data from NASA's *Mars Global Surveyor* spacecraft observed features that suggest there might be current sources of liquid water at or near the Martian surface. Detailed images show the smallest features ever observed from Martian orbit, and they are similar to features on Earth left by flash floods. For example, there are places resembling gullies formed by flowing water; another area shows deposits of rock and soil transported by the flows. Although the water does not flow like Earth's rivers, they might be responsible for many riverine features observed on the red planet.

HOW EARTH DIFFERS FROM OTHER PLANETS

How does **Mercury's surface** differ from Earth?

Mercury's surface differs greatly from the Earth. Many areas are heavily cratered. It also rotates very slowly—a Mercury day is equal to 59 Earth days; a Mercury year is equal to 88 Earth days. The small planet is alternately baked by the sun, then frozen as it turns away from the sun. Temperatures range from 723°F (402°C) on the sunlit side at "noon" to –96°F (–173°C) on the dark side at "midnight."

How does **Venus differ** from the Earth?

Although Venus is called our "sister" planet—because it is relatively the same size, its surface differs greatly from the Earth's. Venus is a very hostile planet, with a run-away greenhouse effect in place: Thick clouds form a sea of carbon dioxide gas that traps the radiation from the sun. This causes the surface temperatures to reach 890°F (477°C), which is hot enough to melt lead and even make some rocks glow red. The heat wears away at the planet's rock, as does the sulfuric acid "rain" from the atmosphere. Impact craters dot the planet's surface, and long lava rivers flow from recently active volcanoes.

Could **Earth** eventually experience a **greenhouse effect like Venus**?

Venus's atmosphere resembles a greenhouse: As sunlight penetrates the gaseous carbon dioxide atmosphere, part of it is absorbed by the surface and the rest reradiated off the surface in the form of heat. The carbon dioxide absorbs the heat, stopping the heat from reaching space.

Currently, Earth does not have a Venus-like accumulation of carbon dioxide in its atmosphere, and it is not as close to the sun. But recent measurements of the Earth's average annual surface temperature show that the atmosphere is increasing in temperature. The reason for this global warming—whether it is a natural cycle or caused by human interference—remains unclear, but many people believe humans are the main culprits. They blame the increase not only on carbon dioxide (mainly from industrial and commercial emissions), but other human-produced greenhouse gases. (For more information about global warming and greenhouse gases, see "Geology and the Oceans.")

How do **Martian features differ** from the Earth?

There are many geological differences between Mars and Earth, and most of them involve size. For example, the interior of Mars (inferred only by data from spacecraft such as the *Mars Global Surveyor*) is probably composed of a dense core about 1,056 miles (1,700 kilometers) in radius, and a molten rocky mantle somewhat denser than the Earth's. Over this lies a thin crust—about 50 miles (80 kilometers) thick in the southern hemisphere and only about —23 miles (35 kilometers) thick in the north.

Mars has a huge canyon, Valles Marineris, that cuts across about one quarter of the planet. It measures about 2,485 miles (4,000 kilometers) in length and has a depth of 1.2 to 4.3 miles (2 to 7 kilometers). Superimposed on the United States, it would run from New York to Los Angeles.

Mars also has the largest known volcano in the solar system: Olympus Mons. But there has been a discrepancy in the actual measurement of this huge feature. Most of the measurements are based on estimates from *Mariner 9* pictures, resulting in an estimated height of between 14 to 16 miles (22 to 29 kilometers). In addition, some of the measurements were made from the base of the mountain, while others were from the crater surrounding it, Nix Olympica.

Thanks to the *Mars Global Surveyor,* scientists now believe that Olympus Mons is roughly the height of three Mount Everests (the Earth's tallest mountain that stands 29,035 feet [8,850 meters] high) and is nearly 340 miles (550 kilometers) across, which is about as wide as the entire Hawaiian Islands chain.

How do **Mars and Earth** differ in **temperatures**?

Mars and Earth differ greatly in temperature. The Earth's greenhouse effect keeps the planet's average surface temperature at about 60°F (15°C). If no greenhouse gases existed in the atmosphere, our planet's temperature would be 0°F (–18°C), which is too cold for most life. Plus, with no greenhouse gases to hold it in, most of the heat radiated from the Earth's surface would be lost directly into outer space.

On Mars, the temperatures fluctuate much more radically than on Earth. The average temperature on the red planet is about –67°F (–55°C). Martian surface temperatures range widely from –207°F (–133°C) at the winter pole (during the Martian winter solstice, when the north pole of Mars is pointing away from the sun) to almost

80°F (27°C) on the day side during summer. The wild fluctuations in the Martian temperatures are due to the planet's highly elliptical (oval-like) orbit around the sun. The Earth does not experience such extreme temperature changes because our orbit is less elliptical than Mars: We are only about 3 percent closer to the sun in January (our closest approach to the sun) than in July (our farthest distance from the sun).

How do **Jupiter, Saturn, Uranus, and Neptune differ** from the Earth?

Jupiter, Saturn, Uranus, and Neptune differ from the Earth in many ways, with the main differences being size, distance from the sun, and composition. In addition, all the gas planets do not have solid surfaces like the Earth; their gaseous material simply gets denser with depth. And all the outer planets have rings that vary in thickness and brightness (Saturn's are the largest and most famous). Although no one knows the origin of the rings, they might have been there since the planets' formation, perhaps resulting from the breakup of former, larger moons. The following lists a brief summary of some other major differences:

Jupiter—Jupiter is more than twice as massive as all the other planets combined. It is 318 times as massive as the Earth, and the gas giant has an equatorial diameter of about 88,846 miles (142,984 kilometers). Jupiter is 483,631,840 miles (778,330,000 kilometers, or 5.2 AU, from the sun). It also has the Great Red Spot (a huge storm system about two Earths in diameter) that has been seen by Earth observers for more than 300 years. Jupiter furthermore has a huge magnetic field, much stronger than Earth's, which extends past the orbit of Saturn.

Saturn—Saturn is the sixth planet in the solar system at an average distance of about 888,187,982 miles (1,429,400,000 kilometers or 9.54 AU) from the sun, and it is the second largest planet. It is the least dense of the planets, with a specific gravity of 0.7, which is less than that of water. (It is said if you could find a tub large enough, Saturn would float!) Its two prominent A and B rings, and the fainter C ring, can be seen from Earth.

Uranus—Uranus is the seventh planet from the sun, at an average distance of about 1,783,950,479 miles (2,870,990,000 kilometers or 19.22 AU). It is the third largest planet (note that Uranus is larger in diameter but smaller in mass than Neptune). Most of the planets spin on an axis nearly perpendicular to the flat plane (ecliptic) of the solar system, but Uranus's axis is almost parallel to the plane. At the time of the spacecraft *Voyager 2*'s passage through the outer solar system, Uranus' pole was pointed almost directly at the sun.

Neptune—Neptune is the eighth planet from the sun, at an average distance of 2,821,025,213 miles (4,540,000,000 kilometers or 30.06 AU), and the fourth largest planet. This planet is smaller in diameter, but larger in mass than Uranus. Neptune has the fastest winds in the solar system, reaching 1,243 miles (2,000 kilometers) per hour. This planet also radiates more than twice as much energy as it receives from the sun.

How do **Pluto and the Earth differ**?

Though Pluto may be a rocky, terrestrial planet similar to the Earth—the only such planet in the outer solar system—that is where the similarities end. Pluto is very different than the Earth, and it isn't exactly a great place for humans to live. During the Pluto day, only 1/1500th of the intensity of sunlight we receive on Earth reaches the tiny planet. Nevertheless, it is not "dark" there. The sunlight seen on Pluto is about 250 times brighter than the light we see from a full Moon.

With the sun so distant, Pluto's surface temperature is thought to be less than –328°F (–200°C), which is so cold that it would make your exposed skin as brittle as glass—it would shatter upon being touched. The surface is not inviting, either: In 1994, the Hubble Space Telescope imaged 85 percent of Pluto's surface, revealing bright and dark regions. The bright areas are thought to be shifting fields of frozen nitrogen; the dark areas are thought to be valleys, fresh impact craters, and/or methane ice colored by interaction with sunlight. These images also supported the idea that Pluto has extensive polar ice caps, especially when the planet is farthest from the sun. This is the main similarity between this planet and the Earth.

If it's so small, how did astronomers **discover Pluto**?

American astronomer Percival Lowell (1855–1916) postulated the existence of a distant planet beyond Neptune as the cause of slight perturbations in the motions of Uranus. By 1905, Lowell had built one of the most advanced observatories in order to find this planet—the Lowell Observatory in Flagstaff, Arizona. (Lowell was also searching for possible Martians on Mars to explain his observations of what looked like "canals" on the red planet.)

After his death, members of the Lowell Observatory staff continued Lowell's work, and they also sought someone who would be dedicated to the search. In 1928, then-22-year-old Clyde William Tombaugh (1906–1997) answered the call, spending hours pouring over photographic images taken by the telescope. Tombaugh studied pairs of photo plates, many of which contained anywhere from 50,000 to 400,000 stars, galaxies, and asteroids on a single image. By using a machine called a blink comparitor—in which two images are alternated, or blinked, in a single viewer to detect any movement—Tombaugh finally saw a small dot of light "move" in the sky. On February 18, 1930, he discovered Pluto in the constellation of Gemini, which is near the position Lowell originally predicted.

Interestingly enough, scientists knew (as did Tombaugh) that Pluto was not large enough to cause Uranus and Neptune to be perturbed in their orbits. Tombaugh and other astronomers searched further for another possible planet, often called Planet X.

But such a planet was never found and probably never will be. When the true mass of Neptune was determined by data from the *Voyager 2* spacecraft, all the orbital discrepancies vanished.

Do any **moons** in our solar system have an **atmosphere** like the Earth?

Yes, several of the larger moons in our solar system have atmospheres, but they are much different than the Earth's. For example, Saturn's moon Titan—a satellite larger than Mercury and more massive than Pluto—has a significant atmosphere. At its surface, the pressure is 50 percent higher than Earth's; it is composed mainly of nitrogen (as is Earth's atmosphere), about 6 percent argon, several percent methane, and trace amounts of at least a dozen other organic compounds, such as hydrogen cyanide and carbon dioxide, along with water. The organics are formed as methane (the major compound found in Titan's upper atmosphere) is destroyed by the sunlight that reaches the planet. This creates a smog similar to what we find around are larger cities, but on Titan the smog is much thicker. Some scientists believe this might be what the Earth's early atmosphere was like just before life started on our planet. But in the Earth's case, the sun's warmth and light was closer, giving life a better chance to exist.

Another moon with an atmosphere—albeit a tenuous one—is Triton, the largest moon around Neptune. Composed mostly of nitrogen with a small amount of methane, the atmosphere is a thin haze extending about 3 to 6 miles (5 to 10 kilometers) above the surface.

Still another small moon with what is often called an atmospheres is Jupiter's volcanically active moon Io. Its atmosphere is composed of sulfur dioxide and other gases from the satellite's numerous volcanic eruptions. Jupiter's ice-coated moons Ganymede and Europa also have thin atmospheres containing oxygen. But unlike on Earth, the atmospheric oxygen on these moons is almost certainly not of biological origin. Most likely, it was generated by sunlight and charged particles hitting the icy surfaces, producing water vapor. This vapor subsequently split into hydrogen and oxygen, with the hydrogen escaping into space and leaving the oxygen.

PROMINENT GEOLOGICAL FEATURES IN THE EASTERN UNITED STATES

(Note: There are many more prominent geologic features in the eastern United States that are not mentioned in this text. Such a list would entail another book. Therefore, this is merely a representative sampling of geologic features. To find out more, explore the Web sites and books listed in the last chapter.)

GLACIAL PHENOMENA

Where is there evidence of the **last Ice Age** in the **eastern United States**?

There are several places that show glacial evidence of the last Ice Age in the eastern United States, including in New York, Vermont, New Hampshire, Maine, Massachusetts, Rhode Island, Connecticut, Ohio, Minnesota, Indiana, Wisconsin, and Michigan, with touches into Pennsylvania and New Jersey. Evidence includes strange depositional and erosional features, including high waterfalls and drumlins.

There are even "leftover" features caused by melting ice found in the states below the glacier's edge. For example, melting water from the retreating glaciers created deep valleys and gorges in some place in Pennsylvania, such as Letchworth State Park. "Potholes" called kettle holes dot the park, creating depressions (some large enough to contain lakes) caused by the melting of ice chunks buried by glacial gravel deposits.

What happened in **New England** during the last **Ice Age**?

Scientists estimate that during the last two to three million years, New England experienced about 20 to 30 advances and retreats of the Ice Age ice sheets. Because the last ice advances and retreats carved away signs of earlier glaciers, more is known about the latter part of the Ice Age. Mountains were rounded, river valleys widened, and land

371

How did the ice sheets from the ice ages form?

For thousands of years, as the Earth's climate became colder, the huge Arctic ice sheets moved southward, covering large parts of several continents. Snow did not have the chance to melt, so it continued to accumulate over the years. As more snow was added, the increasing weight turned the snow to ice. In some spots, the ice was up to 2 miles (3.2 kilometers) thick, and its mere weight caused the ice mass to move even farther across the land.

The ice sheets acted like natural bulldozers, moving soil and debris, and eroding everything in their paths. Material became part of the glaciers, adding to their weight and their subsequent spread. The amount of material moved was immense—measured in the trillions of tons. Mile by mile, the ice moved over the land, leveling the surface and tearing apart the landscape. This is why many northern regions have little evidence of their past geologic history: It was completely eradicated by the glaciers. And it all happened with a simple drop in temperature of about 9°F (5°C).

scoured. Like a snowplow clearing off ice and snow from a highway, the toe of the ice sheet tore away material and exposed bedrock below, which is why so many outcroppings of bedrock are visible in today's New England.

This last ice sheet was between 3,000 and 9,000 feet (914 and 2,743 meters) thick, reaching from Canada, down into New England, and to the Atlantic Ocean. The leading edge of the ice sheet reached many places, including the highlands of Mount Desert in Maine, near Cape Cod, and to the current upstate New York border (although glacial surges—when a glacier moves quickly in one direction usually because of slight melting—occurred just over the Pennsylvania border from central New York). By around 22,000 years ago, when the ice sheet was at its peak, sea level was more than 300 feet (91 meters) lower than it is today. That meant the Atlantic Ocean shoreline was about 80 miles (129 kilometers) south of present-day Fire Island.

About 18,000 years ago, the ice seemed to stop its southward advance when a warming trend occurred. By the time the ice stopped growing, it had reached the continental shelf off today's New England coast. Quickly, in geologic terms, it retreated, reaching central Maine about 4,000 years later. As the meltwater from the glacier flowed toward the sea, the modern coastline was flooded to a depth of about 300 feet (91 meters). Today, off the coast of Maine we can see the effect of that sea level rise: islands that were once the tops of mountains, and peninsulas that were once tall ridges. By about 10,000 years ago, the climate seemed to stabilize, as did the rise in sea level.

Glaciers clung to the slopes of these mountains just 75 miles east of Los Angeles in Indio, California, as recently as 5,000 years ago, according to one study. The glaciers on San Gorgonio Mountain were likely the southwestern-most in what is now the United States during the end of the last Ice Age. *AP/Wide World Photos.*

What is the **highest free-falling waterfall** in the **northeastern United States** that resulted from **Ice Age glaciation**?

Taughannock Falls, just outside of Ithaca, New York, is about 215 feet (65.5 meters) in height, which is about 33 feet (10 meters) taller than Niagara Falls. This makes it one of the highest free-falling waterfalls in the northeastern United States and one of the highest falls of any kind east of the Rocky Mountains. Like most falls and gorges in New York, Taughannock Falls owes its origin to ice age glaciation.

How did most gorges in the **Finger Lakes of New York** form?

Most of the Finger Lakes gorges—deep, steep-sided ravines—are the result of interaction between glaciers and melting water. As the north-to-south river valleys were continually carved out by numerous glacial advances and retreats over 2 million years, streams flowed into the valleys from the east and west. As the ice melted and glacial sediment dammed the river valleys, deep lakes formed; eventually, the lakes' water levels lowered, exposing steep hillsides. The narrow gorges formed as water drained from the adjacent hillside streams, cutting the steep valley walls as water flowed into the lakes. The majority of gorges formed in this way; others were likely cut during earlier interglacial times, filled with easily eroded, loose glacial sediment during ice advances, and then re-cut since the last glacial retreat.

The erosion of the Finger Lakes gorges continues today, but at a much slower and gradual pace than during the Ice Age. The gorge rocks are mostly sandstone and shale, rocks that are easily eroded by moving water. In addition, straight cracks in rock layers (called joints) are weak. As water flows into the cracks, it freezes and expands, causing even more erosion. For example, Watkins Glen may be famous for car racing, but it is also known for its spectacular glacial geology. Located at the south end of Seneca Lake, Watkins Glen has 19 waterfalls all within 2 miles (3.2 kilometers) of each other. The stream cutting the huge glen descends 400 feet (122 meters) through 200-foot (61-meter) cliffs.

What is known as the "Grand Canyon of the East"?

Letchworth State Park's Letchworth Gorge is often called the "Grand Canyon of the East," even though it is much smaller in scale than Arizona's Grand Canyon. This site, located in Castile, New York, is where the Genesee River runs through Devonian shale and sandstone gorges. There are several major waterfalls there; one of these is 107 feet (33 meters) high and situated between cliffs as high as 600 feet (183 meters) in some places.

Most of the gorge's geology is due to intensive glacial erosion during and after the Ice Age. Initially, rivers flowed in this area; then the ice sheets buried certain parts with huge amounts of sand and gravel. The park's three deep canyons were created as the Genesee River detoured around the blocked sections of these earlier riverbeds.

Where is the "great drumlin field"?

The "great drumlin field" in the east is found in upstate New York and covers an area roughly from Syracuse to Batavia and from the Lake Ontario shore down to the Finger Lakes region. These "upside-down spoon-shaped" deposits left by the last retreating ice sheet were elongated by the ice sheet flow to the southeast, south, and southwest from Lake Ontario.

Where can glacially formed umlaufbergs be seen?

Umlaufbergs are found in many places, especially in upstate New York. These stand-alone, rounded hills formed when glacial meltwater flowed around more resistant bedrock in a river. Some of the best examples can be seen along the floodplain of the Susquehanna River near Windsor and Endicott, New York. Umlaufberg can be double, single, or multiple depending on the number of glaciations.

How did glaciers affect Wisconsin during the ice ages?

Wisconsin has a great many connections to the ice ages. The most recent series of glacial advances and retreats is called the Wisconsin Glaciation, which lasted from about 100,000 to 10,000 years ago. The most recent glacier advanced about 25,000

Are any of the states affected by the Ice Age's ice sheets literally moving?

Yes, in a way, you could say that the states affected by the Ice Age are moving, thanks to the ice sheets that covered the areas in the past 2 million years. This movement, called *isostasy,* is the upward motion after a great weight has been released and the crust tries to achieve a state of equilibrium. Perfect conditions for isostasy occurred after the Ice Age: Ice sheets—some over 2 miles (3.2 kilometers) thick—pushed down on the crust; after the glaciers retreated and melted, the pressure was released, allowing the land to spring back. Since the Ice Age glaciers receded, the Earth's crust in Wisconsin has rebounded about 160 feet (50 meters), rising about a half inch (1.3 centimeters) per year; Newfoundland has rebounded about 600 feet (183 meters) since the end of the Ice Age; and many small earthquakes felt in the Adirondacks Mountains of New York State are thought to be related to isostasy.

years ago, covering about two-thirds of the state; about 14,000 to 16,000 years ago, it began to have a succession of advances and retreats, and the last one touched northwestern Wisconsin about 10,000 years ago.

But the Wisconsin Glaciation did not arrive as one large ice sheet. Instead, the advancing ice was channeled into the lowlands now occupied by Lakes Superior and Michigan and was stopped by nearby uplands. This caused the ice to split into six major lobes as it crossed the state, including the Superior, Chippewa, Green Bay, and Lake Michigan lobes.

The ice formed several of the usual glacial features, such as striations (or scratches) in the local bedrock and long, streamlined drumlins (the state capitol sits on a drumlin). At the edges of the lobes, ridges called moraines developed; gorges formed as glacial lakes drained. These glacially formed lakes are the most impressive—over 14,000 dot the state. Many were formed when blocks of ice from the glacier were buried, then melted, forming the depressions.

What is so special about **Long Island**, New York?

Long Island is special in many ways. For one thing, it is the fourth largest island in the United States, and the largest outside of Alaska and Hawaii. Long Island is 118 miles (190 kilometers) long, and from 12 to 20 miles (19 to 32 kilometers) wide, for a total of 1,723 square miles (4,463 square kilometers) in area. It is separated from Staten Island by the Narrows, from Manhattan and the Bronx by the East River, and from Connecticut by the Long Island Sound; to the south is the Atlantic Ocean. And, most amazingly, the Long Island we know today can actually be attributed to the Ice Age.

Once located where a chain of volcanic islands stood some 300 million years ago, the place that would become Long Island was once a swampland at the edge of a towering mountain range (formed by the slow-motion collision of continents). And only 20,000 years ago, it was dotted with glacial lakes and woolly mammoths, as the continental ice sheets began to retreat.

Just how many times the ice sheets reached Long Island is still a mystery. Some scientists speculate this happened twice; others say more than five times. Either way, the glaciers moved through Connecticut, widening and deepening an ancient river valley that would eventually become Long Island Sound; widened streams would become the many harbors of Long Island's North Shore.

As the glacier finally retreated, it left behind sand and gravel in the form of a moraine (glacial ridge). This ridge is the central spine of the island, extending from Brooklyn to Amagansett and then curving offshore. By about 10,000 years ago, the leftover material finally became a true island, as the rising sea entered the valley and eventually broke through the western edge. Water finally reached the Long Island Sound, surrounding and creating the island.

Long Island continues to change today. Rising sea levels and natural erosion by the ocean are gradually wearing down the island that is really an Ice Age sandbar remnant.

MOUNTAINS

What are the **highest mountain peaks** in the **eastern United States**?

The highest mountain peak in the eastern United States is Mt. Mitchell, which rises 6,684 feet (2,037 meters) in the Blue Ridge Mountains of North Carolina. The second highest peak, Clingmans Dome, is located in the Great Smoky Mountains National Park and is 6,643 feet (2,025 meters) high. The highest mountain peak in the northeastern United States is Mt. Washington in New Hampshire. This 6,288-foot (1,917-meter) mountain is notorious for its extreme winter (and summer) weather: the

What was Laurentia?

Laurentia was the ancient continent that incorporated much of what is now our North American continent. (Note: Several of the dates and names differ from one source to another; these time periods are only approximate, and the events are based on sparse data from the field.) Sometime around 750 million years ago, the huge continent of *Rodinia* (once the single, largest continent on Earth) began to split apart into smaller continents; one part formed Laurentia, the ancestor of North America.

strongest winds ever recorded on Earth were measured there at 231 miles (371.8 kilometers) per hour on April 12, 1934.

What is the **Appalachian Province**?

The Appalachian Province is one of the major features that show evidence of ancient continental collisions in the eastern United States. (The word Appalachian is often seen as "Appalchain" or "Apalachin," but the correct spelling is "Appalachian.") This wide belt of mountains and valleys stretches 1,600 miles (2,600 kilometers) from Newfoundland to Alabama and on to southwest Texas.

Skiers and snowboarders make the final climb up Tuckerman Ravine on Mt. Washington in New Hampshire in May. The location of some of the most extreme weather in the United States, temperatures were in the 30s with wind gusts up to 40 miles per hour on Mt. Washington on this particular late-spring day. *AP/Wide World Photos.*

How did the **three major zones** in the **Appalachian province** form?

The Appalachian Province can be divided into three major zones: *Valley and Ridge,* or sedimentary rocks that represent ocean material at the edge of the old Laurentia continent (the main portion of this zone is the Appalachian Mountain Range); the *Blue Ridge Mountains,* which form the eastern flank of the Appalachian Mountains (mostly metamorphic or intrusive igneous rocks resulting from the effects of the Taconic Orogeny); and the *Piedmont,* the volcanics and

What is the Appalachian National Scenic Trail (or the "Appalachian Trail")?

The Appalachian National Scenic Trail—also known simply as the Appalachian Trail—is one of the most popular hiking trails in North America. The footpath is about 2,167 miles (3,488 kilometers) long, running along the ridge crests and across major valleys of the Appalachian Mountains.

There is plenty of ground to cover along the trail, which runs from Katahdin, Maine, to Springer Mountain in north Georgia—in other words, the entire area affected by the Appalachian orogeny. It cuts across the states of Maine, New Hampshire, Vermont, Massachusetts, Connecticut, New York, New Jersey, Pennsylvania, Maryland, West Virginia, Virginia, Tennessee, North Carolina, and Georgia. (For more information about orogenies, see "Building Mountains.")

Earl Shaffer hikes the Appalachian Trail in Boiling Springs, Pennsylvania. Shaffer, 79, was the first person to hike the Appalachian Trail from Georgia to Maine in one uninterrupted trip in the summer of 1948. *AP/Wide World Photos.*

sediments from the highlands that probably formed during the Acadian Orogeny.

About 480 million years ago, the first of several plate collisions took place that started the formation of the Appalachians (what is often called the Appalachian Orogeny). Other ways in which the Appalachians grew included collisions of islands and blocks of the continent during the late Ordovician (Taconic Orogeny); when northern Europe collided with North America in the Devonian (Acadian Orogeny); and when the ocean closed again in the late Paleozoic as the North American continent collided with the African and European continents (Alleghanian Orogeny, considered the last phase of Appalachian Mountain building).

During these orogenies, large amounts of material were bulldozed westward, while the already present rock layers were pushed upward, probably making them as high as today's Rocky Mountains. Millions of years of subsequent water, wind, and ice erosion has since worn down these ancient mountains.

What are the **Shawangunk Mountains**?

A part of the Younger Appalachians is found in the Shawangunk Mountains, a chain that stretches about 50 miles (80 kilometers), south and west of New Paltz, New York. These mountains were created during the earliest continental collision of the Appalachians. They also include Sam's Point Dwarf Pine Barren Preserve, south of Minnewaska, which is one of the most extensive examples of tectonic faulting in the United States.

What are the rocks like in **Shenandoah National Park**?

The Shenandoah National Park in Virginia is part of the Blue Ridge Mountains (the mountains that form the eastern rampart of the Appalachian Mountains between Pennsylvania and Georgia). The oldest rocks in Shenandoah National Park were formed between 1 and 1.2 billion years ago: At the peak of Old Rag Mountain is an exposed section of a billion-year-old granitic gneiss called the Old Rag Granite. Another billion-year-old exposure can be seen at Mary's Rock Tunnel. Two other older rocks in the park include 570-million-year-old basalts made from individual lava flows—each 30 to 90 feet (9.1 to 27 meters) deep—and sedimentary rocks, such as sandstone, quartzite and phyllite that formed later. The most enjoyable part of Shenandoah National Park—and a good way to see its unique geology—is Skyline Drive, a 105-mile (169 kilometer) road that winds along the crest of the mountains through the length of the park.

What are the rocks like in the **White Mountains** of New Hampshire?

The White Mountains of New Hampshire are some of the tallest mountains in New England and are composed of igneous rocks (mostly granite). About 50 percent of the entire state is underlain by metamorphosed, recrystallized volcanic rocks. Most of the original igneous rock was formed by eruptions and intrusions of magma from two time periods: the first was from 200 to 165 million years ago, and the second was from 130 to 110 million years ago. Both times the igneous activity occurred when the North American and Eurasian tectonic plates moved apart, opening up the North Atlantic Ocean. (For more information about moving tectonic plates, see "The Earth's Layers.")

Vermont's **Green Mountains** are composed of what **type of rocks**?

Vermont's Green Mountains are mostly metamorphosed sedimentary and volcanic rocks, with slivers of ocean crust throughout. These rocks include marble formed by metamorphism of Cambrian to Ordovician limestones—marble that has been used in many famous buildings, including the Thomas Jefferson Memorial in Washington, D.C., and the United Nations Building in New York. Other rocks include slates (formed during the Taconic Orogeny) and granites (mostly formed in the Devonian).

Where are the **Ozark Mountains**?

The Ozark Mountains encompass most of Missouri south of the Missouri River, and parts of northern Arkansas, northeastern Oklahoma, and southern Illinois. These mountains are thought to have originated as the land slowly uplifted over millions of years. The highest elevation is the 1,772-foot (540-meter) Taum Sauk Mountain—one of several igneous rock peaks made of granite and rhyolite. The Ozarks originated as a result of volcanic eruptions and intrusions about 1.3 to 1.5 billion years ago.

BIG RIVERS, BIG WATERS

What famous **river** has a large **submarine canyon**?

The Hudson River has a large submarine canyon called, of course, the Hudson Canyon (or Hudson Submarine Canyon). It extends over 400 miles (644 kilometers) seaward from the New York–New Jersey Harbor, across the continental margin, to the deep ocean basin.. This canyon is carved into the smooth slope of the continental shelf, and probably formed before the last ice age glacier retreated 10,000 years ago. Before that time, the continental shelf was above sea level, with the young Hudson river running though it for about 120 miles (193 kilometers), and emptying into the Atlantic Ocean. The glaciers retreated and sea levels rose, submerging the canyon. Now it is about 9,000 feet (2,743 meters) deep and is still being eroded by turbidity currents flowing down the deep chasm.

Why is the **Mississippi River** so well-known?

The main reason that the Mississippi River is so well-known is its dominance in the central part of the United States. It is also the third longest river (along with the Missouri, which it joins) in the world, measuring about 3,890 miles (6,260 kilometers) in length. The river begins in Lake Itasca in northern Minnesota and heads in a generally southerly direction, entering the Gulf of Mexico in southeastern Louisiana, where it forms the delta on which New Orleans sits. The river is navigable almost all the way, connecting with the Intracoastal Waterway in the south and with the Great Lakes–St. Lawrence Seaway system in the north by way of the Illinois Waterway. This makes it an ideal river for transporting goods by boat, so the Mississippi has been a major commerce route for the United States for centuries.

The origin of the Mississippi River is complex, but the main activity that produced today's river started at the end of the ice ages. As the ice sheets melted, they released tremendous amounts of water, some forming huge glacial lakes. For example, Lake Agassiz covered northwest Minnesota, parts of North Dakota, and the Canadian provinces of Manitoba, Saskatchewan, and Ontario; the southern outlet to this lake was called Glacial River Warren, which eventually carved the valley now occupied by the Minnesota River. As for the Mississippi, water continued where the glaciers left off, continually carving and deepening the river. As it reached the Gulf of Mexico, erosion by the Mississippi created many large delta lobes of sediment; it has also changed its channel many times over thousands of years, forming overlapping channels within the delta.

Today, although the river has less water than it did at the end of the ice ages, it still exhibits many features that are in a constant state of flux. In the southern part, it meanders in great loops across a broad alluvial plain, with oxbow lakes and marshes representing remnants of the river's former channels. Natural levees border the river for much of its length, built up by sediment from periodic flooding over time. The Mississippi discharges 620,000 feet (17,556 meters) of water per second—the fourth greatest amount in the world.

How did the **Susquehanna River** cause one of the **greatest flood disasters** in United States history?

The 444 mile (715 kilometers) long Susquehanna River seems innocuous enough. It rises in Otsego Lake near Cooperstown, New York, winds through Pennsylvania, then

Why is the Connecticut River Valley so geologically interesting?

The Connecticut River Valley is probably interesting to scientists, geologists, and the public for several reasons. Maybe it's because of the valley's connection to the past ice ages; or that it contains rocks representing the three families (igneous, sedimentary, and metamorphic); or maybe it's the proliferation of fish fossils found in black shales there. But what many people consider most interesting (geologically) is the valley's famous fossils from the age of the dinosaurs.

Most of the dinosaur evidence is found as footprints—not from the famous *Tyrannosaurus rex* or other huge reptiles, but probably from *Dilophosaurus* (made famous as the spitting dinosaur in the movie *Jurassic Park*), *Coelophysis,* or other reptilian relatives. Few, if any, bones have been found with these footprints. Most of the prints range from 12 to 18 inches (30.5 to 45.7 centimeters) and are found in shales, many of which split open to reveal the perfectly preserved footprints. This indicates that the valley was once covered by wide flats of soft sediment that were perfect for dinosaurs to walk through and leave their impressions in the ground.

on to the Chesapeake Bay (the bay is thought to be the drowned lower river, covered after the sea level rose at the end of the ice ages). The river is shallow and the water fast-moving, with plenty of flood control works (dam, levees, etc.) along its path.

But in June 1972, Hurricane Agnes hit. By the time it had reached the shore it had become a slow-moving tropical storm. Because of the heavy rains, the water breached 40-foot (12-meter) dikes in various places, inundating the entire basin and resulting in one of the worst such floods in United States history.

Why is the **St. Lawrence River** so popular?

The St. Lawrence River is popular because it is one of the largest rivers in the United States, is used extensively for hydroelectric power (especially in Canada), is a recreational fisherman and boater's dream, and is the only way to get ships to the Great Lakes from the Atlantic. The waterway flows a distance of 2,300 miles (3,700 kilometers), starting from the northeastern end of Lake Ontario, moving northeast along the United States–Canadian border, into southern Quebec, Canada, past Montreal and Quebec City, and into the Gulf of St. Lawrence. It is also a popular tourist attraction, especially the Thousand Islands region, which is a collection of about 900 islands that dot the river.

How did the **Chesapeake Bay** form?

Chesapeake Bay—an inlet of the Atlantic Ocean—is about 200 miles (320 kilometers) long and from 3 to 30 miles (4.8 to 48 kilometers) wide; it separates the Delmarva Peninsula from the mainlands of Maryland and Virginia.

What's so "great" about the Great Lakes?

The Great Lakes—five freshwater lakes in central North America—collectively form the largest body of freshwater in the world by containing a total of 18 percent of the planet's freshwater. With a combined surface area of 95,000 square miles (246,050 square kilometers), from west to east they are: Lake Superior, Michigan, Huron, Erie, and Ontario. The distance from Duluth, Minnesota, at the western end of Lake Superior to the outlet of Lake Ontario is about 1,160 miles (1,867 kilometers). All the lake bottoms, except that of Lake Erie, extend below sea level.

The lakes were first began to form about a billion years ago, as a fracture in the Earth ran from what is now Oklahoma to Lake Superior. This crack generated a great deal of volcanic activity for almost 20 million years. Lava poured out over various regions, causing the higher ground to sink and forming a huge rock basin that would give Lake Superior a head start as the largest Great Lake.

The area was further affected by millions of years of ice carving during the ice ages, with perhaps close to eight to 12 advances and retreats. As the ice leveled mountains and carved out valleys, some rocks were more affected than others, thus creating more resistant high ridges. At the end of the Pleistocene Period, as the ice sheets retreated, huge lakes formed between the ridges, filling with glacial meltwater and creating today's Great Lakes. The lakes were much larger back then, but as the ice retreated and the St. Lawrence River opened up to the Atlantic Ocean, they eventually dropped to their current levels.

The formation of the 3,237 square mile (8,384 square kilometer) Chesapeake Bay is highly controversial. It appears to be the drowned estuary of the Susquehanna River, the long river that starts in Cooperstown, New York, runs through Pennsylvania, and into the bay at Havre de Grace, Maryland.

But some scientists believe the bay is actually a complex, 50-mile- (85-kilometer-) wide impact crater, which would make it the largest in the United States. The idea of a comet or asteroid striking this area about 35 million years ago was presented in the early 1990s, after a buried crater was detected under the bay about 5 miles (8 kilometers) west of Cape Charles, Maryland. Some of the best pieces of evidence for this theory are the shocked quartz grains and impact glass found within nearby sediments, which seems to be indicative of impact ejecta.

What's so special about **Niagara Falls**?

Besides being marketed as the "honeymoon capital of the world," Niagara Falls, on the Niagara River, is an amazing geological feature. The Niagara River runs between Lake

Niagara Falls as seen from the Canadian side. The massive volume of water that pours over the falls, which span two countries, causes the water to appear green. *AP/Wide World Photos.*

Ontario and Lake Erie, forming a portion of the border between the United States and Canada. The falls are located on the international line between the cities of Niagara Falls, New York, and Niagara Falls, Ontario. Goat Island splits the cataract into the American Falls, which are 167 feet (51 meters) high and 1,060 feet (323 meters) wide), and the Horseshoe, or Canadian, Falls, which are 158 feet (48 meters) high and 2,600 feet (792 meters) wide.

The falls are relatively young, in geologic terms, forming about 10,000 years ago as the retreating Ice Age glaciers exposed the Niagara escarpment—the 600 mile (966 kilometer), cliff-like ridge that runs west and north from Rochester, New York, into Canada. This allowed the waters of Lake Erie to flow north over the scarp and into Lake Ontario, creating Niagara Falls; at the same time, it started the erosional process that has caused the falls to move upstream to its present location.

The rate of erosion of the falls depends on the rock, which includes shales, sandstones, and limestones (including dolomite). The edge of the Niagara escarpment is made of more resistant dolomite. As the softer shales and sandstones underlying the dolomite are eroded, chunks of the dolomite capping rock fall, which is why the falls continue to erode today.

When the falls were flowing naturally, it is estimated that they moved back about 3 feet (1 meter) per year, meaning they are at least 1,000 feet (305 meters) further upstream now than they were in 1678, when they were first discovered by French explorer Louis Hennepin. The process continues today, with the escarpment being eroded back toward Lake Erie and the erosion rate differing depending on the part of the falls. For example, Horseshoe Falls erodes at a faster rate than the American Falls because more water is passing over them. Erosion is also slowing down because of humans, especially since at least half the water of the Niagara River is diverted from the falls for such uses as hydroelectric power.

What is the **Everglades**?

The Everglades in southern Florida is a marshy, low-lying, subtropical savanna area. It is about 4,000 square miles (10,000 square kilometers) in size, extending from Lake Okeechobee south to the Florida Bay. This unique lowland is surrounded by a limestone rim that creates a natural retaining wall. Water comes from the Big Cypress Swamp, Lake Okeechobee, and rainfall that averages about 60 inches (152.4 centimeters) each year.

The rocks beneath the Big Cypress Swamp date to about 6 million years, which is young in geologic terms but old for southern Florida rocks. Part of the Everglades also sits on rock that formed during the ice ages, although there were no glaciers in Florida. As the huge ice sheets trapped much of the world's water, sea level fell by as much as 300 feet (91 meters) below today's level. During the warmer, interglacial times, the water would rise again, depositing marine organisms and sediment in the region and eventually forming (mainly) limestones.

OFF THE COAST

Are **barrier islands** found in the **eastern United States**?

Yes, barrier islands—long, narrow deposits of sand that form parallel to the shore and away from surf action—are found all over the world, but are most well known along the east coast of North America. They are usually separated from the mainland by a lagoon, bay, or sound. Although they are hot spots for vacationing and development, because of their geological past these islands are also susceptible to radical changes, especially from intense weather systems, such as hurricanes and nor'easters that run up the coast. One of the most famous barrier islands lies off the shore of North Carolina and is called the Outer Banks.

The formation of barrier islands is a much debated subject in geology. One theory is that the islands were formed about 18,000 years ago as the ice ages ended. As sea level began to rise, it flooded areas behind beach ridges; the rising water also carried the sediments from the beach, depositing it just off the coast. Ocean currents and waves, along with sediment from nearby rivers and streams, continued to bring in sediment, all of which helped to build up the islands.

How did the **Gulf of Mexico** form?

Today's Gulf of Mexico began as part of the young Atlantic Ocean when the North American and Eurasian lithospheric plates began to drift apart about 175 million to 100 million years ago. Land subsequently shifted: Cuba and Hispañola moved in from the Pacific, forming the border of the Caribbean Ocean; fragments that now form Central America moved in from the west, shutting off the ocean waters between North and South America. This helped form the Gulf of Mexico and nearby Caribbean basins, and it also set up ocean currents and flow that still exist today.

Why is **Cape Cod** so unique?

Cape Cod is unique not only because of its proximity to the ocean and as a tourist attraction, but because of its formation. The arm of the cape that extends into the Atlantic Ocean is actually a remnant of Ice Age glaciers, forming the longest shoreline of sand in New England.

About 15,000 years ago, a huge ice sheet covered all of New England, with the ice managing to reach the continental shelf just off today's shore. One of the lobes—appropriately called the Cape Cod Bay lobe—formed a terminal moraine that is now seen across Martha's Vineyard and Nantucket (the two islands south of the Cape). As the ice sheet retreated, meltwater held back by the moraine formed Glacial Lake Cape Cod. Sediment accumulated, covering about 400 square miles (1,036 square kilometers), while a river carved into the moraine drained water from the lake. At the same time, meltwater from another lobe, called the South Channel lobe, carried sediment into most of the area now known as the Outer Cape.

After the ice sheet retreated, sea level rose, flooding the area. For about 10,000 years, the glacial sediments have been shifted and changed by waves and currents. The sediment moved by near-shore currents sequentially formed the series of sand spits and barrier islands now familiar to those who live on or visit the Cape.

FAULT ZONES

Is there a problem with **earthquakes in New York City**?

Although New York City is not on a plate boundary, that doesn't mean there *aren't* any cracks or faults underlying the area. In fact, the entire state of New York lies on numerous faults, most of which are relatively inactive when compared to California's San Andreas fault. But there are still earthquakes. It is estimated that since 1730 there have been over 400 recorded in the state of New York, making it the third highest state in earthquake activity east of the Mississippi River. For example, a magnitude 5.1 earthquake occurred on April 20, 2002, in Plattsburgh; a magnitude 2.4 earthquake occurred on Jan 17, 2001, in Manhattan; and the largest earthquake in the state occurred in 1944 when a 5.8 magnitude quake hit Massena.

Geologists have found faults under the city, too. One fault system, the 125th Street Fault Zone, runs through New York City from 125th Street, through Central Park, and on to East 96th Street. Although earthquakes don't seem to be too much of a hazard, even a small quake could compromise the integrity of taller skyscrapers or other structures, especially since a large part of the city is underlain by loosely compacted glacial till. That's why, in 1996, New York City added a seismic provision to its building code, calling for homes, office buildings, roadways, and bridges to be constructed to withstand a sizable earthquake.

What is the **New Madrid Fault System**?

The New Madrid Fault System extends 120 miles (193 kilometers) southward from around Charleston, Missouri, to Marked Tree, Arkansas. It crosses five state lines and cuts across the Mississippi River in three places and the Ohio River in two places. It is located within the New Madrid seismic zone, a collection of faults in the continental crust located in a weak spot known as the Reelfoot Rift. This zone lies beneath the surface of Missouri, Arkansas, Tennessee, Kentucky, and Illinois. The cracks occurred in weaknesses in the crust, caused when North America tried—and failed—to split in two about 550 million years ago.

The overall fault is active and averages about 200 events per year (about 20 per month or two per week), including tremors large enough to be felt by the public (usually about magnitudes 2.5 to 3.0). The higher magnitude earthquake occurrences are, as in most quake zones, based on statistics: On the average, every 18 months a tremor of 4.0 or more is felt, creating only local, minor damage; a quake of 5.0 or greater usually occurs once every decade; it can be felt in other states and cause significant local damage. Every 70 to 90 years, an earthquake of magnitude 6.0 or greater occurs, and it is estimated that every 250 to 500 years a magnitude 7.0 or greater quake occurs. (For more information about earthquakes, see "Examining Earthquakes.")

Why are scientists **currently watching** the **New Madrid Fault**?

Many scientists believe that the highest earthquake risk in the United States—besides on the West Coast—is along the New Madrid Fault. A significant earthquake in this central region of the United States would affect the entire country, not to mention the extensive damage it would cause in the New Madrid area.

The main reason for concern is the earthquake itself, of course. But there is another reason to worry: Although quakes don't occur as frequently there as in California, the effect would amount to 20 times more damage because of the underlying geology of the New Madrid area. Underneath much of the region are thick, watery deposits of unconsolidated (loose) sands and muds left by the Mississippi River. During a strong earthquake, these loose materials would intensify the shaking.

In addition, under the loose materials are over-one-billion-year-old, basement rocks that allow seismic energy to be transmitted farther and more efficiently, thus creating more damage. It is estimated that a major quake in this area of about magnitude 7.5 would be felt throughout half the United States, causing damage in 20 states or more. Missouri alone could anticipate losses of at least $6 billion from such an event.

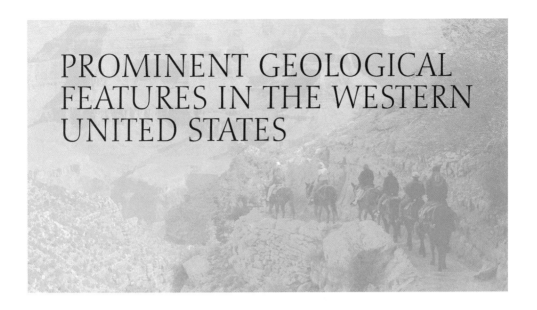

PROMINENT GEOLOGICAL FEATURES IN THE WESTERN UNITED STATES

(Note: The western United States is rich in geologic features, many of which are not mentioned in this text. Such a listing would fill an entire book. Therefore, this is merely a representative sampling of features. To find out more, explore the Web sites and books listed in the last chapter.)

MOUNTAINS

How did the mountains around **Yosemite Valley** in California form?

Yosemite Valley in California is, geologically speaking, a relatively new feature, but the rocks within the valley are about 100 million years old. During several igneous intrusions over millions of years, magma crystallized under the surface in a long, narrow belt that eventually became the Sierra Nevada. Some of the molten rock managed to reach the surface, forming a string of volcanoes on granite roots, most of which were eventually eroded away. By the end of the dinosaur era, about 65 million years ago, the granite was exposed at the surface, then constantly worn down for many millions of years more. When the continental crust east of the Sierra Nevada began to stretch in an east-west direction, the movement also tilted the Yosemite region.

Today, we still see evidence of the base rock in the mountains around Yosemite; it has a salt-and-pepper look because of the light and dark minerals within the granite. But not all the granites are the same, especially in terms of erosion resistance. For example, El Capitan and Cathedral Rocks are made up of a very resistant variety of granite and form two of the most imposing highland features dominating the Yosemite Valley.

What is **Half Dome** in Yosemite?

Half Dome is one of the most well-known features in the Sierra Nevada. It is a smooth chunk of quartz monzonite—a type of igneous rock that forms under the Earth's surface—and is rounded on three sides, with a sheer cliff on the fourth side. Half Dome is

Why are some regions in the Coastal Range dangerous?

Parts of the Coastal Range—especially in California—are particularly danger-ous because of the area's geology. Some spots contain loose sediment over harder bedrock—enough to cause landslides. For example, the knobby hills of the Northern Coast Ranges have many landslides because of a rock layer called the Franciscan formation, which is mostly made of muddy sandstone. The prob-lem is even worse in earthquake-prone California: A relatively mild shaking under the right conditions can cause unconsolidated sediment to turn into something akin to a liquid (solifluction), causing houses, roadways, and other structures to crumble.

In addition, the sea takes its toll on the Coastal Ranges in the form of bluffs composed mainly of easily eroded sedimentary rocks such as sandstone (which crumbles easily) and shale (which can either disintegrate or liquefy when wet), such as the sandstone bluffs at Santa Cruz and shale cliffs at Point Loma. But not all of the Range is made of sedimentary rock—the headlands are made of more resistant igneous rocks such as granites and basalts. For example, Point Reyes Headlands, Morro Rock (near San Luis Obispo), and Point Sur are all highly wave-resistant basalts.

easily seen from the valley floor, towering over 4,800 feet (1,463 meters) high, and standing about 8,800 feet (2,682 meters) above sea level, which gives you a good idea of how high Yosemite Valley sits above sea level.

Where are the **Coastal Ranges** located?

The Coastal Ranges are young granite mountains located along the west coast of the United States. They range from 2,000 to 4,000 feet (610 to 1,219 meters) in height (sometimes as high as 6,000 feet (1,829 meters), have wide valleys, and run from Alaska to southern California. In the Coastal Range section of California, the mountains align in a northwest direction, almost parallel to the San Andreas fault. The majority of the features in this region were formed by rapid uplift along thrust faults.

How did the **Rocky Mountains (or "Rockies") form**?

Geologists believed there were at least three uplifts that formed the Rocky Mountains. The ancestral Rockies began during the Proterozoic, between 1.7 and 1.6 billion years ago. Layers of sedimentary rock—previously deposited in an ancient sea about 2 billion years ago—uplifted during the collision of two tectonic plates. The second uplift took place around 300 million years ago, involving sediment layers formed by a shallow sea.

The third uplift (and the one geologists know most about) occurred about 130 million years ago, as major tectonic plates collided along what was then the west coast of North America. This crushing of rock began to affect the area of the modern Colorado Rocky Mountains about 70 million years ago. Sedimentary rock had previously been deposited in the area from another ancient sea; within a few million years, these rocks had eroded away, and the Proterozoic igneous and metamorphic rocks were again exposed to erosion. More and more cracks occurred as the uplift continued, and between 29 and 24 million years ago, granitic magmas began to rise in certain regions, creating lava flows, ash beds, and volcanic mountains, such as the Never Summer Mountains. Thanks to Ice Age glaciers and the continual erosion by water, wind, and ice, today's Rocky Mountains are now lower and more well-rounded than they were millions of years ago.

How did the **San Juan Mountains** originate?

The basement rocks of the San Juan Mountains—also known as the home of the "million dollar highway," a beautiful stretch of road that runs through these mountains—are a mix of igneous, metamorphic, and sedimentary rocks representing mountain building, volcanics, uplift, glaciers, and ancient seas. On top of these rock layers are visually stunning volcanic mountains.

The San Juans were formed by volcanoes that started erupting about 40 million years ago and continued belching material for another 30 million years. About 28 million years ago, a series of volcanic ash flows covered much of what is now southern Colorado. Lava flows are evident, as are breccias and conglomerates, formed when the lava mixed with the older bedrock. In addition, the tremendous eruptions in the San Juans were characterized by fast-moving, hot, billowy collections of incandescent ash, gases, and small chunks of volcanic glass (often referred to as *nuée ardentes* or "glowing avalanches.") As they settled out, the material fused together, forming an extremely erosion-resistant rock called welded tuff. Today, such tuffs are seen capping softer rock on flat-topped mesas and various other spots throughout these mountains.

What are the **Grand Tetons**?

The Grand Tetons are found in northwestern Wyoming; they form an abrupt wall of bare rocky peaks that rise about 6,000 to 7,000 feet (1,829 to 2,133 meters) high. Although the rocks themselves are over a billion years old, the mountains are very young, rising only about 9 million years ago along the Teton fault zone located at the base of the peaks. As the mountains rose, the valley below tilted westward and sank. Since that time, Ice Age glaciers, wind, water, and ice have eroded the Tetons and deposited sediment in the valleys below.

How did the **Olympic Mountains** form?

The Olympic Mountains of Washington state were actually "pasted to" the North American continent as a result of an ancient ocean floor smashing into the continen-

Is there truly a rain forest on Washington state's Olympic Peninsula?

Yes. Many scientists consider Washington state's western Olympic Peninsula to be a rain forest, albeit a much cooler one than the rain forests of the Amazon River in South America. The reason for this temperate rain forest is simple: The Olympic peaks rise up from the ocean's edge, capturing most of the moisture that flows west from the Pacific Ocean.

Like many mountain scenarios, clouds rise over the Olympic peaks, cooling down as they ascend. This process squeezes out precipitation from the clouds like a sponge, creating rainy conditions in the local forests and mountains. As the air travels up and over the peaks, it becomes much drier, with very little rain occurring on the other side (called a rain shadow). For example, the highest peak, Mt. Olympus, receives about 220 inches (559 centimeters) of precipitation per year; about 30 miles (48 kilometers) over the mountains, the town of Sequim gets only 17 inches (43 centimeters) of rain per year.

Although the area is considered a rain forest, there is quite a bit of precipitation that falls as snow. Mt. Olympus receives mostly snow; over the years, this has helped create just over 260 small glaciers capping the mountaintops.

tal landmass starting about 35 million years ago. Because of the movement of the tectonic plates in this region, the top of the seafloor folded from the pressures, creating what was the beginnings of the Olympic Mountains; the lower layer of the seafloor dove beneath the Olympic area, creating enough heat to form the volcanoes of the Cascade Range. These mountains are not as high as other peaks in the Rockies—the highest, Mt. Olympus, is just under 8,000 feet (2,440 meters)—but they are still large enough to reveal their torturous, twisted past. (For more information about tectonic plate movement, see "Through the Earth's Layers.")

What **mountains** constitute the **Cascade Range**?

The Cascade Range—a belt of volcanoes created by lithospheric plate movements—runs roughly parallel to the Pacific coast. They are the northern extension of the Sierra Nevada, and reach through northern California, Oregon, Washington, and into British Columbia. They range in height from about 4,000 to 14,400 feet (1,200 to 4,400 meters) and includes some well-known peaks, such as Mts. Rainier (the highest peak), St. Helens, Baker, Shasta, and Hood.

The Cascade Range area is also known to be volcanically active. Evidence for this occurred on May 18, 1980, when Mt. St. Helens exploded (for more information about volcanoes, see "Erupting Volcanoes"). The range itself is made up of thousands of

Where is Denali and why is it so special?

Denali is the tallest mountain in Alaska and the tallest mountain on the North American continent. Some people consider the granite and slate mountain to be the tallest land-based peak in the world: The peak is 20,320 feet (6,194 meters) high, but when measured from the 2,000-foot lowlands near Wonder Lake, it actually has a vertical relief of about 18,000 feet, which is more than Mt. Everest in the Himalayas. In other words, Mt. Everest *is* the highest mountain at 29,035 feet (8,850 meters) when measured above sea level, but when its vertical height is measured from its base on the Tibetan Plateau it only stands 11,000 feet (3,353 meters) high.

On the other hand, most people consider the tallest ocean-based mountain in the world to be Mauna Loa, which is 13,448 feet (4,100 meters) above sea level; located on Hawaii, its total height is 33,132 feet (10,099 meters) from the base below the ocean to its peak. However, to be really technical correct, nearby Mauna Kea on the same island of Hawaii is the tallest mountain from its base below the ocean to its peak, this total being about 350 feet (107 meters) more than Mauna Loa. But Mauna Kea doesn't have the huge mass of Mauna Loa, which may be why most people say Mauna Loa is the tallest.

You might know Denali by another name, Mt. McKinley, but the name Denali actually came first. The mountain was called this by the Native Americans (Athabascan) who live in the area, and means "the high one." The mountain was named by whites in the 1800s after then-presidential nominee William McKinley. In 1980, the Mt. McKinley National Park was renamed Denali National Park and Preserve.

short-lived, small volcanoes that have built up the region; the most recent peaks have been forming over the last 1.6 million years. The collision of the North American and Juan de Fuca plates are mostly responsible for the current volcanism.

Did **Crater Lake** form from a **volcano**?

Yes, Crater Lake did form from a volcano. This 1,932-foot-deep (589 meters) lake—one of the world's deepest—formed about 7,700 years ago. Part of the Cascade Range in southern Oregon, the 11,000-foot- (3,353-meter-) high peak of Mt. Mazama erupted, causing an explosion that was probably a hundred times more violent than that generated by Mt. St. Helens in 1980. The top 5,000 feet (1,524 meters) exploded and collapsed, creating a huge caldera (crater) with a rim stretching for 33 miles (53 kilometers).

Ash and volcanic dust covered hundreds of square miles and even reached north into Canada. Over time, two smaller cinder cones erupted within the caldera, the flows

393

Denali ("The High One") is the Native American word for North America's highest peak. It was renamed Mt. McKinley in honor of President William McKinley, by gold prospector William Dickey. *AP/Wide World Photos*.

not only creating the relatively flat floor of the lake, but also Wizard Island. The crater eventually filled with rainfall and snowmelt, creating what we now call Crater Lake.

WESTERN HOT SPOTS

What is the **geologic history** of **Yellowstone**?

The region called Yellowstone, which lies partly in Idaho and Montana, but mainly in northwestern Wyoming, is considered a hot spot because of its volcanic history. The entire area encompasses a volcanic plateau and mountains standing about 8,000 feet (2,400 meters) above sea level in the Rocky Mountains. Huge volcanic eruptions occurred about 600,000 years ago, with at least 1,000 times the power of Mt. St. Helens. Lava spread for thousands of square miles in only minutes as a large underground magma chamber emptied. The chamber eventually collapsed, forming a caldera 30 miles (45 kilometers) wide, 45 miles (75 kilometers) long, and several thousands of feet deep. As the crater filled in with lava over time, it created the central Yellowstone basin we see today.

What creates the **hot springs and geysers** in **Yellowstone**?

Yellowstone is by no means finished with its volcanic activity. This is evident in today's geothermal activity, including geysers, mud pots, hot springs, and fumaroles.

What concerns scientists about Yellowstone today?

Yellowstone may have harmless geysers and hot springs, but it also has some volcanic activity that concerns scientists. In the last decade, scientists have determined that two of the old volcanic areas are bulging (called resurgent domes), rising and falling with an average overall uplift of about 1 inch (2.54 centimeters) per year. One resurging area, called the Sour Creek dome, is causing Yellowstone Lake to tilt in a southerly direction. If such activity continues, scientists are concerned that it will result in a volcanic eruption.

Hot springs and geysers like Old Faithful—along with the fumaroles and mud pots—are the result of water coming in contact with hot magma below Yellowstone National Park. As rain water and snowmelt percolate through the soil and cracks in rock, they eventually reach about 10,000 feet (3,048 meters) underground, the site of the huge magma chamber under the Yellowstone Plateau. The water heats up to above boiling, with the superheated water being much lighter than the cooler water, which sinks around it. This causes the hot water to migrate back toward the surface through cracks in the rock, eventually finding its way out as hot springs, the most abundant thermal feature in the Yellowstone region. The hot water is not "pure," either, but picks up silica and other elements as it travels to the surface, creating chemical deposits on or near the surface, such as mud pots.

Geysers are one of the more amazing features in Yellowstone. They are the result of the rising hot water and expanding steam bubbles trying to squeeze through narrow cracks. Like a pressure cooker, they ultimately "blow up" from the backup of pressure.

Why do scientists believe the area around **Long Valley, California,** should be **monitored**?

Scientists are monitoring the area around Long Valley, California, because of its long, violent volcanic history. The giant Long Valley Caldera is a 9.3 by 18.6 mile (15 by 30 kilometer), oval-shaped depression located some 12.4 miles (20 kilometers) south of Mono Lake along the east side of the Sierra Nevada in east-central California. About 760,000 years ago, a huge volcanic eruption blew out about 150 cubic miles (625 cubic kilometers) of magma from about 4 miles (6.4 kilometers) below the surface. This area eventually sank, creating a caldera, and for the next few hundreds of thousands of years, small eruptions occurred in the central and western parts of the crater. About 35,000 years ago, the activity moved northward to the Mono Lake area, building the Mono-Inyo Craters volcanic chain within the caldera, the most recent eruptions occurring about 250 and 600 years ago.

Today, molten magma still underlies the caldera and heats underground waters, creating hot springs and steam vents. The waters are used to heat three geothermal power plants. But this innocuous activity does not worry scientists. What does worry them is potential trouble: In May 1980, a strong earthquake swarm—including four magnitude-6 earthquakes—struck the southern margin of Long Valley Caldera. Along with the quake, a resurgent dome uplifted on the caldera floor. Thus began the unrest in the Long Valley Caldera that is continuing today. Using devices such as global positioning systems (GPS), scientists are continually monitoring the quakes, uplift, and changes in thermal springs in the region, hoping to understand the potential hazards posed by the activity, and perhaps determining the possibility of a volcanic eruption in the future.

What is the **San Francisco Volcanic Field**?

Arizona's San Francisco Volcanic Field, located in the northern part of the state, is an area of young volcanoes along the southern margin of the Colorado Plateau. For over 6 million years, this field has produced more than 600 volcanoes. The most prominent landmark here is San Francisco Mountain, a stratovolcano that rises to 12,633 feet (3,851 meters) and serves as a scenic backdrop to the city of Flagstaff.

Although it seems strange to have volcanoes so far inside Arizona and away from a lithospheric plate boundary, scientists believe this area is a hot spot, a site of localized melting deep in the mantle. As the North American Plate moved slowly westward over this source of molten magma, eruptions produced a line of volcanoes. The first volcanoes in the San Francisco Volcanic Field began to erupt about 6 million years ago around the town of Williams. From there, they grew eastward, through Flagstaff, and toward the valley of the Little Colorado River. All these volcanoes are in a belt that extends about 50 miles (80 kilometers) in length. Although the last eruption occurred about 1,000 years ago, some scientists believe the volcanic activity in the area is not over. But if it does show signs of an eruption, estimates are that the explosions will be small and result in little damage because of the size and remoteness of the volcanic field.

What is so special about **Arizona's Sunset Crater**?

Arizona's Sunset Crater is one of the youngest scoria cones in the contiguous United States, having first erupted only about 1,000 years ago. The cone is named for the topmost layer capping the volcanic mountain, an oxidized material that gives the cone a reddish glow similar to sunset colors. In the 1920s, the crater was saved from being severely damaged when a Hollywood movie company was stopped from blowing up the mountain to simulate a volcanic eruption. This prompted the establishment of the National Monument at Sunset Crater in 1930. Currently, the mountain is still protected, with trails along Sunset Crater often closed to save the fragile cinder flanks from being trampled by hikers.

BIG RIVERS, BIG WATERS

What are some of the **longest rivers** in the **western United States**?

Some of the longest rivers in the western United States are also some of the longest in the world. The following lists these major rivers:

Columbia—The Columbia River begins in Columbia Lake, British Columbia, and flows more than 1,240 miles (1,996 kilometers), first along the Canadian Rockies, moving west and south to join the Snake River, and into the Pacific Ocean between the Cascade Range of Washington and Oregon. Because of its characteristics, the Columbia is considered by some to have the greatest hydroelectric potential of any other river in the United States.

Rio Grande—The Rio Grande River begins high in the Rocky Mountains of southern Colorado, flowing more than 1,700 miles (2,736 kilometers) through the Chihuahuan Desert of New Mexico, along the Mexico–United States border (marking Texas' southern border), and into the Gulf of Mexico around Brownsville, Texas.

How did the Great Salt Lake become so salty?

The Great Salt Lake is a shallow body of saltwater located in northwestern Utah, and is also the largest salt lake in the United States. Water from the Weber, Jordan, and Bear Rivers feed the lake, which causes the depth and extent of the lake to change with the seasons and extreme weather events. Detouring of spring runoff for reservoirs has also caused water level fluctuations in the lake. On the average, the lake is between 13 to 24 feet (4 to 7.3 meters) deep. Between 1955 and 1975, the lake expanded to its current maximum size of about 2,500 square miles (6,477 square kilometers).

Geologists know that the Great Salt Lake was once part of prehistoric Lake Bonneville, an Ice Age lake that covered most of the region. As the glaciers retreated, the 1,000-foot- (305-meter-) deep lake eventually breached, releasing much of its water. One of the remnants of this huge lake is the Great Salt Lake. The land surrounding the lake contained a great deal of salts, including magnesium chloride, potash, and common table salt, making it natural for the lake to become more briny. And with no large influx of freshwater (such as meltwater from the glaciers), the waters evaporated over time, causing the lake to become even more salty.

Colorado—The Colorado River also originates in the Rocky Mountains, starting at Grand Lake, Colorado, and flowing over 1,400 miles (2,320 kilometers). It winds through the Colorado Plateau (carving out Arizona's Grand Canyon), along the desert of the Arizona-Nevada border, and ends in the Gulf of California, the body of water between Mexico and Baja. However, because so much of its water is either dammed or used for irrigation, the river often dries up before reaching the sea.

Arkansas—The Arkansas River begins in the Rocky Mountains (further north than the Rio Grande's and south of the Colorado River's origins), traveling 1,500 miles (2,414 kilometers) through the nation's central states. It eventually meets the Mississippi River along the Arkansas-Mississippi border.

Fisher Towers, surrounded by the Colorado River (front) and the La Sal mountains (rear) near Moab, Utah. Carved from a 2,000-foot-tall mesa top, the towers stand over the Colorado River, below the 12,000-foot peaks of the La Sal Mountains. *AP/Wide World Photos.*

How did **Lake Tahoe** form?

Lake Tahoe is a 193-square-mile (500-square-kilometer) lake that lies on the California-Nevada state line in the Sierra Nevada mountains. Situated at a height of 6,228 feet (1,898 meters) above sea level, the lake's depth of 1,645 feet (501 meters) prevents it from freezing. Within the last several million years, the Tahoe Basin formed as it dropped between two uplifted blocks—the Sierra crest on the west and the Carson Range on the east. Because of volcanic activity in the region, there was a great deal of uplifting, faulting, and magma welling up through the faults. The Truckee River was connected to the lake, eventually cutting through the lava flows and winding its way to its current course. During the ice ages (2 million to 10,000 years ago), the river was alternately dammed and released, with the lake level fluctuating in response. The maximum lake level during glaciation was nearly 800 feet (244 meters) higher than its present level.

Currently, there appears to be no volcanic, faulting, or mountain building processes occurring at Lake Tahoe. Erosion of the surrounding area continues, however, including sedimentation of the lake. Scientists estimate that Lake Tahoe will continue to fill in at the rate of one foot (one third of a meter) for every 3,200 years, becoming a meadow in about 3,158,400 years.

What is so **special** about **Mono Lake**?

Mono Lake is one of the oldest continuous lakes in North America; the original depression was created by movement along the Mono-Sierra fault four million years ago. Located on the western edge of the Basin and Range province—east of the Sierra Nevada and north of the White/Inyo Mountains in Owens Valley—it sits at 6,392 feet (1,948 meters) in elevation.

During the last Pleistocene glacial advance, the lake was five times bigger and six times deeper than today because of glacial meltwaters. Today's Mono Lake—although smaller—has no outlet, allowing minerals and salts to accumulate over thousands of years, including sulfates leached from volcanic and granitic rocks within the basin, salt and chlorides from desert saline soils, and carbonates from the breakdown of local igneous rocks. Because it has such a high amount of carbonate, the water is very alkaline (basic), with a pH close to 10 (neutral water—neither acidic or basic—has a pH of 7).

The carbonates in particular make Mono Lake a unique place that is filled with giant towers of a sedimentary rock called tufa. This rock forms when fresh calcium-saturated spring water bubbles up through the carbonate-rich lake water. As water evaporates, it produces the huge calcium carbonate chunks of tufa. (Don't confuse this rock with tuff—a compacted deposit of volcanic dust and ash.)

Unfortunately, Mono Lake has been in trouble since the 1940s, when the Los Angeles Department of Water and Power built an aqueduct to divert water from the streams feeding Mono Lake. Since that time, the lake level has dropped, exposing the grand peaks and towers of tuff that give Mono Lake its uniqueness. In 1994, Los Angeles was required to reduce water diversion enough to sustain a lake level at 6,392 feet (1,948 meters) above sea level. Although the lake is still below that level, it is improving, but not enough to save the natural habitat and wildlife along the shores of the lake. In addition, the tufa towers that make the lake unique were all underwater before the diversion; exposed to the air, they are now eroding and not reforming.

OPEN SPACES

Where is the **Columbia Plateau**?

The Columbia Plateau is a large, fertile region in the northwestern United States that lies between the Cascade Range and the Rocky Mountains. It is drained by the Columbia River and its tributary, the Snake River. The Columbia Plateau originated as a depression that was filled by at least 20 basaltic lava flows (mainly from fissure volcanoes) with thicknesses in excess of 5,000 feet (1,500 meters) in many parts. Not only does it show evidence of past volcanism, but this plateau also contains the Blue and Wallowa Mountains and Craters of the Moon (see below for details).

Where is the **Colorado Plateau**?

The Colorado Plateau is a huge expanse of colorful rocks centered in much of Utah, western Colorado, northern Arizona, and northern New Mexico. This plateau is relatively stable compared to the geologic upheaval to the west (the Basin and Range region) and east (the Rocky Mountains), and is composed mostly of sediments (limestone) and lava (on a local scale). The area is also home to many famous geologic features, including the Grand, Zion, and Bryce Canyons.

What is the **Painted Desert**?

The Painted Desert is a wide expanse of arid land in Arizona and has only sparse vegetation. A good deal of wind and water erosion has filled the region with badland hills and flat-topped mesas and buttes. The desert runs in a narrow, crescent-shaped arc about 160 miles (258 kilometers) long, starting near the Grand Canyon and swinging southeast just beyond Petrified Forest National Park. It is named after the many sedimentary rocks—mostly colorful sandstones and mudstones—that cover its surface. The reds, oranges, yellows, and pinks within the rocks are mainly from oxides of iron (hemitite) and aluminum concentrated within the soil; the gray, blue, and light purples are from sediments that were deposited without oxygen (anaerobic conditions).

Why is **Death Valley** so interesting?

Even though the name Death Valley connotes doom and gloom, the site is of great geologic and climatic interest. The valley is almost all below sea level—in fact, it is the lowest point in the Western Hemisphere—yet it contains badlands, snow-covered peaks, sand dunes, intermittent streams (that get very little water), and canyons. But most importantly, Death Valley is the hottest and driest spot on record in North America.

Geologically, Death Valley is a 156-mile- (251-kilometer-) long depression (trough) created between two major mountain ranges—the Amargosa on the east and Panamint on the west. The lowest point in the valley is located in the Badwater Basin

A runner makes his way along Highway 136 during the annual Death Valley Marathon in Death Valley National Park, California. Runners cover 135 miles in 127°F temperatures over some of the most unforgiving terrain anywhere. *AP/Wide World Photos.*

salt pan, which is about 282 feet (86 meters) below sea level. Interestingly enough, only 15 miles (24 kilometers) away lies Telescope Peak, measuring 11,049 feet (3,368 meters) above sea level.

What is the area known as the **Basin and Range**?

The Basin and Range is a structural province covering large parts of Nevada and Utah. It formed when the crust in the western United States stretched in an east-west direction, creating a series of faults lying in a north-south direction. Movements along these faults created many alternating valleys (basins) and mountain ranges—thus the name for this region.

FASCINATING MORPHOLOGY

Are there **Craters of the Moon** on Earth?

Yes, there are Craters of the Moon on Earth, but they did not originate on or anywhere near our natural satellite. Craters of the Moon is a lava field in Idaho that covers 618 square miles (1,601 square kilometers), making it the largest young basaltic lava field in the lower 48 states. Sixty distinct flows form the Craters of the Moon lava field, all

A mule train winds its way down the Bright Angel trail in Grand Canyon National Park, Arizona. The trail winds for 10 miles down to the canyon's bottom. *AP/Wide World Photos.*

ranging in age from 15,000 to just 2,000 years old and containing such lava types as aa, pahoehoe, and blocky pahoehoe.

In 1969, four Apollo astronauts visited Craters of the Moon to better understand what they would find on their visit to our natural satellite. This is because the Moon is mainly volcanic rock that is very similar to the terrain found at Craters of the Moon.

What is **Ship Rock**?

Ship Rock—or TseBitai, meaning "the winged rock" in the Navajo language—is the huge chunk of a volcanic neck that towers above the middle of the flatlands of New Mexico. Originally formed underground about 30 million years ago, this volcanic core was exposed by erosion of the surrounding rock. The main part of Ship Rock is 1,969 feet (600 meters) high and 1,640 feet (500 meters) in diameter, and is criss-crossed by lava dikes that radiate away from the central neck. It is also composed of an unusually high potassium magma called minette that is thought to have formed in the Earth's mantle.

What makes the **Grand Canyon** geologically interesting?

The Grand Canyon is one of the most well-known and geologically interesting features in the United States. The canyon itself is relatively young, forming only over the last five or six million years as the Colorado River cut its way through the Colorado Plateau.

What is Meteor Crater?

Meteor Crater (also once called the Barringer Meteorite Crater), located just outside of Winslow, Arizona, is a huge, almost one-mile- (1.6-kilometer-) wide and 700-foot- (213-meter-) deep hole thought to have been formed by a stony asteroid striking the Earth. Although the area is now desert, the space body struck the surface about 49,000 years ago when the surrounding landscape was covered with forests. The impact ripped up and overturned the surrounding sandstone, creating a 150-foot-high (46 meter) rim of jumbled and smashed boulders, some of them larger than a house. Scientists estimate that the asteroid was about 150 feet (46 meters) in diameter, weighed about 300,000 tons, and was traveling about 40,000 miles (64,374 kilometers) per hour.

When the crater was first discovered, most scientists believed it was volcanic in origin, even though over 30 tons of iron meteorites were found within a radius of 8 to 10 miles (13 to 16 kilometers). It took until about 1963—in a landmark paper presented by geologist Eugene Shoemaker—to present evidence that proved the crater was formed by an impact. Since that time, other tell-tale signs have been uncovered, including rock structures called shattercones (cone-shaped features created by intense pressure on local rock); shock lamellae, or minute quartz grain cracks caused by an impact; and coesite and stishovite, two materials that do not form naturally on Earth, but can form from the intense pressures and heat from an impact.

Although many of its rock types can be found elsewhere in the world, the canyon's uniqueness comes from their variety and beautifully displayed cross-section. There is almost a complete record of the Paleozoic Era (about 540 to 250 million years ago), but less evidence of the Mesozoic and Cenozoic times (250 million years to the present), which is thought to have either eroded away or else it was never deposited in this region.

The bottom layer dates back two billion years in geologic time and contains rocks called the Vishnu schist, interspersed by granites (Zoroaster granite). Geologists believe that these rocks were once uplifted into mountains about 6 miles (10 kilometers) high. Between 1.7 and 1.2 billion years ago, the mountains eroded, leaving only the roots—the rock layers we see today.

How are the **Grand Canyon, Bryce Canyon,** and **Zion** connected?

The Grand Canyon, Bryce Canyon, and Zion are connected by locality and geology: Rock layers in the region have been uplifted, tilted, and eroded, forming a feature called the Grand Staircase, which includes colorful cliffs that stretch between Bryce   **403**

Canyon and the Grand Canyon. The connection is simple: The bottom rock layer at Bryce Canyon is the top layer at Zion, while the bottom layer at Zion (Kaibab limestone) is the top layer at the Grand Canyon.

Bryce and Zion, both in Utah, are carved into the Navaho sandstone formation; Zion in the sandstone of the White Cliffs, Bryce in the formation's Pink Cliffs. The major features in Bryce are called *hoodoos*—pinnacles, spires, and pillars of rock that have often been weirdly shaped by erosion. In Zion, the (mostly) white and red Navaho sandstone is more than 2,000 feet (610 meters) thick, forming pinnacles and other features above the narrow Zion Canyon.

What are the **Channeled Scablands**?

The Channeled Scablands are "giant-sized" landforms created when Ice Age glaciers retreated. A huge ice sheet blocked the flow of water from what is now the Clark Fork River, eventually creating a large lake. Rising water behind the barrier and melting of ice caused the dam to fail, releasing a 2,000 foot (610 meter) high wall of water that rushed into rivers, then overflowed onto the flatlands of what is now northern Idaho and eastern Washington state. Such catastrophic floods occurred across this south-dipping plateau many times, collectively creating the Channeled Scablands.

As great amounts of water eroded the region, landforms similar to those seen in rivers and streams, albeit much larger, were created. For example, the Dry Falls (located at Grand Coulee, Washington) is a "fossil" of one of the greatest waterfalls in geologic history. Although since eroded, today the falls stand 3.5 miles (5.6 kilometers) wide, with a drop of more than 400 feet (122 meters). In comparison, the American side of Niagara Falls has a drop of about 167 feet (51 meters). Giant features in the region include ripple marks and waterless canyons—all caused by the colossal floods.

FAULT ZONES

Where is the **San Andreas fault** located?

The San Andreas fault is a continuous, narrow crack in the Earth. In general, the "San Andreas fault" is the term geologists use to describe this northern section of the fault; several other right lateral faults branch from Cajon Pass, all of which usually move in response to the same crustal motions as the San Andreas.

When all the associated faults are taken into account, the entire fault system stretches for more than 800 miles (1,200 kilometers). From south to north, it travels from the Gulf of California and through central California. It goes out to sea at Mussel Rock in Daly City, reappears onshore southeast of the Point Reyes Peninsula, then threads its way on- and offshore to Point Arena before finally heading into the ocean to the Mendocino Triple Junction, where there is a meeting of three plates—the North American, Pacific, and southern part of the Juan de Fuca plates. The main and associ-

What event is nicknamed the "World Series earthquake"?

Geologists call the "World's Series earthquake" by another name: the Loma Prieta earthquake, which measured at a magnitude of 7.1 on the Richter Scale. The quake occurred at 5:04 p.m. Pacific time on October 17, 1989, just before the World Series was to be played. Some 63 people died and more than 3,000 were injured. Property damage reached almost $6 billion, making it the most costly natural disaster in the United States up to that time.

The epicenter was located in the Santa Cruz Mountains of central California, and it was the first major event along the San Andreas fault since the 1906 San Francisco earthquake. The average movement of the strike-slip fault was 4 feet (1.2 meters) in one direction and 5 feet (1.6 meters) in the other direction, which is not typical for movement along this fault. Some scientists believe the quake actually occurred on a sub-parallel fault to the San Andreas and not on the main fault itself.

ated faults of the San Andreas were (and continue to be) created as the North American and Pacific tectonic plates pass by one another. (For more information about the San Andreas, faulting, and plate movement, see "Examining Earthquakes.")

Where can **evidence** of the **San Andreas Fault** be seen at the surface?

There are several places where the San Andreas fault can be seen at the surface, or where there is evidence of its presence. For example, San Andreas Lake and Crystal Springs reservoir, which parallel California Interstate 280, are both in the valley of the San Andreas fault. The San Andreas Lake is a "sag pond" that naturally formed in the area in which the fault created a low spot; it has also been expanded with a dam, with two additional lakes in the valley—the Upper and Lower Crystal Springs reservoirs that provide water to the city of San Francisco.

Evidence of the fault itself can be seen south of San Francisco, at Daly City's Mussel Rock, which is noticeable on a rock ledge that leads out into the Pacific Ocean. The rift that formed from the 1906 San Francisco earthquake can be seen on the rock face; in addition, the 1957 San Francisco earthquake epicenter was located here. Cajon Pass is also a feature created by the movement of the San Andreas, and is one of only three gaps in the Transverse Ranges that allows access into the city of Los Angeles (the other two are the San Gorgonio and Tejon Passes).

Why is **Owens Valley** famous?

Owens Valley is famous for being associated with one of the largest faults in the western United States. The valley (or basin) is a fault-bounded trough where the valley

405

floor has dropped down relative to the mountain blocks on either side, a feature geologists call a graben. Owens Valley lies between the Sierra Nevada Mountains to the west and the White-Inyo Mountains to the east.

Owens Valley also had a large lake at one time that was a remnant of Ice Age glacial melting. Although it had been shrinking since the end of the ice ages, it had human help: In the 1800s, it was diverted for agriculture in the valley, and in 1913, the water was diverted by the Los Angeles Department of Water and Power. Within 11 years, the lake had completely dried up, and it remains that way today.

Why is the **Hayward fault** so interesting to geologists?

The Hayward fault runs for 74 miles (120 kilometers) from San Jose, California, to the base of the East Bay Hills, and on to San Pablo Bay. A huge magnitude-7 quake occurred along this fault in 1868, and was known as the "great quake of the Bay area"—at least until the 1906 San Francisco earthquake took over the title.

This fault is interesting to geologists for two reasons: Not only is the right lateral fault often responsible for catastrophic ruptures that generate large earthquakes, but it also shows evidence of tectonic creep. The slow movement—only a few fractions of an inch per year—does not generate earthquakes in itself, but it does cause a deformation of structures that are easily noticed. For example, buildings, curbs, streets, and other structures that straddle the fault are often bent in specific directions, which is a good way of identifying the precise location of the fault.

PROMINENT GEOLOGICAL FEATURES IN THE NORTHERN HEMISPHERE

GLACIAL FEATURES

What is the **largest island** in the world?

The largest island in the world is Greenland, which covers 840,000 square miles (2,175,590 square kilometers), and lies largely within the Arctic Circle. It is bounded by the Greenland Sea to the east, the Davis Strait and Baffin Bay to the west, the Arctic Ocean to the north, and the Atlantic Ocean to the south.

Geologically, Greenland is part of the Canadian Shield—the Precambrian core of the North American continent—and thus is often considered part of North America. Along the east and west coast are mountain ranges, including the highest peak, Mt. Gunnbjorn (12,139 feet [3,700 meters]), which is located in southeast Greenland. Only about 158,430 square miles (410,450 square kilometers) of Greenland, mainly the coast and coastal islands, are ice free. More than half of these ice-free regions contain Precambrian rocks. Mostly granites and gneisses, they are evidence of how much the land has been affected by volcanics and lithospheric plate movements. A huge ice sheet and many minor ice caps and glaciers cover the island.

How do **Greenland glaciers** affect **North Atlantic shipping**?

The glaciers of Greenland affect North Atlantic shipping because of the icebergs they create. Because the island has so much ice—especially the major tidewater glaciers in western Greenland—chunks often calve (break away and collapse) into the ocean. It is estimated that between 10,000 and 15,000 icebergs form each year, mostly from 20 major glaciers located between the Jacobshaven and Humboldt glaciers. These icy chunks directly impact shipping lanes in the North Atlantic Ocean.

What is the **Humboldt glacier**?

The Humboldt glacier is the largest glacier in Greenland, as well as the largest known glacier in the Northern Hemisphere.

What is the **largest glacier** in continental **Europe**?

The largest glacier in continental Europe is the Aletsch glacier in southern Switzerland, which is a huge sheet of ice that covers more than 45 square miles (120 square kilometers) of land. It is bounded to the east by Lake Marjelensee (a glacial lake), to the west by the Aletschhorn (a mountain), and to the south by the Rhone River.

Why is **Iceland** so **geologically interesting**?

Iceland is very geologically interesting because of its landforms created by ice and volcanics there, many of which are still being produced, since the island lies on the boundary between two lithospheric plates (for more information about plate tectonics, see "Through the Earth's Layers"). Glaciers cover about 12 percent of the total area of the country. To the north and west, the island's landscape is mostly sculpted by ice in the form of deep fjords.

Iceland is geologically young. For thousands of years, lava from volcanic eruptions has created a basaltic plateau that averages 2,000 feet (610 meters) in height. This activity also formed (and continues to form) features such as volcanoes, geothermal vents, and hot springs. All the ice and volcanic activity makes only about one-fourth of the island inhabitable, so most of the cities are located along the coast.

What is the **Vatnajokull glacier**?

The Vatnajokull glacier, located in the central part of southeast Iceland, is the largest glacier (also often called an ice cap) on the island. Of the over 4,600 square miles (11,922 square kilometers) of glaciers that cover Iceland, the Vatnajokull covers 3,240 square miles (8,400 square kilometers), which is equal to the size of all the continen-

---

What happened to the Vatnajokull glacier in 1996?

On September 30, 1996, an approximately 2.3-mile- (4-kilometer-) wide volcanic fissure (crack in the ground) began to melt the ice under the Vatnajokull glacier. By October 14, the eruption was over, but the ice trapped the meltwater from the glacier, causing nearby Grimsvotn Lake to rise by some 330 feet (100 meters). By early November, the water pressure was too much: It breached the southern part of the lake, sending a cascade of water into the Skeidarasandur region. Bridges, roads, and levees in the surrounding towns were washed away, as the torrents of water flowed toward the Atlantic Ocean. A few days after the peak flow, the water finally subsided.

---

tal European glaciers combined. The ice sheet also reaches a thickness of 3,300 feet (1,000 meters, or 1 kilometer) in some places.

MOUNTAINS

What are the **Himalayas**?

The Himalayan mountain range, which includes the Karakoram mountains, stretches for more than 1,500 miles (2,414 kilometers). This mountain range cuts across the northern edge of the Indian subcontinent, separating India from Asia (the Chinese regions of Tibet and Turkestan). The Himalayas formed over tens of millions of years, as the subcontinent of India moved northward and collided with Eurasia, a process that still continues today.

What **geologic features** are found in the **Himalayas**?

There are many outstanding geologic features in the Himalayas. The following lists only a few:

Mt. Everest—As most of us learn in grade school, the tallest mountain in the world is Mt. Everest in the Himalayas, which is named after Colonel Sir George Everest, the former Surveyor-General of India. The most recent satellite measurement (taken with a global positioning system device in 1999) showed that the mountain is 29,035 feet (8,850 meters) high. The first known people to climb Mt. Everest were New Zealander Sir Edmund Hillary and the Nepalese sherpa Tenzing Norgay.

K2 and Kanchenjunga—Although Mt. Everest gets all the publicity, there are other high mountains in the Himalayas: K2 measures 28,251 feet (8,611 meters) in height, and the Kanchenjunga is 28,169 feet (8,586 meters) high.

Yes. Thanks to the movement of two main lithospheric plates (Indian and Eurasian), the Himalayas are one of the fastest-growing mountain chains in the world. They are rising by more than a half inch (1 centimeter) a year, a growth rate of 6 miles (10 kilometers) every million years. (For more information about the Himalayas and plate tectonics, see "Building Mountains" and "Through the Earth's Layers.")

Siachen glacier—The Siachen glacier is the longest glacier in the world outside the polar regions, measuring about 47 miles (76 kilometers) in length.

Mt. Rakaposhi—Mt. Rakaposhi is 25,550 feet (7,788 meters) tall and has the sheerest mountain wall known. It rises 3.27 vertical miles (6 kilometers) from the Hunza Valley in only 6.2 horizontal miles (10 kilometers), having an overall gradient of 31 degrees.

What are the **Canadian Rockies**?

The Canadian Rockies are an "extension" of the United States' Rocky Mountains; both are part of the Rocky Mountain system. The Canadian Rockies are divided into five major sections: the Front Ranges, Eastern Main Ranges (with the tallest mountains), Western Main Ranges, Western Range, and Rocky Mountain Trench (the western boundary of the Rockies and the main valley that contains the Columbia River). The highest peak in the Canadian Rockies is Mt. Robson at 12,972 feet (3,954 meters).

What **famous volcano** is the symbol of **Japan**?

The Mt. Fuji volcano is the symbol of Japan—a classic basaltic stratovolcano (build up in layers) and perpetually snow-covered mountain of almost perfect symmetry. Called Fuji-san by the Japanese, the volcano lies about 60 miles (97 kilometers) southwest of Tokyo, on the main island of Honshu. It is the highest mountain in Japan, measuring about 12,460 feet (3,798 meters), and is extremely important to the Japanese culture.

Geologists have found that Mt. Fuji is based on two older volcanoes: The Kofuji volcano, which was active from about 50,000 to 9,000 years ago and is now covered up by the lavas from Mt. Fuji, and a geologically younger volcano (sometimes called Younger Fuji) that is exposed on the northern side of the volcano. Mt. Fuji's last eruption occurred in 1707, forming Hoei-zan crater on the southeastern flank.

What famous fossil site is in the Canadian Rockies?

One of the most famous fossil sites in the world is found in the Canadian Rockies: The Burgess shale rock formation on Fossil Ridge between Mt. Field and Wapta Mountain in British Columbia. The first evidence of the fossil bed was uncovered in 1909, and it has since been described as the most important fossil deposit ever found.

The Burgess shale rock unit is not huge, measuring only about 200 feet (61 meters) long and up to 8 feet (2.4 meters) thick. Amazingly, all of the fossils are invertebrates (animals without backbones). Also, they are all from the Cambrian Period about 515 million years ago and are immaculately preserved. The exceptional quality (and quantity) of these fossils—some known and others that have no modern descendants—has made this rock layer a major paleontological discovery.

What are the **European Alps**?

The European Alps are the large mountain system located in southcentral Europe and measuring about 750 miles (1,207 kilometers) long and 100 miles (160 kilometers) wide. A great deal of scientific alpine terminology originated with the Alps, the first mountain system to be extensively studied by geologists.

From west to east, the mountains run through southeastern France, all of Switzerland, southwest Germany, northern Italy, through Austria, and into Slovenia. The mountains are the source of many European rivers, including the Rhine, Rhone, and Danube. Lakes carved during the last Ice Age also lie within the mountains, including Lucerne and Maggiore. The tallest mountain in the Alps, Mont Blanc, is found in France and is 13,100 feet (3,993 meters) high. The Eiger (13,024 feet [3,970 meters] high) and Jungfrau (13,641 feet [4,158 meters] high) mountains in Switzerland are two other famous peaks in the Alps.

How did the **Alps form**?

The European Alps formed in the same way the Himalayas did, but in this case, it was the African lithospheric plate that moved northward and collided into the Eurasian plate to create the mountains. About 180 million years ago, the African plate began moving north; about 40 million years ago, the crust began to buckle and rise, forming the Alps.

The majority of rocks that form the mountains are either igneous or metamorphic (granite, mica schist, and gneiss), evidence of the violent collision. At the foot of the mountains are sedimentary rocks (sandstones and shales), which provides evidence of erosion in the Alps.

411

Why is the **Matterhorn** famous?

Although it is not the highest mountain in the Alps, the Matterhorn in Switzerland is one of the most dramatic and distinctive in appearance. Situated on the Italian-Swiss border, the mountain rises about 14,690 feet (4,480 meters) and is composed of hard crystalline rock. It is actually a remnant of the Ice Age; the pyramid peak was formed by glacial activity over the past two million years. The distinctive shape was created by a series of cirques that converged until they formed a point, which is similar to how Mt. Everest's peak formed.

The mountain was first conquered by a team led by English climber Edward Whymper, who climbed the Swiss side of the mountain on July 14, 1865. Three days later, an Italian team climbed the Italian side under the leadership of Giovanni Carrel. The south face remained untouched by humans until 1931.

What **African mountains** are an extension of the **European Alpine system**?

The African mountains known as the Atlas Mountains (Haut Atlas) are an extension of the European Alpine system. These mountains, which lie in northwestern Africa, cut across Tunisia, Algeria, and Morocco, and are approximately 1,500 miles (2,400 kilometers) long. The highest range is called the High or Grand Atlas, located in southern Morocco; the highest peak is Toubkal, measuring over 13,665 feet (4,165 meters) in height.

The deformed rocks of the Atlas mountain system and the neighboring mountain ranges are evidence of the Africa and Eurasia lithospheric plate collision, marking the final stage that closed the Tethys Sea. (These mountains are also one of the most exposed folded belts in the world, showing the result of the plate collision that began during the Cenozoic Era.)

VOLCANICS

Where is Italy's **Mt. Vesuvius**?

Mt. Vesuvius is an active volcano located on the Bay of Naples, towering above the city of Naples in southern Italy. At 2,190 feet (1,277 meters) high, the mountain seems (most of the time) to be benign, occasionally spewing out smoke and steam. But it wasn't always calm.

What happened to **Mt. Vesuvius** in **79 C.E.**?

On August 24, 79 C.E., about one o'clock in the afternoon, the Roman residents in Herculaneum (below Mt. Vesuvius) heard a huge roar. The sound coming from the mountain—thought to be "extinct"—was a volcanic explosion that caused molten pumice and ash to rain down on the city. Most citizens believed they could wait out the eruption in their homes, but by midnight, many began to flee to the sea, which was their only escape route. Suddenly, a great surge of ash, and then molten rock, exploded from the volcano, both racing down the slope toward the town. The massive amounts of material engulfed most of the fleeing residents, covering them where they fell outside or in their homes.

But Vesuvius wasn't finished. Residents of the nearby town of Pompei believed the worst was over and that they had been spared. But the next day, Vesuvius erupted again, sending out rapidly moving ash and gases, and burying the Roman city under about 20 feet (6 meters) of ash in a single day. There was no time to try to escape to the sea or into homes this time. People died where they fell, asphyxiated and covered with a thick layer of ash. An estimated 2,000 people died, with more still being discovered today as scientists continue to excavate the site.

This was the first recorded eruption of the volcano. It caused the collapse of an older crater, and a new one formed in its place; within the new caldera, a smaller cone grew. At the time, Vesuvius was thought to be an extinct volcano, but it was very active indeed! Almost 2,000 years later, the city was uncovered, providing a time-capsule view of Roman art, architecture, and artifacts. One of the more well-known discoveries was the remains of people who were "frozen" in place in the volcanic ash, lying where they fell after suffocating.

The most recent major eruption of Vesuvius was in 1944. Since that time, scientists continue to watch the volcano around the clock. Not only are they monitoring steam and smoke emissions, but they are also recording subsidence (sinking) of towns surrounding the volcano, showing that no one can be complacent when it comes to Vesuvius.

What is **Italy's highest** and most voluminous **volcano**?

Mt. Etna, above the city of Catania (Sicily's second largest city), is Italy's most voluminous and active volcano. It is also the largest volcano in Europe, being more than twice as high as Vesuvius and towering 10,900 feet (3,322 meters) above the seabed on Sicily's eastern coast. This mountain has one of the longest documented records of volcanism in the world, beginning many thousands of years ago.

Smoke rises from the central craters of Mt. Etna volcano in Sicily. Mt. Etna is Europe's tallest active volcano. *AP/Wide World Photos.*

This basaltic stratovolcano began forming about 600,000 years ago, issuing explosive eruptions and minor lava flows. Unlike Vesuvius, the magma under Etna is very fluid, allowing gases to escape (Vesuvius traps gases with its sticky, thick lava). Etna has a complex growth history, with scores of secondary cones, vents, and craters within craters. For example, it has a 3-mile (5-kilometer) by 6-mile (10-kilometer), horseshoe-shaped crater open to the east, and two other prominent summit craters (the southeastern crater formed in 1979 after a massive eruption).

What are the **Deccan Traps**?

The Deccan Traps are an immense field of lava located in west-central India, covering about 250,000 square miles (650,000 square kilometers), and roughly the size of Washington state and Oregon combined. It is often described as the plateau of central peninsular India. About 65 million years ago, at the end of the Cretaceous Period (and beginning of the Tertiary Period), a huge outpouring of lava greater than 6,500 feet (2,000 meters) thick occurred called a flood (or plateau) basalt. Geologists do not agree about the origin of the molten material: Some believe it came from just beneath the crust (upper mantle); others believe it came all the way from the boundary between the Earth's lower mantle and molten core, mainly due to the presence of high amounts of magnesium.

Some scientists believe huge outpourings of lava often coincide with the biggest extinctions in Earth's history. For example, extensive volcanism in Siberia

Is there a relationship between the Deccan Traps and the Chicxulub impact?

Scientists are trying to determine whether there is a relationship between the volcanic activity that formed the Deccan Traps and the Chicxulub impact. Did the two events occur at the same time, or did the impact exacerbate the flow of magma from the Deccan Traps?

Most scientists believe the 112-mile- (180-kilometer-) wide Chicxulub crater—a buried depression located in the Gulf of Mexico and partially on the Yucatan Peninsula—formed when a space body struck the Earth. Such a natural catastrophe would cause havoc, and is thought to be partially responsible for the extinction of the dinosaurs. Both the crater and the Deccan Traps occurred around the same time, and both together would have caused major changes in the Earth's climate. Such sudden (in geologic time) variations in the world's climate would also cause extinctions, especially for organisms that failed to adapt to the changes.

But could the impacting body also have helped the flow of magma? Scientists continue to debate such a theory. In general, many do believe that the excess heat and shock waves caused by the impact (or several impacts) could have strengthened the melting and movement of the magma, causing even more material to rise to the surface.

occurred about 248 million years ago, which was the time one of the largest extinctions in Earth's history: the Permian extinctions. About 65 million years ago, during the eruption of the Deccan Traps, another mass extinction was taking place—that of the dinosaurs.

What is the **Giant's Causeway**?

The Giant's Causeway, located in Northern Ireland, is composed of huge basalt columns formed when volcanoes released tons of lava in a flood basalt (similar to, but not as large as, the Deccan Traps of India). About 20 million years ago, as the southeastern coast of Greenland separated from the northwestern coast of the British Isles, major volcanoes developed, producing the flood basalts.

The fluid lava spread across a wide area; as it cooled, the lava crystallized and cracked in a regular pattern—in this case, a hexagonal design. In most places around the world, the lava cracks do not extend from the top to the bottom of the flow. But at the Giant's Causeway, huge basalt columns formed, all packed relatively close together and forming what geologists call columnar basalts.

What **Mexican volcano** is precariously close to a **major city**?

Popocatépetl is a steep-sided, volcanic cone that towers above Mexico City. Even though the mountain is 34 miles (55 kilometers) east of the city, an explosive eruption would greatly affect the residents there. The volcano is huge, and it has an eruptive past of spewing out ash and steam. A major eruption would endanger the 30 million people living in the city, as well as affect airlines using the international airport.

The last significant activity at Popocatépetl occurred in 1920–1922. Still, the mountain remains very active: For example, on January 2, 1998, Popocatépetl erupted for the second time in two weeks, sending ash 2 miles (3.2 kilometers) into the air. This eruption only lasted for a minute and a half, but it was powerful enough for ash to rain down on the outskirts of Mexico City. In addition, win-

Popocatépetl, which means "smoking mountain" in the Nahuatl language, spews ash and volcanic rock nearly two miles into the sky in July 2003. The large, active volcano is dangerously close to the 30 million residents of Mexico City. *AP/Wide World Photos.*

dows and doors were shaken in nearby communities by the force of the eruption. Since that time, scientists monitoring the volcano have noted plenty of ash, steam, and gases erupting, changes in vents, and minor tremors and earthquakes, all of which are indications that Popocatépetl is by no means dormant.

What happened at **Mt. Pinatubo** in 1991?

Located in the Philippines, Mt. Pinatubo is one of the highest peaks in the Luzon volcanic arc. On June 15, 1991, it stood 5,725 feet (1,745 meters) tall, but after a huge eruption, the highest point on the crater rim was only 4,872 feet (1,485 meters). The explosion created a 1.5-mile- (2.5-kilometer-) wide caldera and filled valleys around Pinatubo with about 1.3 cubic miles (5.5 cubic kilometers) of pyroclastic flow deposits. In the 20th century, the eruption was second in size only to the one in Katmai, Alaska, in 1912; it was also ten times larger than the eruption of Mt. St. Helens (1980).

Before the eruption, more than 30,000 people lived on the volcano's flanks; around the mountain, there were some 500,000 people. But they were ready: As a giant ash cloud rose 22 miles (35 kilometers) into the sky and hot blasts seared the surrounding area, timely, accurate warnings averted disaster. The Philippine authorities evacuated 60,000 people from the slopes and valleys, and the American military evacuated 18,000 personnel and their dependents from Clark Air Base situated just below the mountain. The quick action, along with the ability to monitor the volcano and predict the eruption, saved thousands of lives and an estimated one billion dollars in property damage.

But there were still consequences, and not all were directly related to the volcano. Ash deposits about 2 inches (5 centimeters) or more thick covered about 1,544 square miles (4,000 square kilometers) around Pinatubo, burying crops and structures. Additional problems were caused by wind and rain from an ill-timed typhoon called Yunya, which mixed its water with the volcanic ash. More than 300 people lost their lives as structures and roofs collapsed. The added weight from the rain, buffeting by winds, and deposits of ash mixing with torrential rains increased the death toll and structural damage over what the volcano alone had produced.

EVIDENCE OF EARTH MOVEMENTS

Why is **Loch Ness** geologically important?

Loch Ness may be associated with the Loch Ness Monster, but this area in Scotland is also geologically important. It is the site of the 62-mile- (100-kilometer-) long Great Glen Fault. This strike-slip fault runs from Inverness on the Moray Firth to Fort William at the head of the Loch Linnhe, splitting the Scottish Highlands in two. The Grampian Highlands are to the southeast and the Northern Highlands to the northwest—both composed of metamorphic rocks that reveal their crustal-moving past. It is also the location of Britain's deepest freshwater loch—Loch Ness.

The origin and subsequent history of the fault is still conjecture, but geologists do know it was active by the Devonian Period. It moved again in the middle Jurassic Period, probably in association with volcanic activity in the region.

Why does **Japan** have so many **earthquakes**?

The reason for the profusion of earthquakes in and around the Japanese islands—not to mention volcanoes such as Mt. Fuji—is their proximity to the Pacific Ocean's "Ring

Why is Turkey being squeezed?

The country of Turkey is being squeezed like a tectonic tube of toothpaste. To the north is the huge tectonic plate of Eurasia; to the south, the African plate; and to the east, the Arabian plate. These three huge plates grind against each other, literally squeezing Turkey in between.

As a result, this small area is moving to the west toward the Aegean Sea, which is also shrinking as part of the seafloor is being subducted. Movements are mainly caused by two strike-slip faults—the North Anatolian and East Anatolian faults—creating major earthquakes up to magnitude 8 or higher. The North Anatolian fault exhibits a rare behavior: Over the years, scientists have noticed that rupturing in each earthquake relieves local stress in the fault, but it also causes the stress to build up in other parts, setting up the next earthquake. This pattern of earthquakes traveling down the fault line is now being studied to help predict future quakes.

of Fire." As the giant Pacific lithospheric plate moves steadily westward, its leading edge dives (subducts) below the Eurasian plate. Along this zone, volcanoes are produced, and earthquakes occur in response to the heat and pressures caused by the subduction. Japan is also on the cusp of a triple plate junction—a meeting of the Pacific, Philippine, and Eurasian plates—which makes the region even more active.

What was the **Kobe earthquake**?

The Kobe earthquake was one of the most devastating in Japanese history. On January 17, 1995, a quake measuring 7.2 on the Richter scale (or 6.9 moment magnitude, the new way of measuring quakes) struck. The epicenter was about 12 miles (20 kilometers) southwest of the city, between the northeast tip of Awaji Island and the mainland. More than 5,000 people died in southern Hyogo prefecture, Kobe's most important port, and one fifth of the city's 1.5 million inhabitants were left homeless. Damages were estimated to be close to $100 billion.

But it wasn't only the first quake that rocked the city: Aftershocks occurred for several days afterward. Disaster relief was slow in coming, too, held back by Kobe's location and its narrow transportation route: It is located on a narrow strip of land between Osaka Bay to the southeast and the Rokko mountains to the northwest. This area links western and northeastern Japan, but all major transportation systems were cut off because of the collapse of elevated highways and railways. The Shinkansen route—the high speed rail line between Tokyo and all of western Japan—was closed by the collapse of bridge spans in Kobe, as were two other rail lines. In almost every part of the city, the quake cracked, moved, and sunk road pavement, causing even more congestion.

BIG WATERS, BIG RIVERS

What is the **Mediterranean Sea**?

The Mediterranean Sea is a huge body of water located between continental Europe to the north and Africa to the south. It is connected to the Atlantic Ocean through the Strait of Gibraltar; to the Black Sea via the Dardenelles, Sea of Marmara and the Bosporus; and to the Red Sea through the Suez Canal.

The sea is also considered to be "in between" plate boundary zones. The Mediterranean-Alpine region between the African and Eurasian plates is not well defined; in particular, it contains several smaller fragmented plates, called microplates, in between the major plates. Because of this, the geological structure of the area—even the earthquake patterns that occur in and around the Mediterranean—are extremely complex.

How was the **Black Sea** named?

The Black Sea is an inland sea located between southeast Europe and Asia. It has a maximum depth of 7,364 feet (2,245 meters) and is about 159,600 square miles (413,360 square kilometers) in area. Several countries border the sea: To the north is the Ukraine, Russia is to the northeast, to the east is Georgia, Turkey is to the south, and Bulgaria and Romania are on the west. The sea was named because of its color: it appears to be mostly black. This dark coloration is caused by the composition of the water: The top layer has only a little salt and most of the fish; the lower layer gives the sea its color because it contains hydrogen sulfide, a great deal of salt, and little or no currents or organisms.

Why is the **Aral Sea** in **trouble**?

The Aral Sea, a large inland lake in Russia, is about 15,500 square miles (40,145 square kilometers) in area. But because of overuse for irrigation decades ago, it has gone from the fourth largest to the eighth largest lake in the world. Its area also changes each year; therefore, the measurement listed above is merely an estimate.

The Aral Sea has been in existence since the ice ages, when it was only slightly saline. It has been used and abused for centuries, but the real disaster came in the 1950s, when what was then the Soviet Union decided to cultivate cotton in the region, a crop calling for large-scale irrigation. Thanks to natural evaporation and irrigation that cut into the flow of water by the two major rivers that fed the lake, the Aral shrunk to 40 percent of its former area and 25 percent of its volume.

This caused the quality of the water to deteriorate. It became more saline, killing the fish and thus ending the fishing industry around the lake. In addition, as the lake retreated, it exposed salty, open shorelines; winds whipping across the land and stirred up salt and dust, causing respiratory problems in susceptible people. Local weather has also changed as the lake became smaller; and a former island—the site of a former Soviet germ warfare waste dump—is no longer isolated from the mainland. Time will tell whether the Aral Sea ever rebounds, but a recent study by the United Nations was not hopeful. They estimated that this sea could essentially disappear by 2020 if nothing is done to stop the decline.

What is the **Caspian Sea**?

The Caspian Sea is the largest lake in the world, measuring about 143,630 square miles (372,002 square kilometers) in area, with a maximum depth of 3,264 feet (995 meters), and a surface that lies 92 feet (28 meters) below sea level. It is bordered to the northeast by Kazakhstan, to the south by Iran, to the southeast by Turkmenistan, to the southwest by Azerbaijan, and to the northwest by Russia.

What is the **oldest and deepest lake** in the world?

The oldest and deepest lake in the world is Lake Baikal (Baykal or Bajkal) in Russia. The lake is about 5,315 feet (1,620 meters) deep and is about 25 million years old. It is also considered the world's largest freshwater lake by volume. (For more about Lake Baikal, see "Geology and Water.")

What created the **Red Sea**?

The Red Sea runs from the Gulfs of Suez and Aqaba in the north, to the narrow straits of the Bab al-Mandab (the Gate of Sorrows), where the sea joins the Indian Ocean. It covers about 170,000 square miles (440,300 square kilometers), and is about 1,450 miles (2,330 kilometers) long and 225 miles (362 kilometers) wide.

Geologically speaking, the Red Sea is relatively young at a mere 25 million years. The sea is actually a "baby ocean" that has been forming as the African and Arabian continents move away from each other. The process of rifting creates new crust under the Red Sea as it gradually widens. It also means the sea is deep, measuring over 6,600 feet (2,000 meters) in some spots. The tectonic activity produces volcanic vents in some of the deepest parts of the sea, with seawater percolating through the vents and producing hot, salty waters that are rich in minerals and metals.

What is so special about the **Dead Sea**?

The Dead Sea, which is also considered a lake, forms part of the border between Jordan and Israel. It is known for many unique geological and geographical features. For example, its shoreline is the deepest depression on land in the world, measuring about 1,312 feet (400 meters) below sea level, and is known as the lowest dry point on Earth.

Even though seven million tons of water from the Jordan River and a number of smaller streams feed into the lake each day, the Dead Sea is considered by most to be

Bathers float effortlessly in the healing waters of the salty Dead Sea in Israel. *AP/Wide World Photos.*

the world's saltiest lake: it is about nine times saltier than the oceans. Evaporation from the searing sun is the main culprit; plus, the sea has no outlet to allow for any movement of its waters. Its shoreline reveals an abundance of minerals, with clusters of mineral salts, including potash, bromine, and magnesium, as well as sodium chloride, many of which are commercially extracted.

Geologically, the sea is a remnant of tectonic activity, standing in what is called the Jordan trough of the Great Rift Valley. The Dead Sea is actually two attached basins that together measure around 45 miles (72 kilometers) long and 9 miles (14 kilometers) wide.

What is **Hudson Bay**?

Hudson Bay is a huge inland sea located in Canada. It is about 475,000 square miles (1,230,000 square kilometers) in area, and about 850 miles (1,370 kilometers) long and 650 miles (1,050 kilometers) wide. The Hudson Strait connects the bay with the Atlantic Ocean, while Foxe Channel connects it to the Arctic Ocean. The basin's origins are highly debated, but scientists do know it was deepened during the last Ice Age. Since the miles-thick ice sheets retreated, the floor of the bay—and the area surrounding it—have been slowly rising (istostacy), causing the basin to become more shallow.

How did the **Verdon Gorge** form?

Before the River Verdon reaches the Rhone at Avignon, France, it turns west, entering the Verdon Gorge, a chasm about 12 miles (19 kilometers) long and with canyons that

Is Hudson Bay an impact crater?

Scientists now know the Hudson Bay basin was probably not formed by an impact crater, but some once believed the southeast margin was the result of a space object colliding with the Earth. But after extensive study by the Geological Survey of Canada, scientists found no evidence that an impact had occurred. They found no rocks with shock metamorphism or shatter cones (from the intense heat and pressure of an impact), no central uplift in the center of the suspected crater (which most impact craters have), and no materials usually associated with impacts, such as the mineral coesite. Geologists believe the extremely folded and faulted rocks of the region were shaped by slow tectonic processes, not the sudden events that accompany an impact.

can reach a maximum depth of 2,300 feet (701 meters). The river has cut into this French gorge's many limestone layers, creating the deepest and longest gorge in the country. And the river continues to carve deeper into the rock layers, especially since the same uplift that is going on in the European Alps is taking place in the gorge. This is similar to what is happening in many other deep gorges around the world, including the Grand Canyon.

Why does the **Yellow River** look **yellow**?

The Yellow River—or the Huang he or Hwang ho—is the second longest river in China and the fourth longest in Asia, measuring about 3,395 miles (5,463 kilometers) in length. The river gets its name from the silt it carries as it cuts its way through a yellowish-colored loess (a fine-grained deposit thought to be dust from ancient deserts) in the upper part of the channel. Because of this suspended material, the Yellow River is one of the world's muddiest, carrying about 57 pounds (26 kilograms) of silt per cubic yard of water. In comparison, the Nile, the longest river in the world, carries about 2 pounds (0.9 kilograms) and the Colorado River some 17 pounds (7.7 kilograms) of silt per cubic yard of water. One reason for the high volume of sediment in the Yellow River is the consistently high speed of the water, which can carry millions of tons of silt to the ocean each year.

What is the **highest waterfall** in the world?

The highest waterfall in the world is Angel Falls in southeastern Venezuela, South America (located a mere six degrees north of the equator, it just qualifies for being a Northern Hemisphere waterfall). This waterfall is located along the Rio Churun, a tributary of the Caroni River, which winds over a sandstone plateau until it reaches a cliff on the northern edge. There it shoots over the cliff edge, dropping about 2,648 feet (807 meters), hits an obstruction, then takes a second dive of 564 feet (172 meters) to the jungle floor for a total of 3,212 feet (979 meters).

PLATEAUS AND DESERTS

What is nicknamed the **"roof of the world"**?

The "roof of the world" is the nickname for the Tibetan Plateau, a huge upland area that measures some 2,175 by 932 miles (3,500 by 1,500 kilometers) and is north of the Indian subcontinent. It averages 1,640 feet (5,000 meters) in elevation, making it the largest, highest area on Earth—and maybe the largest and highest ever in the history of the planet. At its southern rim, called the Himalaya-Karakoram complex—there are 14 peaks higher than 26,247 feet (8,000 meters), including Mt. Everest.

The Tibetean Plateau first rose about 13.5 million years ago as the Indian lithospheric plate (also called the Indian subcontinent) slammed into the Eurasian plate. When the two continents met, their rocks were too light to subduct (dive) under one another, so the rocks piled up instead. Because of this, the Tibetan Plateau has the thickest continental crust known, averaging about 43 miles (70 kilometers) thick. In the northwest part of the plateau, under the Pamir Mountains, the continental crust can be up to 62 miles (100 kilometers) thick.

What is the world's **largest desert**?

The largest desert in the world is the Sahara in Africa, measuring about 3,500,000 square miles (9,065,000 square kilometers) and covering most of northern Africa. The eastern Sahara is usually divided into three regions that are most familiar to everyone: the Libyan Desert (west from the Nile Valley through western Egypt and eastern Libya), the Arabian Desert (or Eastern Desert, between the Nile valley and the Red Sea in Egypt), and the Nubian Desert (in northeast Sudan).

Will the Tibetan Plateau become any taller?

Geologists believe that the Tibetan Plateau will not get any higher, mainly because there is a balance between the thickness of the plateau and how high it can get. When the Earth's crust reaches a certain maximum height, it will begin to spread outward and create a flat-stopped plateau. In the case of the Tibetan Plateau, the maximum height can rise 3.1 miles (5 kilometers) before the crust begins to spread out due to the force of gravity. And that is currently the average height of the plateau.

The Sahara has less than 10 inches (25 centimeters) of rain each year, with the prevailing winds drying out any moisture before it reaches the interior. The best-known geologic features are the sand dunes, though they only cover about 15 percent of the desert. But the areas the dunes do cover (collectively called ergs) can reach 39,000 square miles (101,010 square kilometers). Some dunes travel rapidly because of winds, sometimes moving at a rate of 36 feet (11 meters) per year; others appear not to have moved for thousands of years.

OTHER FEATURES

What is the **Rock of Gibraltar**?

The Rock of Gibraltar is located in Gibraltar, a British Crown colony only 2.5 square miles (6.5 square kilometers) in area. Most of the colony is sitting on a peninsula that extends into the Mediterranean Sea, and most of that peninsula is made up of the Rock. It lies at the Strait of Gibraltar, the narrow neck that separates Europe from Africa and provides the only link between the Atlantic Ocean and the Mediterranean Sea.

The Rock of Gibraltar is composed of Jurassic limestone. It is geologically different from the surrounding landscape—being much older—and the limestone is riddled with more than 140 caves. Some geologists believe the Rock formed as the African lithospheric plate crashed into the Eurasian plate, creating mountain chains such as the Alps. Larger chunks of rock were pushed out of the way, including a piece thrust westward that formed the Rock.

Is there a fossil forest in the Arctic?

Yes, there is a fossil forest found in the Arctic, in the islands of the Canadian High Arctic. Although little now grows in this region except lichen and moss—the temperature averages about freezing—it once held a temperate forest. In several places the wood from the old forest has been mineralized by calcite; in other sections, the wood has been mummified. One of the best examples is found on Axel Heiberg Island, in which stumps and roots from 45 million old trees are still found in their growth position (not fallen over like most trees). Scientists exploring the island, which is only 680 miles (1,094 kilometers) from the North Pole, do not need a geologist's hammer to gather samples—the wood is still soft, so a saw can be used. Amazingly, digging around some of the roots also exposes the ancient forest floor leaf litter, complete with cones and fossilized remains of insects (and even animals) that lived in the region long ago.

What are the **Guilin Hills** of China?

The Guilin Hills of China are the famous limestone pillars that rise abruptly from the valley floor (they are seen in many Chinese artworks). These hills cluster together on the shores of the River Li, about 800 miles (1,289 kilometers) from Shanghai. Limestone was laid down by an ancient sea in this area. Millions of years later, the land uplifted, and acidic rain began to dissolve the rock. Cracks and fractures weakened the limestone, and when a second uplift occurred, some of the weaker rocks collapsed, creating what is called a tower karst landscape. These hills—riddled with caves, as are most karst areas—average about 325 feet (100 meters) high.

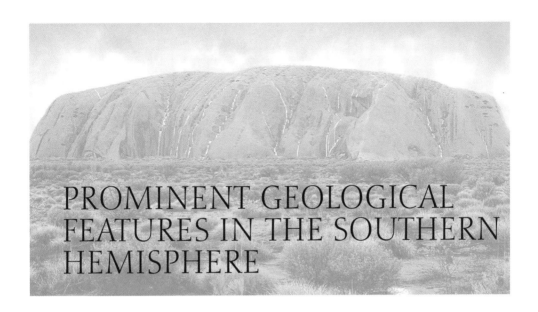

PROMINENT GEOLOGICAL FEATURES IN THE SOUTHERN HEMISPHERE

MOUNTAINS

Where are the **Andes Mountains** and how did they form?

The Andes are located mainly along the west coast of South America, cutting into the countries of Chile, Argentina, Peru, Columbia, Bolivia, Equador, and Venezuela. They continue in some of the islands of the West Indies. In Panama, a spur of the northern Andes connects with the mountains of Central America.

The Andes formed as two lithospheric plates collided and subducted. The major subduction zone is located along the west coast of Central and South America; this is mostly due to the Nazca plate subducting beneath the South American and Caribbean plates, and, to a lesser extent in the north, to the Cocos plate subducting under the Caribbean plate. Typical of most subduction zones, this area is marked by deep trenches and overlying chains of volcanoes. The mountains are relatively young, in geologic terms, originally uplifting in the Cretaceous and Tertiary Periods. They are still rising today, and active volcanoes and earthquakes are common in the region. The Andes are loftier than any other mountain ranges except the Himalayas, with many snowcapped peaks more than 22,000 feet (6,700 meters) high. These mountains also feed many of the great South American rivers, including the Orinoco, Amazon, and Río de la Plata.

What collection of mountains are called the **Southern Alps**?

The Southern Alps are found on the western side of South Island, New Zealand. They are a tall mountain range topped with glaciers and formed by tectonic activity. About 500 million years of tectonic plate movement (New Zealand was once attached not only to Australia, but also to Antarctica) created mountains and numerous faults on South Island. The more than 311 mile (500 kilometer) Alpine Fault defines part of the

Indo-Australian and Pacific plates boundary. This crack in the Earth's crust runs between the Southern Alps (to the east of the fault) and the low coastal plain bordering the Tasman Sea (to the west of the fault).

VOLCANICS

Are there any **major volcanoes** in **Antarctica**?

There are several volcanoes in Antarctica, but they are not as much of a concern as in other places. After all, the population is sparse and there are no indigenous inhabitants in Antarctica, unless you count seals, penguins, and other oceanic birds.

One volcano is near McMurdo Station, the largest United States base. The 12,448-foot- (3,794-meter-) high Mt. Erebus, is the world's southernmost historically active volcano. Erebus is the largest of three major volcanoes that form the roughly triangular Ross Island. The cone represents an intraplate volcano, situated on the southern end of the Terror Rift, which itself sits within the Victoria Land basin, a major depression of Antarctica underlain by a 13-mile (21-kilometer) crust.

This volcano is actually a cone within a cone: the older Fang volcano (partly destroyed sometime in the past) and the modern cone of Mt. Erebus. With an inner crater that measures 328 feet (100 meters) deep, the volcano can often be seen spewing lava and steam from its fissures. This glacier-covered volcano was erupting when first sighted by Captain James Ross in 1841. Since 1972, scientists have recorded continuous lava-lake activity that is occasionally accompanied by strombolian explosions that eject bombs onto the crater rim.

What **rift in Africa** has created the continent's highest and lowest **volcanoes**?

The East African Rift has created the continent's highest and lowest volcanoes: the massive Mt. Kilimanjaro, and below sea level, the volcanic vents in Ethiopia's Danakil Depression, which is located just a few degrees north of the equator in the Northern Hemisphere. Mt. Kilimanjaro, which was made famous by writer Ernest Hemingway, is about 19,340 feet (5,895 meters) high and is located in Tanzania, right next to the border of Kenya. There are actually three volcanoes that "blend in" with one another to create a single, complex volcanic structure: the Kibo, the center volcano with the highest, permanent glacier-covered peak; Shira, the oldest and most westerly; and Mawenzi, the most easterly peak, reaching just over 17,500 feet (5,350 meters) high.

How did the **Galapagos Islands** form?

The Galapagos volcanic islands (which are part of Ecuador) straddle the equator about 600 miles (1,000 kilometers) west of the South American coast. Like the Hawaiian and Reunion Islands, this collection of islands is thought to have formed from a mantle plume,

Mt. Kilimanjaro looms in the distance in Tanzania. *AP/Wide World Photos.*

a huge hot spot of magma that rose and erupted at the surface of the ocean. (For more information about hot spots, see "The Earth's Layers" and "Erupting Volcanoes").

Española is the oldest of the Galapagos Island, which are located above the Nazca Plate. The Galapagos plumes have not produced a linear chain of islands similar to the Hawaiian Islands, but rather a loose collection of islands; they have also produced a chain of seamounts known as the Carnegie Ridge. Some geologists believe that the Galapagos mantle plume could be about 90 million years old.

When did the famous **Krakatau volcano** erupt?

The Krakatau volcano in Indonesia has erupted many times in the past, but the most famous explosion occurred on August 27, 1883. The island group that is home to this volcano lies in the Sunda Strait between Java (Jawa) and Sumatra (Sumatera). It formed along the Sunda Fault in an area where the Indo-Australian lithospheric plate is being recycled back into the mantle below the Eurasian plate. Material subsequently melted with the molten magma, eventually reaching the surface to form the island arc. It is this geologic setting that caused the catastrophic eruption of Krakatau, destroying the nearby volcanoes that had formed over the years—the Danan and Perbuwatan volcanoes—and leaving only a remnant of the Rakata volcano.

The eruption was reportedly heard as far away as 3,000 miles (4,800 kilometers), but the real havoc occurred nearby. More than 40,000 people died as the shock from the eruption caused a tsunami—a seismic sea wave that wiped out more than 160 villages along the coastlines of Java and Sumatra. The waves—probably generated from the explosion's shock waves, crater collapse, and a huge amount of material entering the ocean—exceeded 130 feet (40 meters) in height in some places. Like ripples in a pond, the waves (although in a dissipated form) reached as far away as South Africa and India, as far south as southern Australia, and as far north as Japan. And for almost a year afterward, ash and pumice rained down on the islands.

This site has also experienced frequent eruptions since 1927, forming the "newest" cone, Anak Krakatau, or Child of Krakatau, which is located between the former Danan and Perbuwatan volcanoes. This approximately 700-foot- (213-meter-) high volcano is still very active, a fact that concerns many geologists. After all, the heavily populated coasts of Java and Sumatra are very close to Anak Krakatau, and another huge eruption could cause even more death and destruction.

What is so special about a **volcano on Réunion Island**?

The massive Piton de la Fournaise is an active basaltic shield volcano on the French island of Réunion, which is about 435 miles (700 kilometers) east of Madagascar in the Indian Ocean. It is also one of the world's most active volcanoes: More than 150 eruptions—most of which have produced fluid, basaltic lava flows—have occurred there since the 17th century. As with the Galapagos and Hawaiian Islands, geologists believe a hot spot is responsible for the volcanic activity on Réunion Island.

Where is the **Ngorongoro crater**?

The Ngorongoro crater is located in northern Tanzania and sits on the edge of the Great Rift Valley escarpment to the north of Lake Manyara. The crater is immense, measuring between 10 and 12 miles (16 to 19 kilometers) in diameter, and ranging from 1,312 to 2,000 feet (400 to 610 meters) deep. Within the rim is a sanctuary for several diverse habitats that harbor an abundance of wildlife.

This huge hole in the ground is the world's largest unbroken caldera and is often called the second largest extinct volcano. It is a leftover from great volcanic outpourings of lava that once occurred in this geologically active region starting about 20 million years ago. As the lava mounted and volcanoes around the area grew, it is believed that Ngorongoro also grew; it may have even rivaled Mt. Kilimanjaro, which is to the east of Ngorongoro, in size. The lava that filled the volcano formed a solid cap, which subsequently collapsed when the molten rock subsided, forming the caldera that we see today.

GLACIERS AND ICE

How **big is Antarctica** and where is it **located**?

Antarctica is one of the 7 global continents and the second smallest continent after Australia. Most of it is located south of the Antarctic Circle. The continent totals

5,405,430 square miles (14 million square kilometers) in area, with an estimated 108,109 square miles (280,000 square kilometers) that are ice-free. It is slightly less than 1.5 times the size of the United States—or larger than the United States and Mexico combined.

The Antarctic coastline is about 11,165 miles (17,968 kilometers) long, and there are no political boundaries on the continent. Technically, it belongs to all nations. In reality, there are several areas that have been "claimed" by the 12 countries that signed the original Antarctic Treaty on December 1, 1959 (although now about 18 countries have stations or temporary camps around the continent). The treaty came into force on June 23, 1961.

With a view of the Ross Ice Shelf extending out in front of them, visitors make their way back to the vehicles that carried them across the snow and ice surrounding Scott Base, Antarctica. *AP/Wide World Photos.*

What are some **other physical characteristics** of **Antarctica**?

Antarctica is considered to be the coldest, windiest, highest, and driest continent in the world. It also holds 70 percent of the Earth's freshwater and 91 percent of the Earth's total ice. During the summer, more solar radiation reaches the surface of the South Pole than is received at the equator in an equivalent amount of time. But that

What does the Antarctic terrain resemble?

Antarctica's terrain resembles an alien world: The flat, white landscape at places like the South Pole and Ronne and Ross Ice Shelves seems to go on forever; dry and barren rock with snowcapped mountains surround the Dry Valleys (the place that scientists believe resembles the planet Mars); the Transantarctic Mountains cut through the continent like a great divide; and the cinder-ridden areas of places like McMurdo make one realize that the continent has been very volcanically active in the recent past.

Antarctica is comprised of about 98 percent thick continental ice sheets and around 2 percent barren rock. Elevations vary greatly from sea level at the coast and up to 16,863 feet (5,140 meters) at Vinson Massif (although the exact height varies depending on the reference source). The continent averages about 1.5 miles (2.4 kilometers) in height above sea level, making it about a mile (1.5 kilometers) higher than the average global land height.

24-hour-a-day sunshine does not heat up the area; instead, the snow reflects back the majority of the sun's radiation into space. That is why, at the height of summer, the South Pole rarely, if ever, reaches above-freezing temperatures.

What is the **Southern Patagonia Icefield**?

The third largest ice field in the world is the Southern Patagonia Ice Field in Chile, measuring about 5,019 square miles (13,000 square kilometers) in area and about 3,280 feet (1,000 meters) thick in some parts. The 223-mile- (360-kilometer-) long by 25-mile- (40-kilometer-) wide ice field not only feeds many of the fjords in western Chile, but also the lakes on the inland side. Because of its location, the ice field also receives more than 276 inches (700 centimeters) of precipitation a year.

What is the **largest valley glacier** in the world?

The largest valley glacier in the world is in East Antarctica. Called the Lambert Glacier, it is located within the Lambert graben and measures up to 50 miles (80 kilometers) in width and more than 311 miles (500 kilometers) in length. This huge glacier drains one-fifth of the East Antarctic Ice Sheet and contributes over 14 square miles (35 square kilometers) of ice to Prydz Bay (via the Amery Ice Shelf) each year.

What is **another large valley glacier** located in **Antarctica**?

The Beardmore Glacier, located in the Transantarctic Mountains in central Antarctica, is one of the world's largest valley glaciers, measuring 125 miles (200 kilometers) long and 25 miles (40 kilometers) wide. This glacier is wedged between the Queen Maud and Queen Alexandra mountains and descends about 7,200 feet (2,200 meters) from the South Polar Plateau to the Ross Ice Shelf. The Beardmore was first discovered in

1908 by Ernest Shackleton (1874–1922) during the British Antarctic expedition that began in 1907; he referred to it as "The Great Glacier," or just "The Glacier." It was later used by Robert Scott (1868–1912), who crossed the glacier to get over the Transantarctic Mountains on his ill-fated trip to the South Pole.

BIG WATERS, BIG RIVERS

What is the world's **second largest freshwater lake**?

The world's second largest freshwater lake is Lake Victoria, a 26,830 square mile (69,490 square kilometer) lake in east central Africa on the Uganda-Tanzania-Kenya borders. This lake sits between two branches of the Great Rift Valley and occupies an approximately 250-foot- (75-meter-) deep basin on the Equatorial Plateau. It is thought that the eastern flank of the western Great Rift Valley branch uplifted, causing the water in the local rivers to pond and form the lake.

Where is the **Amazon Basin** located?

The Amazon Basin extends from the Andes Mountains to the coast of Brazil (at the Atlantic Ocean), the main river and its tributaries cutting through nine surrounding South American countries. The main artery of the Amazon River is the second longest

Abandoned huts stand in the floodplain of the Zambezi River in northern Mozambique. Villages such as this one rely on the river for food and transportation but also risk being destroyed by the same body of water. *AP/Wide World Photos.*

A rainbow forms in a cloud of mist from the Zambezi River, Africa's fourth largest river, at Victoria Falls. *AP/Wide World Photos.*

river in the world, running more than 3,900 miles (6,276 kilometers) in length and covering an area of about 2,368,000 square miles (6,133,091 square kilometers). Combined with its tributaries, it contains about 25 percent of all the water carried by the world's rivers, and is also home to the world's largest rainforest.

Why are there so many **falls** along the **Zambezi River**?

There are numerous falls along the Zambezi River in Africa for a good geological reason: The river zigzags back and forth as it cuts into very faulted rock, carving into the region's sandstone and basalt plateau. In most cases, where the different rocks meet, a fault occurs, as does a waterfall.

One of the most famous waterfalls along this course is Victoria Falls, the 350-foot- (106-meter-) tall and over a mile- (1.6-kilometer-) wide falls located near the Zambia-Zimbabwe borders. In full flood, the spray from the immense amount of water pouring over the falls can reach a height of 1,000 feet (305 meters).

OFF THE COAST

How did the **Great Barrier Reef** form?

The Great Barrier Reef extends for more than 1,250 miles (2,011 kilometers) off
Australia's east coast, and is actually a combination of living and dead (fossil) organ-

Is the Great Barrier Reef in trouble?

The Great Barrier Reef may be in trouble, not only from humans, but also from natural causes. Tourism is one problem caused by humans. People often break corals by reef walking or dropping anchors, and boat fuel and oil dumping also damages the coral. It is estimated that about 500 commercial vessels carry tourists to the reef system each year. Because the ideal environment for a coral is shallow warm water, plenty of sunshine, water movement, and a proper level of nutrients, human-induced changes such as building moorings, groins, or breakwaters can change all of the above factors. This, in turn, can cause an imbalance in the reef environment, contributing to widespread die-outs of the corals and loss of marine life.

But natural events can also affect the reef. For example, since the 1960s, Crown of Thorns starfish have been eating the corals, with outbreaks appearing every 1 to 15 years. Other problems also plague the reef, including such natural events as the El Niño and La Niña, which cause the waters off western South America to warm or cool, affecting the global climate. Some scientists believe that bleaching (a great dying off of corals) at the Great Barrier Reef and other reefs around the world is caused by rising ocean water temperatures, although evidence for this is still inconclusive. No one can agree on what causes these changes in ocean water temperatures; they may be related to the El Niño, La Niña, other anomalous ocean waters, or overall global warming.

isms. Although there are hundreds of different species of corals there—generally classified as either hard or soft coral—only the hard corals are responsible for building the reefs.

Although we don't think of them as such, reefs are true geological marine features. They are built by living corals, which consist of individual coral polyps—tiny live creatures that join together to form colonies. Polyps resemble minute, jelly-like blobs crowned by tentacles that allow them to feed on passing particles in the water. These organisms live inside a hardened shell of aragonite, a form of calcium carbonate the polyp extracts from the ocean waters. When polyps die, other polyps use the shells as a foundation for building new skeletons on top of the old.

Collections of these occupied and abandoned shells are what we recognize as a coral reef. These organisms create the huge reefs, cays (reef shingle islands), and high islands seen at the Great Barrier Reef and hundreds of other reefs around the world. The Great Barrier Reef, which is made up of about 3,000 individual reefs, is a haven for marine wildlife, including several varieties of dolphins and whales, more than 1,500 species of fish, 4,000 species of mollusk, and 200 species of birds.

Where are the **ice-free and ice shelf** areas around **Antarctica**?

The ice-free coastal areas around Antarctica include parts of southern Victoria Land, Wilkes Land, the Antarctica Peninsula area, and parts of Ross Island on McMurdo Sound (the location of the largest United States base in the Antarctic).

Ice shelves are present along about half the coastline, and floating ice shelves constitute 11 percent of the continent's area. These shelves form where glaciers and ice streams flow into a bay, losing ice as the underside melts. This is also the source of the cold bottom water that flows north on the ocean floor and eventually oxygenates the tropical waters close to the equator.

The largest ice shelf in Antarctica is the Ross Ice Shelf, a somewhat triangular piece of ice that covers about 209,400 square miles (542,344 square

One of the hundreds of thousands of divers who come to view the rich diversity of marine life on Australia's Great Barrier Reef, the largest reef on the planet. *AP/Wide World Photos.*

kilometers). Fed by the many glaciers of the Transantarctic Mountains (such as the Beardmore glacier), the shelf's thickness ranges from about 600 feet (183 meters) at the seaward edge to more than 4,264 feet (1,300 meters) on the landward edge. This shelf, which is about the size of France, actually sits on the water and rises and falls with the tide. Large icebergs break off each year, some larger than the state of Rhode Island.

How did **Madagascar** form?

Madagascar, the large island off the coast of eastern Africa in the Indian Ocean, was once part of the African continent. It is the world's fourth largest island, about half as

What present-day islands were once at the center of the Gondwana supercontinent?

About 400 million years ago, the Falkland Islands (now located off the coast of Argentina) were at the center of the supercontinent Gondwana. As the lithospheric plates eventually separated into South America, Africa, Antarctica, India and Australia, the Falkland Islands drifted about 6,500 miles (10,461 kilometers) and rotated so that what was once pointing north now points south.

The most common rocks on the island are quartzites, sandstones, and mudstone; some people suggest that diamonds and gold could be found in areas once connected to South Africa. Evidence of the supercontinent's breakup is also evident in the igneous dolerite dikes that cut through the island's rock layers.

large as California and measuring approximately 1,000 miles (1,609 kilometers) long. Although geologists don't know all the details, they do know that about 165 million years ago this chunk of land ripped away from Africa, creating an island with dramatic wildlife and plant diversity.

What is the world's most **remote island**?

The world's most remote island is Bouvetoya, located in the Southern Ocean. This volcanic island, which is about 1,590 miles (2,558 kilometers) southwest of Cape Town, South Africa (at the southern extremity of the Mid-Atlantic Ridge), measures about 4 miles (6.4 kilometers) long and 3 miles (4.8 kilometers) wide, and has glaciers covering most of its surface. It was first discovered in 1739, lost, then rediscovered in 1822. In 1825, the island was found yet again by a British sealing expedition. The sailors also noted a volcanic island nearby called Thompson Island, but to this day this island has never been seen again and is thought to have been a victim of a huge eruption between 1895 and 1896. Bouvetoya became a Norwegian Territory in 1929; in 1971, it was declared a nature preserve. As of this writing, no one has ever wintered over on Bouvetoya.

OTHER FEATURES

What is **Ayers Rock**?

Ayers Rock—also known as Uluru and considered the largest monolith in the world—is located in the southwest corner of Australia's Northern Territory. This huge, orange-brown rock (and nearby Mt. Olga) are in the middle of the flat Australian

437

Ayers Rock in central Australia takes on an unusual color as waterfalls cascade down its walls. The rock, which is usually a reddish color, appears almost purple due to drenching rains and flooding in the area. *AP/Wide World Photos*

desert. The roughly oval monolith is about 1,132 feet (345 meters) high, 2.2 miles (3.6 kilometers) long, and just over 1 mile (2 kilometers) wide.

Nearby Mt. Olga is about 1,500 feet (457 meters) high and is part of a collection of about 30 rounded hills called the Olgas. Both features are made of Precambrian rock, but their origins are still unknown. Many geologists believe the rocks are the denuded remnants of a mountain range: An ancient ocean deposited layers of sediment that eventually compressed to create arkosic sandstone. At some point, tectonic movement pushed up the rock layers, tilting some of the strata to about 85 degrees, which is why the layers of Ayers Rock are approximately vertical (those at Mt. Olga are still horizontal). Subsequent erosion by rain and desert winds has continued for hundreds of millions of years.

Why is **Tierra del Fuego** so geologically interesting?

The Tierra del Fuego region, an archipelago at the southern tip of South America, extends toward Antarctica and is separated from the southern continent by about 600 miles (965 kilometers). It was at one time attached to Antarctica until the continents began to move apart about 25 million years ago.

The region was formed by the 373-mile- (600-kilometer-) long Magallanes-Fagnano fault system. This is also one of the few places on Earth where geologists can

Is all of Antarctica covered with ice?

No. The McMurdo Dry Valleys of Antarctica are one of the few places that are not covered by ice. In these areas—sometimes called Antarctic oases—extensive outcrops of rocks are exposed. This allows geologists to collect rock samples and fossils of ancient life, and is one of the reasons why we know about the some of Antarctica's geologic history.

Lakes are also features of the Dry Valleys. For a few weeks each summer, temperatures warm enough to melt glacial ice, creating streams that feed freshwater lakes at the bottom of the valleys. Under a protective cap of ice, the lake water remains unfrozen all year, supporting colonies of bacteria and phytoplankton.

Just over the past decade, summers have been colder than usual in Antarctica, which is in great contrast with the rest of the Earth, which appears to be warming. This has caused more freezing in the lakes, a possible detriment to the organisms living in the water. No one knows if the trend is related to global climate change, or just a quirk in the weather of this icy continent.

examine surface and subsurface geological features of a continental transform plate boundary. But because of the remoteness and difficulty hiking into the area, it has not been easy to study.

GEOLOGY RESOURCES

GETTING STARTED

How can I **get started** in geology?

There are a number of ways to get started in geology. First and foremost—and the most fun—is to get outside and look at the local rocks and landscapes near you. The list goes on from there: Observe the geology around you on trips. Start your own rock, mineral, fossil, and/or gem collection. Join a local geology or gem and mineral club. Take an introductory Earth science course. Visit geology exhibits at natural history museums, colleges, and universities. Read some of the many books and magazines on the subject. Watch nature programs on television that feature geology that take you to remote places and feature exotic phenomena. And finally, surf the Internet for the almost overwhelming amount of information on geology, ranging from live webcams of active volcanoes to virtual tours of natural history museums around the world.

Where is a **good place** to start looking for **rocks and fossils**?

The best place to look for rocks and fossils is in your own backyard and immediate area, especially if you live in the country. For city dwellers, a short jaunt to local stream beds, river banks, and road cuts will yield fascinating and unusual rocks and fossils. Just make sure you know where you are; if you know you are on private property, *always ask permission to go on someone's land;* and, if they say no, respect their request.

What **other outdoor precautions** should a rock and fossil collector heed?

Some of the other precautions that a collector might take before going into the field include being aware of weather conditions (such as predicted storms and warnings of

The first, and most important, precaution anyone should take before searching for rocks and fossils is to obtain permission to search the area. This permission may be from the organization, government agency, or person that owns the land. Also, collectors must verify that they can collect and keep any specimens that they find. Failure to do so may result in arrest and charges of trespassing, destruction of property, or even theft. For example, in almost all cases it is illegal to collect rocks and minerals from parks, monuments, and sundry other sites that are under the jurisdiction of the United States' National Park Service.

severe conditions), finding out if the area is known for quick weather changes and being prepared (remember that in certain areas weather can change in a matter of minutes), knowing what types, if any, of dangerous animals inhabit the collecting area (such as venomous snakes or grizzly bears), and learning what kinds of stinging insects are present (such as wasps and poisonous spiders, making sure to carry any necessary medications if you are severely allergic to such insects).

COLLECTING ROCKS, MINERALS, AND FOSSILS

How do you **start** a rock, mineral, and/or fossil **collection**?

Most collectors define a good rock collection as a carefully selected, representative, well-labeled group of specimens. With time, study, and patience, you, too, can have such a collection. The rock, mineral, or fossil collection must fit your goals: Do you want a large or small collection? Do you want your rocks, minerals, and/or fossils to have a local or international scope? Or do you want your collection to be based on one, two, or all of the rock families (igneous, sedimentary, or metamorphic)? Some people only collect crystals; others deal only in micromounts; and still others collect only fossils. As a beginner, you might not know where your collection is heading, but eventually you will see a pattern emerging.

As time goes on, your collection will improve as more good specimens are added, or when lesser-quality rocks, minerals, or fossils are replaced with better ones. The best way to begin is to train yourself to analyze rocks by their appearance—from texture and size to clarity of crystals and color. One way to better understand good specimens is to attend gem and mineral shows, go to rock museums, and talk with other rockhounds and collectors.

How can you **identify rocks, minerals, and fossils**?

The best way to identify rocks is to educate yourself. This can be in the form of taking geology courses at a local university, reading basic books about rock identification and classification, or even volunteering to help geologists in the field (usually through a major university with a geology department). You will not learn all the rocks, minerals, and fossils overnight; as with any science, it takes time to learn the basics.

A geologist uses tools as well as his hands to collect rock samples. *AP/Wide World Photos.*

As a last resort—usually to identify a single specimen—take a rock you are having problems identifying to your local university geology department. Although their time is valuable (they are usually doing time-consuming geologic research), some professors or technical personnel might be able to help you identify a particular specimen.

What are the **best places** to collect **rock specimens**?

If you are searching for rocks, flat country is not the place to try, since there are few areas of exposed rock there. Instead, try places such as quarries, road cuts or natural cliffs, and outcrops (but again, *make sure you have permission to collect—and avoid overhanging rock and the edges of steep natural or quarried walls*). The best rock surfaces are those that are fresh and unweathered—places that have just been cut to put in roads or other structures. If possible, visit several exposures of the same rock type to be sure you collect a good representative sample.

What kind of **clothing and equipment** are needed for **geology field work**?

Clothing and equipment that are suitable for outdoor activities, such as for hiking or backpacking, will do very nicely for geology field work. Naturally, they should be customized to meet the needs of the local environment. An expedition into a remote area of the world for several months would have much different requirements than a day dig at a local site. If you don't have any experience with the great outdoors, there are numerous sources of information concerning clothing and gear, including experienced local guides, backpackers, hiking books, and outdoor recreation store personnel.

What are some tools used for rock or fossil collecting?

Collecting rocks, minerals, gems, or fossils often requires a number of different tools, with each situation having its own unique needs. But the most indispensable, general purpose tool to have for all situations is the rock hammer (called by many names, including geology hammer, geologist's hammer, or geohammer), which has a square knob at one end and a chisel or pick at the other.

Other tools include: a rock saw, stonemason's chisels (with an assortment of blade widths in order to remove layers of hard rock), a club hammer (for hitting heavy chisels), old knives or trowels (to scrape away soft rock), lightweight spades or shovels (for removing loose material), sieves and strainers (for sorting and washing samples), and all types of brushes—from paint brushes to toothbrushes (for removing fine loose rock and dirt from delicate specimens). Since collecting sites can be miles from roads—and everything has to be carried in—lightweight tools with multiple uses are preferable.

The following are only general suggestions for clothing and equipment (it is the responsibility of the collector to make sure his or her own personal requirements are met):

1. Most importantly, the clothing should be appropriate to the local conditions and weather. A dig in Antarctica's Dry Valleys would require much different clothing than one in the Gobi Desert, or even at your local rock outcrop. Look for synthetics (microfleece works nicely) for cold weather and control your temperature through the use of layers. In warmer weather, wear light-colored clothing that wicks away perspiration.

2. Sturdy boots, especially those with good ankle support, are a must for hiking to and around a dig site, where rocks are usually sharp and uneven. Some geologists prefer boots that cover the ankle as protection from hazards in tall, thick grasses or even snake bites in some regions.

3. To protect your hands, and keep warm if needed, choose appropriate gloves. Thicker gloves can be used for tougher rocks, such as pounding the heavy, rough dolomite (a hard limestone) in which Herkimer diamonds are found.

4. Rain gear can be used not only to stay dry, but also to keep warm.

5. The sun's rays are always a concern: Sunblock is essential for protection from ultraviolet rays. A good hat will keep off the sunlight or a light rain—and can make you look jaunty. However, if there is a danger of falling rocks, substitute a hard safety helmet for the soft hat.

6. Eyes can be protected by well-made sunglasses, or safety glasses/goggles if there is a danger of flying rock chips.

7. Other equipment that might be useful includes a flashlight, water bottles or canteens, a first aid kit, compass and maps, a global positioning system (GPS)

device, cameras (regular and/or video), and a backpack to carry everything. Depending on the length of stay, a sleeping bag, tent, cooking gear, and food may also be required.

What **other items** can a **rock collector use** in the field?

There are several other items to carry along in your backpack. Don't forget to bring either cloth or sturdy paper bags for your collected rocks. Many geologists wrap specimens in newspaper to protect the rock, write a brief description on a label with a few rock details (where it was collected, the type of rock, etc.), and then pack them in cloth or paper bags. A notebook is also handy for keeping field notes until more permanent records of the rock or fossil specimen can be made. Finally, one other indispensable tool is the hand lens—sometimes called a pocket magnifier—which is used to identify mineral grains. A six- to ten-power magnification works best. These can be purchased in jewelry stores, optical shops, or scientific and geology supply houses.

Where do I start **searching** for **specific rock types**?

The simplest way to locate specific rocks is by using geologic maps that show the local distribution and extent of particular rock types (sometimes they include brief descriptions of the rock). Large or small areas are covered depending on the map's size and scale.

Where can **geologic maps** be obtained?

There are many sources for geologic maps. The following lists a few good sources:

- The United States Geological Survey (USGS) is the premier federal government publisher of geologic maps, in addition to hydrological and topographical maps, aerial photos, digital data, and satellite images. To contact: USGS Information Services, Box 25286, Denver, Colorado 80225, phone: 1-888-275-8747.

- Your state (or the state where you want to collect) government is another source to contact. Try under the state's own geological survey: for example, the New York State Geological Survey for map information about the Adirondack Mountains. Various countries' geological surveys also often publish geologic maps of their region.

- There are also many organizations that produce and sell their own geologic maps, including the American Association of Petroleum Geologists, P.O. Box 979, Tulsa, OK 74101-0979, phone: 1-800-364-2274; and the Geological Society of America, P.O. Box 9140, Boulder, CO 80301-9140, phone: 303-447-2020.

- There are also a plethora of commercial dealers that sell geologic maps in the United States and overseas, including East View Cartographic, Inc., 3020 Harbor Lane N., Minneapolis, MN 55447, phone: 763-550-0965; Geopubs, 4 Glebe Crescent, Minehead, Somerset, TA24 5SN, England, phone 01643 709001; and UNESCO Publishing, 7 place de Fontenoy, 75352 Paris 07 SP, France.

- Certain commercial retail stores dealing in outdoor equipment for hikers, climbers, and bicyclists often carry topographic maps of the local region, and sometimes they have maps of popular hiking sites in the state.

- And finally, you can often find local and sometimes national geologic maps at public or university libraries.

How do I **catalog and store** my collection?

Although cataloging and storing a rock, mineral, and/or fossil collection is a matter of personal preference, here are a few suggestions:

Storing—Because most rocks are durable and do not require special treatment, they can be stored wrapped in newspaper, then packed in ordinary shoe boxes, discarded cardboard boxes (from the grocery store), or even plastic containers. Smaller or more fragile specimens can be stored in egg cartons; small plastic containers lined with soft materials such as cotton can also be used for more delicate specimens. Even old jewelry boxes can be used to store smaller rocks, since most are lined with soft fabrics or materials.

Labeling—Labeling your collection is important, mainly so specimens do not get mixed up. Many people paint a very small dot of white paint (lacquer) on a specimen, then number the spot with black paint. Each number corresponds to the most important items about the rock written on a separate sheet of paper: the date of collection, rock name, collector's name, description of collection site, geologic age and formation (if known), and other important data. Still other collectors merely put a sturdy numbered card with the specimen in a small box or wrapped individually in newspaper.

Displaying—If you want to display some of your collection, you can use a table, an old bookcase, or even wooden soda bottle crates. There are also a variety of wall shelves and ornate bookcases commercially available—some complete with glass doors—depending on how much you want to spend.

VISITING GEOLOGY

What are **gem and mineral shows**?

Gem and mineral (also called rock and mineral) shows are usually held by local, regional, or national gem and mineral organizations. They are a good source of information for those starting out in geology, as well as for those who chose geology as their profession. The shows usually include rock, gem, mineral, and fossil dealers who either collect their own specimens or buy from brokers around the world. Some dealers also sell geological equipment, tools, books, and magazines—both used and new.

The sponsoring organization usually displays member's collections, or has a booth where novices and professionals can ask questions about rocks, gems, minerals, and fossils. For those interested in making jewelry out of rocks and gems, there are often lapidarists who either show their wares or give demonstrations.

What are some major **cave or mine tours** in the United States?

There are a large number of commercial tours available that will take you into spectacular caves or former working mines. It sometimes seems that there are almost as many tours as there are caves and mines! Good sources of information about these tours are local tourist information offices; another is to use any search engine on the Internet, with the keywords "cave tours" or "mine tours." The following is just a sampling of what is available in the United States:

A group of travelers are guided through crevasses on a cave tour hundreds of feet underground at Blanchard Springs Cavern in the Ozark National Forests near Mountain View, Arkansas. *AP/Wide World Photos*

Howe Caverns—Howe Caverns is located in upstate New York. Its cavern walls consist of two types of limestone and several other types of rock. Contact: Howe Caverns, Inc., 255 Discovery Dr., Howes Cave, NY 12092, phone: 518-296-8900.

Carlsbad Caverns National Park—Carlsbad Caverns National Park is located near Carlsbad, New Mexico. It is actually situated within a Permian Period fossil reef. There are a hundred known caves within the park, including one cave that is the third longest in the United States, and another that has one of

447

the world's largest underground chambers. Contact: Carlsbad Caverns National Park, 3225 National Parks Highway, Carlsbad, NM 88220, phone: 1-800-967-2283.

Mollie Kathleen Gold Mine—Located in Colorado, the Mollie Kathleen bills itself as the country's only 1,000 foot (305 meter) vertical shaft gold mine tour. Contact: Mollie Kathleen Gold Mine, 1 Mile N. Hwy. 67—P.O. Box 339, Cripple Creek, CO 80813, phone: 719-689-2466.

Lackawanna Coal Mine Tour—The Lackawanna Coal Mine is located in Pennsylvania; the tour takes you through one of the places where anthracite, a rare form of coal, was once mined. Contact: Lackawanna Coal Mine Tour, McDade Park, Scranton, PA, phone: 1-717-963-MINE.

Are there places where you can **search for gems and minerals**?

There are a large number of places that allow amateur geologists (and professionals, too) to search for gems and minerals in relative safety and comfort. Again, the best sources of information about this activity are the local tourist information office, the information desk at your local library, or the Internet. The following are just some examples to whet the collector's appetite (In most places, you can bring water and a box lunch and make a day of it. But also ask if the site is in the sunlight or not, and bring your suntan lotion if the sieving or collection area is not covered.):

Franklin, North Carolina—There are numerous mines located in the Franklin area that allow gem and mineral hunting. Some allow collectors to search the streambeds, while others sell buckets of stream sediment for washing and sieving at a flume. Contact: Franklin Area Chamber of Commerce, Inc., 180 Porter St., Franklin, NC 28734, phone 704-524-3161.

Herkimer Diamond Mines—Located in upstate New York, Herkimer diamonds are mostly doubly terminated quartz crystals that come already faceted by nature. They are dug out of dolomite bedrock that formed approximately a half billion years ago in a shallow Cambrian Period sea. Contact: Herkimer Diamond Mines, 4601 Route 28N, Herkimer, NY 13350, phone: 1-800-562-0897.

Are there any organized **geology tours and expeditions**?

There are a large number of geocentric tours available, both in the United States and around the world. Sources for these tours include your local travel agency, advertisements found in geology-related and science magazines, and the Internet. The following is a small sampling of what's available for someone who wants to go beyond the backyard:

Earthwatch Institute
Earthwatch International
3 Clock Tower Place, Suite 100
Box 75
Maynard, MA 01754
Phone (United States/Canada): 1-800-776-0188

A wide range of international expeditions where tour members actively assist scientists doing research in the field.

Himalayan Folkways & Chandertal Tours
20, The Fridays
East Dean
Near Eastbourn
East Sussex BN20 0DH
England
Phone: 00 44 (0)1323 422213

Geology tours of the Himalayas with guest lecturers; 15 days exploring the Chandra and Upper Spiti valleys in the mountains, or at the western edge of the Tibetan Plateau.

Mineral Search Safaris
Walter S. Bowser
4884 Agate St.
Las Cruces, NM 88012
Phone: 505-382-9018

Collecting tours for geologists and rockhounds in remote areas of Mexico, China, Tanzania, and the Peruvian Andes.

Sierra Club
National Headquarters
85 Second St., 2nd Floor
San Francisco, CA 94105
Phone: 415-977-5500

Over 60 international excursions ranging from volcanoes in Ecuador to trekking in the shadow of Mt. Everest.

Western Paleo Safaris
P.O. Box 1042
Laramie, WY 82073
Phone: 307-742-4651

Week-long outings into the American West in search of fossils, including dinosaur fossils.

Wisconsin Dells Nature Safaris
Time Travel Tours
14 East Geneva Cir.
Madison, WI 53717
Phone: 1-800-328-0995

Guided tours of Devils Lake, Parfrey's Glen, or Wisconsin Dells by local geologists.

GEOLOGICAL ORGANIZATIONS AND SOCIETIES

What is the **major governmental organization** that deals with geology?

The United States Geological Survey (USGS) is the federal agency that deals exclusively with geology. According to the Survey, its mission is to provide reliable scientific data to scientists and the public; describe and understand the Earth; minimize loss of life and property from natural disasters; manage natural resources (water, biological, energy, and mineral); and enhance and protect the quality of life in the United States.

Currently, there are three major regional offices:

Eastern Region and Headquarters—United States Geological Survey National Center, 12201 Sunrise Valley Dr., Reston, VA 20192, phone: 703-648-4000.

Central Region—United States Geological Survey, Box 25046 Denver Federal Center, Denver, CO 80225, phone: 303-236-5900.

Western Region—United States Geological Survey, 345 Middlefield Rd., Menlo Park, CA 94025, phone: 650-853-8300.

What **other U.S. government organizations** deal with geology?

There are many other United States government organizations that have some dealings with geology including, but not limited to, the following:

- Bureau of Reclamation (USBR)
- Department of Energy (DOE)
- Environmental Protection Agency (EPA)
- Federal Emergency Management Agency (FEMA)
- Fish and Wildlife Service (USFWS)
- Forest Service (USFS)
- Minerals Management Service (MMS)
- National Park Service (NPS)
- Office of Surface Mining (OSM)

Are there any **state geologic surveys**?

Yes, every state has its own geologic survey, although they are sometimes called by different names. For example, Massachusetts has its state geologist listed under the Executive Office of Environmental Affairs, not the geological survey.

The best way to find a state's geological survey is to contact the United States Geological Survey (see the address above). A list of state geological survey offices is also on the Internet at http://www.umr.edu/~library/geol/geoloff.html.

What are some **nongovernmental U.S. organizations and societies** devoted to geology?

The following are some examples of U.S. geology-oriented organizations and societies that are independent of the government:

- American Association of Petroleum Geologists (AAPG)
- American Geological Institute (AGI)
- American Geophysical Union (AGU)
- American Institute of Professional Geologists (AIPG)
- Association for Women Geoscientists (AWG)
- Association of American State Geologists (AASG)
- Association of Engineering Geologists (AEG)
- Geological Society of America (GSA)
- Geoscience Information Society (GIS)
- Mineralogical Society of America (MSA)
- Society for Mining, Metallurgy, and Exploration (SME)
- Society of Economic Geologists (SEG)
- Society of Exploration Geophysicists (SEG)

What are some **international organizations and societies** devoted to geology?

There are a huge number of organizations and societies devoted to geology throughout the world. The following are a few representative examples:

- Canadian Geophysical Union (CGU, or UGC if written in French)
- Deutsche Geologische Gesellschaft (German Geological Society)
- Edinburgh Geological Society
- Geological Society of Australia
- Geological Society of New Zealand
- International Association of Hydrogeologists
- International Association of Seismology and Physics of the Earth's Interior
- International Association of Volcanology and Chemistry of the Earth's Interior
- International Meteor Organization
- International Union of Geological Sciences

451

GEOLOGY CAREERS

What kind of **careers** are there in **geology**?

In general, geologists study the Earth to increase their understanding of this planet. They are curious about the history of the Earth, the processes going on within it, and the consequences of those processes, such as earthquakes, volcanoes, and mineral depositions. Through this study, they can also help others by discovering life-improving natural resources, mitigating the effects of natural disasters, and educating people about the planet we call home.

Geologists, like all scientists, often specialize in fields that are of special interest to them. They can become one or a combination of the following:

- *Economic geologists* explore natural resources and their development.
- *Engineering geologists* are concerned with geologic factors that affect structures such as buildings, dams, and bridges.
- *Environmental geologists* solve problems associated with waste disposal, pollution, urban development, and natural hazards such as flooding.
- *Geochemists* investigate the chemical makeup of rocks and minerals.
- *Geochronologists* use radioactive dating to determine the age of rocks and reconstruct the geologic evolution of the Earth (and sometimes its organisms).
- *Geodynamacists* study tectonic plate motions and interactions between the plates.
- *Geomorphologists* analyze the effects of Earth processes on the origin and nature of landforms.
- *Geophysicists* interpret data to uncover the planet's interior, and investigate phenomena such as the Earth's electric and magnetic fields.
- *Glaciologists* investigate the movement and properties of ice sheets and glaciers—and how the ice affects each other and the land beneath.
- *Hydrogeologists* deal with all aspects of groundwater, including how water creates karst topography or how groundwater pollution moves in underground realms.
- *Marine geologists* explore the continental shelves and ocean floors; some use this information to determine the movement of the tectonic plates around the planet.
- *Mineralogists* discover the composition, properties, and formation of minerals.
- *Paleontologists* look for and study fossils to uncover ancient animals and plants, determine organisms' evolution, and understand the environment in which the organisms lived.

- *Petroleum geologists* explore for oil and natural gas; they are also often involved in the production of petroleum products.
- *Planetary geologists* (also called *planetologists*) analyze data from the Moon and other planets and satellites to gain an understanding of our planet's long geologic history, as well as the formation, impact history, and current morphology of other planetary bodies.
- *Sedimentologists* deal with sedimentary processes and the resulting rocks.
- *Seismologists* investigate all aspects of earthquakes; they use earthquake waves to discover the Earth's interior structure—even using artificial (humanmade) "seismic" waves to better understand structures immediately below the surface, such as faults.

A geologist measures the productivity level of a new water well using satellite imagery and a unique, efficient drilling technique. *AP/Wide World Photos.*

- *Stratigraphers* study the relationships (both time and spatial) between rocks, minerals, and fossils contained within rock layers.
- *Structural geologists* explore phenomena that affect the Earth's crust, such as folding, faulting, and deformation.
- *Volcanologists* examine volcanoes and volcanic phenomena, including the formation and types of gases emitted; some even attempt to predict possible eruptions.

What is the best strategy for **obtaining a degree** (or two) in geology?

Although many students opt for the geology programs at "big name" colleges and universities, there is only limited space available, competition is keen, and undergraduate classes are typically large. A much better way is to attend an academically strong smaller college that focuses on undergraduates. Many of these do not even have graduate programs, so the focus becomes more on teaching than research.

Find a mentor/advisor who will work with you, and take as many courses dealing directly, or even peripherally, with geology as possible. After you've finished your junior year, discuss with your mentor what graduate programs are best for you, taking

What is the best way to start a career in geology?

As with most sciences, the best time to start is when you are young, though this does not preclude those who wish to change careers later in life. Take as many science courses as possible in high school, especially "Earth science." Take any continuing education courses on geology. Make geology your hobby: go to gem and mineral shows; explore local creek beds and road cuts; take vacations that turn into field trips; and read as much as you can about geology and related fields. Also, get some practical experience, which might mean doing lapidary work yourself, taking summer field trips sponsored by your local college or university, or even volunteering at a local natural history or science museum.

The point is to immerse yourself in the field to see if you really like it and what areas of geology fascinate you. It might be volcanoes, earthquakes, minerals, or mountains. In any case, make believe you are a geologist—act like one, make it your lifestyle, and see what happens.

into consideration the strength of the department, the current research, and the reputations and personalities of the individual professors. Ignore the media hype and rankings of "hot" schools. Shop around and be sure to visit each place on your "short list."

Perhaps the best place to start shopping for a degree is your local library—just peruse the many "guides to college" books now available. And for additional information, check out college Web sites on the Internet for the most updated information.

GEOLOGY MAGAZINES AND BOOKS

What are some **popular** national and international **geology magazines and newsletters**?

The following are popular national and international magazines and newsletters dealing exclusively with geology:

Geology Today—The publishers of this magazine is the Geologists' Association, in conjunction with the Geological Society of London. It is a bimonthly periodical, providing an instructive read for all earth scientists, both amateur and professional.

Geotimes—The American Geological Institute puts out this publication that reports geoscience news, research developments, geologic events, professional meetings, and trends in government policy, education, and industry.

Lapidary Journal—This magazine is a source of information about the gem, bead, lapidary, jewelry-making and mineral fields (with readers in 106 countries).

Rock & Gem—This magazine covers gem and mineral rockhounding, along with lapidary and jewelry work.

EOS—Published by the American Geophysical Union (AGU), this weekly newsletter is filled with information—detailed geology articles, news (research, government activities, and education), book reviews, job postings, and conference listings—on almost every facet of geophysics from space to the oceans. You have to join the AGU to get *EOS*, but many science libraries carry the newsletter.

What are some **popular national magazines** that often feature articles on geology?

Although the following magazines do not exclusively deal with geology, they often have articles about this field: *Discover, Popular Science, Astronomy, Scientific American, Science, Nature, Mercury, New Scientist, National Geographic, Natural History Magazine, Science Digest,* and *Science News.*

What are some **popular books** on geology?

A listing of popular books dealing with geology would be as large as *this* book (probably larger). The best strategy to find geology books is to look through the science and nature sections of your local bookstores, pull out a few titles, and browse them to see if they interest you.

But we have found the following books to be good geology "reads" for the layperson. They are engaging, educational, and fun, the types of books you will want to read outside, preferably by a rock outcrop:

The Map That Changed the World: William Smith and the Birth of Modern Geology by Simon Winchester (HarperCollins; August 2001; ISBN: 0060193611). This is the fascinating (and heroic) story of William Smith, the surveyor who discovered that there were different layers of rock in the Earth—layers that appeared in the same order no matter where they were found. In 1815, after twenty years of work, he published his famous map that changed the way we study our planet.

Annals of the Former World by John A. McPhee (Farrar Straus & Giroux; June 1998; ISBN: 0374105200). This John McPhee book is an entertaining look at the geologic history of North America, centered on the 40th parallel, as told by geologists themselves. It is essentially four of his previously published books in one: *Basin and Range* (1981), covering the land between the Sierra Nevada and the Rocky Mountains; *In Suspect Terrain* (1983), which is an overview of the Appalachian Mountains; *Rising from the Plains* (1986), which is about the Rocky Mountains; and *Assembling California* (1993), which is about the tectonics and volcanics of the Rockies. Also included in this book is *Crossing the Craton,* a previously unpublished new volume for the series that is about North America's ancient core that lies underneath Iowa, Nebraska, and Illinois.

The Practical Geologist by Dougal Dixon (Fireside; August 1992; ISBN: 0671746979). This is a great introduction to geology, starting with the history of the Earth's formation and its subsequent development, then covering everything from the composition of rocks and minerals to continental movements, erosion, and the effects of water and wind. There are also sections on mapping, searching for rocks and minerals, identification, preparing samples, and preserving your collection.

Volcanoes by Robert Decker and Barbara Decker (W H Freeman & Co.; 3rd edition, August 1997; ISBN: 0716724405). This is a wonderful introduction into the world of volcanoes. The information is at a level that the layperson can understand, but has enough depth that the serious student will appreciate it. The Deckers use examples and descriptions of real eruptions to illustrate volcanic processes.

Are there any **field guides** to geology?

Yes. For the person interested in getting out in the field, there are a plethora of field guides available written by people who have been "out there and done that." Again, a trip to your local bookstores is in order, especially if you are interested in your immediate locality. The authors swear by (not at) the following books when they are in the field:

Roadside Geology Series—Written and edited by various professional geologists, this Mountain Press Publishing Company series (P.O. Box 2399, Missoula, MT 59806) began in 1972 and now includes 24 titles to date. Covered are the geology of such states as Alaska, Colorado, Hawaii, Maine, New York, California, Utah, and Wyoming. Each books offers information (in the form of text, photos, and illustrations) about the geology, landforms, and places to see sites of geological interest.

Geology Underfoot Series—Various authors also write books for this series, also published by Mountain Press Publishing Company. These books are written for general readers with an interest in geology, but in particular for those with little or no formal training in this area. Each examines 20 to 40 sites of geologic interest in detail, with more in-depth excursions into the geology of an area. Titles cover such places as central Nevada, Death Valley and Owens Valley, Illinois, and southern California.

A Field Guide to Geology: Eastern North America, Peterson Field Guides— Written by David C. Roberts; illustrated by W. Grant Hodsdon; edited by Roger Tory Peterson. These guides are published by Houghton Mifflin Company (Trade & Reference Division, 222 Berkeley St., Boston, MA 02116). This 2001 guide describes each geologic region of eastern North America in detail, including the geologic history and processes that shaped the land, and contains numerous maps and cross-sections.

A Field Guide to Rocks and Minerals, Peterson Field Guides—Written by Frederick H. Pough; photographs by Jeffrey Scovil; edited by Roger Tory Peterson. This book was also published by Houghton Mifflin Company in 1998. It includes 385 color photographs illustrating rocks, minerals, and geologic formations in the front section of the book. The geographic distribution, physical properties, chemical composition, and crystalline structures of hundreds of minerals are cross-referenced with the photos.

Are there any **children's or young adult publications** that deal with geology?

There are no children's or young adult magazine publications that deal strictly with geology. However, as with magazines for adults, the following publications often include articles on geology: *National Geographic World, Ranger Rick, Kids Discover, MUSE, Boys' Life,* and *Odyssey.*

GEOLOGY MUSEUMS

Where can I see **geology exhibits**?

Most of the major natural history museums in the United States have extensive geology exhibits. Examples include the Smithsonian Institution's National Museum of Natural History in Washington, D.C.; the Field Museum of Natural History in Chicago, Illinois; the Carnegie Museum of Natural History in Pittsburgh, Pennsylvania; the American Museum of Natural History in New York City; and the Natural History Museum of Los Angeles County in Los Angeles, California.

Are there any **smaller museums** that are devoted to **geology**?

Yes. In addition to the larger museums mentioned above, there are many smaller museums that are exclusively devoted to geology, minerals, and gems. Here is just a small sampling, arranged by state:

The sphere of the Hayden Planetarium sits within the Rose Center for Earth and Science at New York's American Museum of Natural History. *AP/Wide World Photos.*

Delaware

Mineralogical Museum
University of Delaware
114 Old College
Newark, DE 19716
Phone: 302-831-8242

This mineral collection numbers over 5,000 specimens—600 on display in the Irénée du Pont Mineral Room (her personal collection forms the core of the museum).

New Jersey

Rutgers University Geology Museum
Geology Hall, Old Queens Section
College Avenue Campus
New Brunswick, NJ 08901
Phone: 732-932-7243

This museum emphasizes New Jersey geology and anthropology. The museum includes a 30-foot (10-meter) geologic cross section of New Jersey (from the Delaware Water Gap to the southern New Jersey coastal plain), mineral exhibits, and a small rock shop.

New Mexico

Mineralogical Museum
New Mexico Bureau of Geology and Mineral Resources
New Mexico Tech
801 Leroy Place
Socorro, NM 87801
Phone: 505-835-5140

The main exhibit hall highlights minerals from New Mexico, with over 2,000 specimens on display. There are also minerals from around the southwestern United States and the world, in addition to mining memorabilia, fossils, and an ultraviolet mineral exhibit.

Oregon

Rice Northwest Museum of Rocks and Minerals
26385 NW Groveland Dr.
Hillsboro, OR 97124
Phone: 503-647-2418

This museum collection includes thousands of specimens from around the world, one of the two finest red rhodochrosite specimens in the world, and rare and beautiful gem crystals.

Pennsylvania

Earth & Mineral Sciences Museum and Art Gallery
Penn State University
112 Steidle Bldg.
Pollock Road
University Park, PA 16802
Phone: 814-865-6427

The museum has more than 22,000 specimens of rocks, minerals, and fossils, and the world's most extensive collection of mineral properties exhibits.

South Carolina

Bob Campbell Geology Museum
Clemson University
103 Garden Trail
Clemson, SC 29634-0130
Phone: 864-656-4600

There are over 5,000 specimens in this museum's collection, along with fossils from around the world, meteorites, a display of fluorescent minerals, "hands-on" specimens, a 2,000-volume library, and displays of historic mining artifacts.

South Dakota

Museum of Geology
South Dakota School of Mines & Technology
O'Harra Bldg.
501 East St. Joseph
Rapid City, SD 57701
Phone: 605-394-2467

There are nearly 300,000 specimens of fossils, rocks, and minerals in this museum, with an emphasis the Black Hills region of South Dakota.

Virginia

Virginia Tech Geological Sciences Museum
Department of Geological Sciences
Virginia Polytechnic Institute and State University
4044 Derring Hall
Blacksburg, VA 24061-0420

This museum contains displays of fossils, an extensive mineral collection, and exhibits of local mining history.

Mineral Museum
James Madison University
Department of Geology and Environmental Sciences
MSC 7703
Harrisonburg, VA 22807
Phone: 540-568-6421

Features over 500 crystals and gemstones from around the world.

OTHER GEOLOGY RESOURCES

Are there any **television shows** that emphasize geology?

The Discovery Channel and the National Geographic Channel often have shows and specials that deal with different aspects of geology. The Public Broadcast Service (PBS) also has a number of series that sometimes deal with geology, such as *Nature, Nova, Savage Earth,* and *Scientific American Frontiers.*

Are there any **videos or DVDs** that deal with geology?

There are many videos and DVDs available that deal with geology. One source is your local library; sometimes titles can also be found in a video store. Searching online at sites such as Amazon.com and Barnes and Noble will yield a treasure trove of geology-related titles. Here are a few examples of what's out there:

Glaciers: Alaska's Rivers of Ice—Release date: May 5, 2003; format: DVD

Living Rock: An Introduction to Earth's Geology—Release date: August 13, 2002; format: DVD

Physical Geography: Introduction to Rocks and Minerals—Release date: January 1, 2000; format: VHS

World of Volcanoes: 20 Years of Climbing Volcanoes—Release date: November 20, 2000; format: VHS

What are some **companies** that feature **geologic supplies**?

The following are just a very few of the numerous companies that can supply the amateur and/or professional geologist with all the equipment his or her heart may desire (the addresses and phone numbers provided below were accurate at press time):

ASC Scientific
2075 Corte del Nogal, Suite G
Carlsbad, CA 9200
Phone: 1-800-272-4327

CR-Scientific
232 Cokesbury Rd.
Lebanon, NJ 08833
Phone: 908-236-0369

Dad's Rock Shop
P.O. Box 6124
Mohave Valley, AZ 86446
Phone: 1-800-844-3237

Edmund Scientific (limited field equipment)
60 Pearce Ave.
Tonawanda, NY 14150
Phone: 1-800-728-6999

Geologic Resources
21 Pointe Rok Dr.
Worcester, MA 01604
Fax: 508-755-1047

Kooter's Geology Tools
P.O. Box 310
1018 Wells Fargo Rd.
Lindsborg, KS 67456
Phone/Fax: 1-888-383-5219

Are there any stamps that represent geology?

Yes, there are stamps that represent geology. They mostly represent natural hazards, gems and minerals, or some special features, such as caves. Gems and minerals are not a favorite subject of the United States Post Office; only about eight such stamps have been issued to date. But they have issued other stamps representing the Earth sciences, including those depicting volcanoes, mountains, the Grand Canyon, Old Faithful, prehistoric mammals and dinosaurs, and others.

If you want gem, mineral, fossil, mining, and other such stamps, you can turn to those from other countries, which issue such stamps in the hundreds, if not thousands. For example, African governments have issued many stamps depicting minerals and mining, probably because these areas are very important to their economic well-being.

If you are interested in geology-oriented stamps, you might want to look into the American Topical Association (founded in 1949). The ATA includes a Gems, Minerals, and Jewelry Study Unit whose members are interested in stamps depicting gems, minerals, fossils, jewelry, mining, and other earth science related topics and subjects. You can contact the ATA at P.O. Box 57, Arlington, TX 76004-0057, phone 817-274-1181.

U.S. Geosupply, Inc.
P.O. Box 40217
Grand Junction, CO 81504
Phone: 970-434-3708

SURFING THE INTERNET

(Note: These Web sites were active at the time of this writing. Because content on the Internet can change so rapidly, some of these sites might no longer be functional, though they were active at press time. We apologize for any inconvenience this may cause.)

What kind of **geology resources** are available on the **Internet**?

The Internet can be thought of as a world-wide geology (virtual) library, and its plethora of Web sites on the subject reflects that fact. There is, literally, everything about geology online: academic course notes written at middle school, high school, college, and graduate school level; scientific papers from international conferences; Web-rings devoted to one specific subject such as minerals; live Web cams (Web sites with live cameras or camcorders) of everything from erupting volcanoes to moving ice; virtual tours of geologically significant areas around the world; and on and on. You can spend endless hours online, immersing yourself in this field. But realize that, just as with a library, there comes a time when you also have to get outside and experience geology in the real world.

Is there **one good Web site** to start learning about **geology**?

If you don't feel like using search engines or surfing around the Internet for hours, but you want to get started with one, comprehensive, informative site for geology, then go to the United States Geological Survey's (USGS) Web page at http://www.usgs.gov/. There you'll find everything about volcanoes, earthquakes, landslides, floods, and more. It is an absolute must resource site, one that you'll return to again and again.

Are there any **e-zines** that feature geology?

Yes, there are numerous e-zines (or magazines found on the Internet's World Wide Web) that not only feature articles on geology, but also specialize in certain fields within geology, such as paleontology. The following lists some of the most well-known ones:

Bob's Rock Shop—For rock hunters everywhere, this site is touted as the Internet's first e-zine for rockhounds; it is found at http://www.rockhounds.com.

Dinosaur Illustrated Magazine—This is another free, online dinosaur magazine; it is located at http://illustrissimus.virtualave.net/dimcont.html.

Dinosaur Interplanetary Gazette—This is touted as the ultimate online dinosaur magazine—and it is!—at http://www.dinosaur.org/frontpage.html.

Discovery.com—Discovery.com is found on the Internet at http://www.discovery.com; it often has articles and updates on geology news around the world.

Are there any **virtual Internet exhibits** that highlight geology?

Yes. Many museums offer, in addition to their regular exhibits, virtual tours of their rock, mineral, and gem collections. Some examples include:

- New York State Museum's online Mineral Exhibit includes 353 specimens with high quality images: http://www.nysm.nysed.gov/minerals/database.html.
- The Smithsonian's National Museum of Natural History is a multimedia online presentation of "The Dynamic Earth": http://www.mnh.si.edu/earth/.
- The University of California, Berkeley, Museum of Paleontology's Geology Wing has extensive online geology exhibits that take you on a journey through the history of the Earth, emphasizing stratigraphy and the fossil record: http://www.ucmp.berkeley.edu/exhibit/geology.html.

463

What is the best way to search for online geology resources?

The best way to find geology resources online is to use a good search engine, such as Google, Lycos, Yahoo, etc. Typing "geology" into the search field will result in an overwhelming number of hits, many of which may be of little interest to you; try to be more specific to limit the number of Web pages you get in your results. For example, if you're interested in information about erupting volcanoes, use the key words "volcanoes current eruptions." You'll still get a large number of Web sites, but many of them will be relevant. Another strategy is to peruse the Web site of college and university geology departments. They usually contain information about research areas and often provide geology-related links of interest. And sometimes the best strategy is to have no strategy at all. Just search on a whim, surf at leisure—and see what pops up.

Are there any **virtual geology field trips**?

Yes, there are a large number of virtual geology field trips available on the Internet, and that number seems to be growing every day. The following are just a few examples of what's out there. Typing the keywords "geology virtual field trip" in any search engine is guaranteed to have you glued to the computer screen for hours:

- Southern California Geology and the Significance of the San Andreas Fault Zone; Department of Geological Sciences at California State University, Long Beach: http://seis.natsci.csulb.edu/VIRTUAL_FIELD/vfmain.htm.
- Rocks, fossils, and landscapes of the American Southwest, Indiana University of Pennsylvania: http://www.iup.edu/fieldtrip.
- Indian Peaks, Colorado Front Range; Department of Geography and Geology, University of Wisconsin, Stevens Point: http://www.uwsp.edu/geo/projects/virtdept/ipvft/start.html.
- Mount Rodgers, Virginia; the Geology Field School, Radford University: http://www.runet.edu/~fldsch/RUFieldschool/fieldtrips/MountRodgers/MtRogers Index.html.

And when you are done with those, there is a large listing of virtual geology field trips to be found at Santa Monica College at http://homepage.smc.edu/robinson_richard/regionalUS.htm.

What are some **Web sites** devoted to **rocks and minerals**?

Here are only a few sample Web sites, but they will stimulate your appetite for rocks and minerals:

- Atlas of Igneous, Metamorphic Rocks, Minerals, and Textures (photographs of many rock and mineral thin sections from the University of North Carolina): http://www.geolab.unc.edu/Petunia/IgMetAtlas/mainmenu.html.
- The United States Geological Survey's State Minerals Statistics and Information (information about minerals divided by state): http://minerals.usgs.gov/minerals/pubs/state.
- The United States Geological Survey's site about collecting rocks (everything you need to know about finding rocks and starting a collection): http://pubs.usgs.gov/gip/collect1/collectgip.html.
- The Mineral Information Institute's page of common minerals and their uses: http://www.mii.org/commonminerals.php.
- American Federation of Mineral Societies (a good resource to discover the mineral societies near your region): http://www.amfed.org.
- Rocks and minerals slide show from the University of North Dakota's "Volcano World" (a good site to help identify various rocks and minerals): http://volcano.und.nodak.edu/vwdocs/vwlessons/lessons/Slideshow/Slideindex.html.
- A mineral collection site with many links to information, aptly called: http://www.mineralcollecting.org.

What are some **Web sites** devoted to **fossils**?

Here are some good introductory Web sites that deal with fossils:

- CretaceousFossils.com (contains information about all kinds of Cretaceous fossils found in the United States and around the world): http://www.cretaceousfossils.com.
- Finding Fossils from the San Diego Natural History Museum (contains general fossil information—not just for kids): http://www.sdnhm.org/kids/fossils.
- United States Geological Survey's Fossils, Rocks, and Time (a good introduction to geologic history and fossil formation): http://pubs.usgs.gov/gip/fossils/contents.html.

What are some **Web sites** dealing with **earthquakes**?

The following Web sites will get you started learning about earthquakes:

- How Earthquakes Work (a good, general introduction to earthquakes and their associated phenomena): http://www.howstuffworks.com/earthquake.htm.
- Nevada Seismological Laboratory (general earthquake information with an emphasis on earthquakes in Nevada and California): http://www.seismo.unr.edu.
- United States Geological Survey Earthquake Hazards Program (with information on earthquake science, worldwide earthquake activity, and hazard reduction): http://earthquake.usgs.gov.

What are some **volcano Web sites**?

The following are Web sites selected for their great content and educational value (not to mention that they're really cool, too):

- Cascades Volcano Observatory (contains all sorts of information about volcanoes, has lots of great links, and emphasizes the Cascade Range): http://vulcan.wr.usgs.gov.
- How Volcanoes Work (an educational Web site that describes the science behind volcanoes and their processes): http://www.geology.sdsu.edu/how_volcanoes_work.
- Volcanoes.com (a huge amount of information about volcanoes around the world): http://www.volcanoes.com.

What are some **Web sites** dealing with the geology of the **oceans**?

The following sites serve as a good survey of marine geology:

- Looking at the Sea: Physical Features of the Ocean (very informative site about the geological features on the ocean floor from Boston's Museum of Science): http://www.mos.org/oceans/planet/features.html.
- Marine Geology: Research beneath the Sea (a good introduction to the history, methods, and phenomena associated with underwater geology): http://walrus.wr.usgs.gov/pubinfo/margeol.html.
- NOAA Ocean Explorer: Explorations (a fascinating look into how scientists explore the ocean depths, including visits to volcanoes, seamounts, and trenches; contains voyage logs, images, and descriptions): http://oceanexplorer.noaa.gov/explorations/explorations.html.

What are some interesting **oil, gas, and mining Web sites** on the Internet?

The following lists only a few of the interesting petroleum Web sites, emphasizing public education about oil and natural gas, and a site on mining:

- American Petroleum Institute: http://api-ec.api.org/frontpage.cfm.
- History of the Oil Industry (emphasis on California): http://www.sjgs.com/history.html.
- Institute of Petroleum (United Kingdom): http://www.schoolscience.co.uk/petroleum/index.html.
- Natural Gas Supply Association: http://www.naturalgas.org/index.asp.
- Oil and Gas Museum, Parkersburg, West Virginia: http://www.little-mountain.com/oilandgasmuseum.
- Petroleum Museum, Midland, Texas: http://www.petroleummuseum.org.
- Samuel Pees' site about the history of oil: http://www.oilhistory.com.
- Society of Petroleum Engineers: http://www.spe.org/suitcase/default.html.

rhyolite A highly acid volcanic rock that pours forth as lava from mountain and island arc volcanoes; it is the extrusive form of granite.

scoria cone Scoria cone is also the name used for a cinder cone, the most common type of volcano.

sediment Unconsolidated mud or sand deposited by water, wind, or glaciers; when consolidated, such materials are usually referred to as sedimentary rocks.

slickenlines The striations on a rock surface formed by friction as the rock moves, usually along fault surfaces; a surface with many slickenlines is called a slickenside.

specific gravity A measurement of weight per volume. It is the ratio of a subtance's density to the density of another substance.

strata The plural form of stratum (a single layer).

stratum A bed or layer of rock that lies between beds of another kind of rock.

stream Streams are often used as a catch-all term—usually including any body of running water such as a river, creek, or brook. In geology, streams are usually smaller than rivers.

strombolian eruption Volcanic activity characterized by short-lived, explosive outbursts of pasty lava ejected a few tens or hundreds of feet into the air.

structural province A large region characterized by certain geologic features and structures.

subduction The process in which one edge of a crustal plate descends below the edge of another plate; most subduction zones lead to large mountain and volcanic chains.

subsidence The sudden collapse of the surface, creating a depression in the land.

swell A long, large wave or succession of waves in the ocean, usually caused by a disturbance, such as an offshore gale or hurricane.

talus slope A steep slope formed by the accumulation of rock fragments, most often at the base of a cliff.

tributary A stream that feeds a larger river or lake.

turbidite Sedimentary rocks deposited by currents; they are mixtures of water and sediment. Because the current is more dense than water alone, it will flow downslope under the pull of gravity.

uplift The upward push of rock layers caused by geological forces, such as mountain building and volcanic magma.

Index

Note: (ill.) indicates photos and illustrations.